DYNAMICS OF SELF-ORGANIZED AND SELF-ASSEMBLED STRUCTURES

Physical and biological systems driven out of equilibrium may spontaneously evolve to form spatial structures. In some systems molecular constituents may self-assemble to produce complex ordered structures. This book describes how such pattern formation processes occur and how they can be modeled.

Experimental observations are used to introduce the diverse systems and phenomena leading to pattern formation. The physical origins of various spatial structures are discussed, and models for their formation are constructed. In contrast to many treatments, pattern-forming processes in nonequilibrium systems are treated in a coherent fashion. The book shows how near-equilibrium and far-from-equilibrium modeling concepts are often combined to describe physical systems.

This interdisciplinary book can form the basis of graduate courses in pattern formation and self-assembly. It is a useful reference for graduate students and researchers in a number of disciplines, including condensed matter science, nonequilibrium statistical mechanics, nonlinear dynamics, chemical biophysics, materials science, and engineering.

Rashmi C. Desai is Professor Emeritus of Physics at the University of Toronto, Canada.

Raymond Kapral is Professor of Chemistry at the University of Toronto, Canada.

DYNAMICS OF SELF-ORGANIZED AND SELF-ASSEMBLED STRUCTURES

RASHMI C. DESAI AND RAYMOND KAPRAL

University of Toronto

CAMBRIDGE
UNIVERSITY PRESS

CAMBRIDGE
UNIVERSITY PRESS

University Printing House, Cambridge CB2 8BS, United Kingdom

One Liberty Plaza, 20th Floor, New York, NY 10006, USA

477 Williamstown Road, Port Melbourne, VIC 3207, Australia

314-321, 3rd Floor, Plot 3, Splendor Forum, Jasola District Centre, New Delhi - 110025, India

103 Penang Road, #05-06/07, Visioncrest Commercial, Singapore 238467

Cambridge University Press is part of the University of Cambridge.

It furthers the University's mission by disseminating knowledge in the pursuit of education, learning and research at the highest international levels of excellence.

www.cambridge.org
Information on this title: www.cambridge.org/9780521883610

First published 2009

A catalogue record for this publication is available from the British Library

Library of Congress Cataloging in Publication data
Desai, Rashmi C.
Dynamics of self-organized and self-assembled structures /
Rashmi C. Desai and Raymond Kapral.
p.   cm.
ISBN 978-0-521-88361-0 (hardback)
1. Pattern formation (Physical sciences)   2. Phase rule and equilibrium.
3. Dynamics.   I. Kapral, Raymond.   II. Title.
Q172.5.C45D47 2009
500.201´185–dc22

2008045219

ISBN 978-0-521-88361-0 Hardback

To our families

The dimmed outlines of phenomenal things all merge into one another unless we put on the focusing glass of theory, and screw it up sometimes to one pitch of definition and sometimes to another, so as to see down into different depths through the great millstone of the world.

Analogies, James Clerk Maxwell

Contents

Preface

The idea for this book arose from the observation that similar-looking patterns occur in widely different systems under a variety of conditions. In many cases the patterns are familiar and have been studied for many years. This is true for phase-segregating mixtures where domains of two phases form and coarsen in time. A large spectrum of liquid crystal phases is known to arise from the organization of rod-like molecules to form spatial patterns. The self-assembly of molecular groups into complex structures is the basis for many of the developments in nano-material technology. If systems are studied in far-from-equilibrium conditions, in addition to spatial structures that are similar to those in equilibrium systems, new structures with distinctive properties are seen. Since systems driven out of equilibrium by flows of matter or energy are commonly encountered in nature, the study of these systems takes on added importance. Many biological systems fall into this far-from-equilibrium category.

In an attempt to understand physical phenomena or design materials with new properties, researchers often combine elements from the descriptions of equilibrium and nonequilibrium systems. Typically, pattern formation in equilibrium systems is studied through evolution equations that involve a free energy functional. In far-from-equilibrium conditions such a description is often not possible. However, amplitude equations for the time evolution of the slow modes of the system play the role that free-energy-based equations take in equilibrium systems. Many systems can be modeled by utilizing both equilibrium and nonequilibrium concepts.

Currently, a wide variety of methods is being used to analyze self-organization and self-assembly. In particular, microscopic and mesoscopic approaches are being developed to study complex self-assembly in considerable detail. On mesoscales, fluctuations are important and influence the self-organization one sees on small scales, such as in the living cell. Nevertheless, many common aspects of these pattern-forming processes can be modeled in terms of order parameter fields, which describe the dynamics of relevant collective variables of the system. The patterns that are formed and the way they evolve are often controlled by certain common elements that include the presence of interfaces, interfacial curvature, and defects.

In order to present an approach to the study of such self-assembled or self-organized structures that highlights common features, we have intentionally limited the scope of the presentation to descriptions based on equations for order parameter fields. Approaches of this type are able to capture the gross features of pattern formation processes in diverse systems, including those in the equilibrium and far-from-equilibrium domains. We have also intentionally omitted descriptions based on various coarse-gained molecular dynamics methods and a variety of other mesoscopic particle-based methods, which are proving to be powerful tools for the study of such systems. In addition, to sharply focus our presentation we have restricted our discussion to systems where hydrodynamic flows are not important.

A selection of the material in this book formed the basis for a one-semester course entitled "Interface Dynamics and Pattern Formation in Nonequilibrium Systems" given jointly in the Departments of Chemistry and Physics at the University of Toronto. Many of the topics covered in the book have been the subjects of intense investigations, and a large literature exists. In order to make the material as self-contained as possible, in most cases we have provided an introduction to each topic in a form that allows the main ideas to be exposed and derived from basic principles. The final chapters of the book provide some additional examples of applications that combine the two underlying themes that are developed in the book: free-energy-functional and amplitude-equation descriptions. These chapters show how the dynamics of physical and biological systems can be modeled using the concepts developed in the body of the book.

Some of the material presented in the book derives from work with our colleagues and students. In particular we would like to acknowledge the contributions of Augustí Careta, Hugues Chaté, Francisco Chávez, Mario Cosenza, Jörn Davidsen, Ken Elder, Simon Fraser, Leon Glass, Martin Grant, Andrew Goryachev, Daniel Gruner, Christopher Hemming, Zhi-Feng Huang, Anna Lawniczak, François Léonard, Roberto Livi, Anatoly Malevanets, Paul Masiar, Alexander Mikhailov, Gian-Luca Oppo, Antonio Politi, Sanjay Puri, Tim Rogers, Katrin Rohlf, Chris Roland, Guillaume Rousseau, Celeste Sagui, Ken Showalter, Kay Tucci, Mikhail Velikanov, Xiao-Guang Wu, Chuck Yeung, and Meng Zhan. We also owe a special debt of gratitude to our colleagues who read and commented on portions of the book: Markus Bär, Jörn Davidsen, Walter Goldburg, Jim Gunton, Christopher Hemming, Zhi-Feng Huang, Chuck Knobler, Maureen Kapral, Alexander Mikhailov, Steve Morris, Evelyn Sander, Len Sander, Celeste Sagui, Peter Voorhees, Tom Wanner, Chuck Yeung, and Royce Zia. The preparation of this book would have been difficult without the help of Suzy Arbuckle and Raul Cunha, and we would like to express our special gratitude to them for their assistance.

1

Self-organized and self-assembled structures

Almost all systems we encounter in nature possess some sort of form or structure. It is then natural to ask how such structure arises, and how it changes with time. Structures that arise as a result of the interaction of a system with a template that determines the pattern are easy to understand. Lithographic techniques rely on the existence of a template that is used to produce a material with a given spatial pattern. Such pattern-forming methods are used widely, and soft lithographic techniques are being applied on nanoscales to produce new materials with distinctive properties (Xia and Whitesides, 1998). Less easily understood, and more ubiquitous, are self-organized structures that arise from an initially unstructured state without the action of an agent that predetermines the pattern. Such self-organized structures emerge from cooperative interactions among the molecular constituents of the system and often exhibit properties that are distinct from those of their constituent elements. These pattern formation processes are the subject of this book.

Self-organized structures appear in a variety of different contexts, many of which are familiar from daily experience. Consider a binary solution composed of two partially miscible components. For some values of the temperature, the equilibrium solution will exist as a single homogeneous phase. If the temperature is suddenly changed so that the system now lies in the two-phase region of the equilibrium phase diagram, the system will spontaneously form spatial domains composed of the two immiscible solutions with a characteristic morphology that depends on the conditions under which the temperature quench was carried out. The spatial domains will evolve in time until a final two-phase equilibrium state is reached. The evolution of such structures is governed by thermodynamic free energy functions, suitably generalized to account for the heterogeneity of the medium and the existence of interfaces separating the coexisting phases. The spontaneous formation of such structures is the system's response to an initial instability or metastability (Bray, 1994; Debenedetti, 1996; Dattagupta and Puri, 2004).

1

 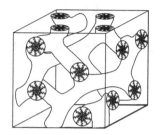 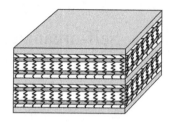

Fig. 1.1. Schematic depictions of hexagonal, gyroid and lamellar nanocomposites that result from the self-assembly of diacetylenic surfactants on silica. From Brinker (2004), p. 631, Figure 6a.

The formation of macroscopic coherent spatiotemporal structures arising from an initial instability or metastability is often a consequence of some inherent symmetry-breaking element. Fluctuations and conservation laws also play an important role in determining the character of the time evolution leading to self-organized structures. As the system evolves, interfaces which delineate the boundaries of local domains also move: thus an understanding of interface dynamics, and more generally of defect dynamics, is a central feature of the evolution of self-organized structures.

Ultimately, self-organized structures have their origin in the nature of the intermolecular forces that govern the dynamics of a system. In some instances, the connection between the macroscopic coherent structure and specific features of the intermolecular forces is rather direct. Self-assembly of molecular constituents in solution is such a process. Self-assembly leads to a variety of three-dimensional structures: strong hydrophobic attraction between hydrocarbon molecules can cause short chain amphiphilic molecules to organize into spherical micelles, cylindrical rod-like micelles, bilayer sheets, and other bicontinuous or tri-continuous structures (Fig. 1.1) (Gelbart *et al.*, 1994; Grosberg and Khokhlov, 1997; Brinker, 2004; Ozin and Arsenault, 2005; Pelesko, 2007). Self-assembly of long-chain block copolymers can also occur through microphase separation as a result of covalent bonds between otherwise immiscible parts of the polymer. This process can lead to three-dimensional structures with topologies similar to those of amphiphilic molecules (Fredrickson and Bates, 1996; Bates, 2005). Similarly, two-dimensional systems, such as Langmuir monolayers at a water–air interface or uniaxial ferromagnetic films, can self-assemble into unidirectional periodic stripes and hexagonally arranged circular drops as a result of the competition between long-range repulsive dipolar interactions and relatively shorter-range attractive van der Waals interactions. Monolayers on a metallic substrate can also self-organize into ordered structures (Fig. 1.2). The most direct way to model such self-assembly is by following the motions of the constituent elements by molecular dynamics. A number of different coarse-grain schemes have been devised in order to extend the size,

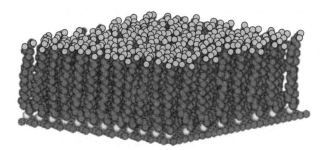

Fig. 1.2. Results of a molecular dynamics simulation of a densely packed assembly of 16-mercapto-hexadecanoic acid molecules tethered to a gold surface. From Lahann and Langer (2005), p. 185, Figure 2.

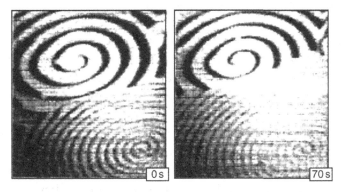

Fig. 1.3. Spiral wave CO oxidation patterns on the surface of a Pt(110) surface. Reprinted with permission from Nettesheim *et al.* (1993). Copyright 1993, American Institute of Physics.

length, and timescales of such simulations (Karttunen *et al.*, 2004; Nielsen *et al.*, 2004; Venturoli *et al.*, 2006). On mesoscopic scales self-assembly can be analyzed and understood through models based on free energy functionals and relaxational dynamics.

Self-organized structures also arise in systems that are forced by external flows of matter or energy to remain far from equilibrium (Nicolis and Prigogine, 1977; Kapral and Showalter, 1995; Walgraef, 1997; Manrubia *et al.*, 2004; Hoyle, 2006; Pismen, 2006). If chemical reagents are continuously supplied to and removed from a container where an oxidation reaction takes place on a catalytic surface, in many circumstances the chemical reaction does not occur homogeneously over the entire surface but instead proceeds by the propagation of chemical waves of oxidation that travel across the catalytic surface. The combination of nonlinear chemical kinetics and conditions that force the reaction to occur in far-from-equilibrium conditions is responsible for the existence of the evolving patterns of chemical waves seen on the surface of the catalyst (Fig. 1.3).

Biological systems almost always operate under far-from-equilibrium conditions since input of chemical and other energy sources is needed to maintain the living state. Consequently, the conditions for the appearance of self-organized structures are present in these systems. Indeed, the nonlinear chemistry associated with biochemical networks, in combination with diffusion of chemical species, can lead to the formation of chemical waves which are often implicated in the mechanisms responsible for biological function (Winfree, 1987, 2001; Murray, 1989; Goldbeter, 1996). Chemical waves are known to play a role in cell signaling processes leading to cell division, aggregation processes in colonies of the amoeba *Dictyostelium discoideum*, and the pumping action of the heart, to name a few examples. Perhaps even more interesting is the fact that chemical patterns have been observed in individual living cells (Petty *et al.*, 2000).

Although applications to fluid dynamics are not considered in this book, fluid flow also provides many examples of self-organized structures (Cross and Hohenberg, 1993; Frisch, 1995; Nicolis, 1995; Walgraef, 1997). The hexagonal patterns arising from Rayleigh–Bénard convection when a fluid is heated from below are familiar, as are the complex spatiotemporal patterns seen in turbulent fluids. In such cases, descriptions of the origins and dynamics of the patterns are usually based on an analysis of the Navier–Stokes equation; the instabilities are seen to emerge as a result of the convective nonlinear terms in this equation.

In contrast to equilibrium systems, in far-from-equilibrium systems free energy functions do not always exist, and the description of the dynamics of self-organized structures must be based on different premises. In the case of chemical and biochemical systems the starting point is usually a reaction–diffusion equation, while, as noted above, for fluid dynamics problems the Navier–Stokes equation is a natural starting point for the analysis.

In spite of the fundamental differences in the origins of diverse self-organized structures, there are often superficial similarities in their forms, and there exist common basic elements which are needed to understand their formation and evolution. At the macroscopic level, one needs a description in terms of suitable field variables or order parameters that account for the existence of spatial structure in the system. Other common elements include the presence of interfaces that separate phases or spatial domains that constitute the self-organized structure, and the existence of defects in the medium. Both of these features often control the dynamical evolution of the structure on certain time scales.

During the second half of the twentieth century, the concept of universality played a major role in our understanding of structural correlations and dynamics in condensed matter systems. Starting with Landau's unifying concept of the order parameter (Landau, 1937) and culminating in the renormalization group theory of critical phenomena (Wilson and Kogut, 1974), these developments demonstrated that a description of the relevant physics does not necessarily lie at the smallest available length or time scales for many fundamental problems. In many instances, the description of the dynamics of self-organized structures in nonequilibrium systems can be examined within a similar context. Consequently, often a macroscopic perspective may be adopted to describe the dynamics of these structures. Even for situations such as self-assembly where the crucial role of the underlying intermolecular forces is evident, the nature and dynamics of the self-assembled structures on long distance and time scales can be captured by approaches based on suitably defined field variables. While the use of such a perspective limits the spatial and temporal scales on which the description is valid, it is general enough to provide a basis for understanding most of the commonly observed structures, even on mesoscopic scales.

In the chapters that follow we describe the dynamics of self-organized structures based on equations of motion for order parameter fields, which provide a description of systems at the mesoscopic and macroscopic levels. Equations of motion for such order parameter fields can be constructed for systems described by free energy functionals, as well as for systems which are constrained to lie far from equilibrium, for which no such functionals exist. Such formulations enable one to identify similarities in both the forms of the self-organized structures and features that determine their evolution in equilibrium and far-from-equilibrium systems.

We begin with an analysis of the familiar phenomenon of phase segregation following a quench of a system into the two-phase region of the phase diagram. An essential ingredient in the dynamics is the behavior of interfaces separating domains of coexisting phases, and we develop a description of such interface dynamics. Domain segregation is modified when long-range repulsive interactions exist: the theoretical description of these systems is considered. In the far-from-equilibrium regime, the order parameter equations are constructed on the basis of weakly nonlinear theory where crucial slow modes are identified in the dynamics. Once again interfaces and fronts play an important role in determining the evolution of the system. Because of the lack of a free energy functional, a much richer variety of self-organized structures is observed, which includes structures with periodic or chaotic temporal behavior. While no truly unified picture of diverse self-organizing structures is possible, the presentation in this book provides the tools needed to analyze and understand the origins of various types of self-organized structure. The material should permit one to see similarities in the structures observed in nonequilibrium and driven systems, and draw parallels in the methods used to describe the phenomena.

2

Order parameter, free energy,
and phase transitions

The kinetics of first-order phase transitions involves the separation of an initially one-phase system into two coexisting phases. The formation and coarsening of domains of the coexisting phases as the system evolves are of central interest. The phase segregation process is usually studied by first preparing the system in a region of the phase diagram where the homogeneous state is stable. The system is then suddenly quenched into the two-phase region, and segregation into domains of the two stable phases takes place. Such phase segregation arises in a variety of physical contexts, including binary alloys and fluid mixtures, ferromagnetic systems, superfluids, polymer mixtures, and chemically reacting fluids. The temperature–composition (T, c) phase diagram for a binary mixture composed of constituents A and B is shown in Fig. 2.1. For low enough temperatures, in the region bounded by the coexistence curve the binary mixture will segregate into A-rich and B-rich phases.

A quench that takes the system from a homogeneous to a two-phase region is often performed by changing temperature suddenly at fixed concentration. Such a quench from the one-phase state at high temperatures may be carried out either along the critical isoconcentration line that passes through the critical point (path a), or along off-critical paths (path b). Phase segregation may be monitored by the changes in the local concentration of the binary mixture. In general, the variable that signals the passage from the one-phase to two-phase regions is called the order parameter ϕ.

The kinetics of the phase separation process in a binary mixture is often discussed in terms of a free energy function $f(c)$. In the one-phase region the free energy function has a simple single minimum, while it is bistable in the two-phase region (Fig. 2.2). The chemical potential is defined as $\mu = (\partial f / \partial c)_{T,\rho}$. In mean field descriptions based on the free energy function, the spinodal line (dashed line in Fig. 2.1) is defined as the locus of points where the derivative of the chemical potential with respect to the concentration is zero. The definition of the spinodal

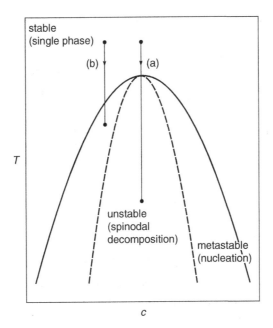

Fig. 2.1. Binary mixture phase diagram in the concentration–temperature plane showing critical (path a) and off-critical (path b) quenches into the two-phase region from the one-phase region. Path a starts at T_i and ends at T_f.

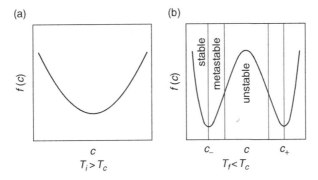

Fig. 2.2. Mean field free energy as a function of concentration for temperatures in the (a) one-phase and (b) two-phase regimes.

line may be generalized to any order parameter. A quench to a state within the spinodal region may lead to a labyrinthine pattern if it is associated with decay from an initial unstable state corresponding to the maximum in the free energy barrier. The evolution to a labyrinthine pattern is called spinodal decomposition. In contrast, an off-critical quench, such as the one shown as path b in Fig. 2.1, corresponds to evolution from an initial metastable state and involves a nucleation mechanism.

Both the morphology and phase separation dynamics depend crucially on the nature of the order parameter field. The order parameter may be conserved as in binary alloys or fluids, or nonconserved as in chemically reacting systems and antiphase domain growth in crystals. Dynamical models for the order parameter field that embody these conditions may be constructed and used to simulate phase separation dynamics. The system may require more than a single order parameter field for its description, and these order parameter fields may have different characters and symmetries. In such more general cases it may not be possible to define a free energy functional, and new phenomena may exist.

2.1 Mean field theory

2.1.1 Binary mixtures: alloys, fluids, and polymer blends

The specification of the thermodynamic state of a binary mixture requires three independent thermodynamic variables, say, the number density ρ, the concentration of the A species c and the temperature T. Only the (c, T) pair is relevant for the phase segregation process: thus, we consider a mixture of N_a molecules of type A and N_b molecules of type B with a fixed total number of molecules $N_o = N_a + N_b$ and fixed volume V. We define $c = N_a/N_o$, which implies $(1 - c) = N_b/N_o$. The differential Helmholtz free energy for a binary mixture may be expressed as

$$
\begin{aligned}
dF &= -SdT + \mu_a dN_a + \mu_b dN_b, \\
&= -SdT + N_o \mu dc,
\end{aligned}
\tag{2.1}
$$

where the fixed N_o constraint is used in the second equality and $\mu = (\mu_a - \mu_b)$. It follows that the differential of the free energy per molecule $f = F/N_o$ is

$$
df = -sdT + \mu dc,
\tag{2.2}
$$

where $s = S/N_o$. From this equation one can deduce that

$$
\mu = \left(\frac{\partial f}{\partial c} \right)_T.
\tag{2.3}
$$

The chemical potential μ is thermodynamically conjugate to c, and the product of such a conjugate pair has dimensions of energy. The equation of state specifies the functional dependence among the three variables $\mu = \mu(c, T)$ and defines a surface in the three-dimensional (c, T, μ) space. A projection of this surface in the (c, T) plane was sketched in Fig. 2.1. The binary mixture critical point (c_c, T_c) is at the apex of the coexistence curve and is also referred to as the consolute point.

There are two other projections of the equation-of-state surface. The (μ, T) projection consists of two regions: a homogeneous A-rich-phase region and a homogeneous B-rich-phase region. The two are separated by a monotonically increasing coexistence line of first-order phase transitions, which ends at the critical/consolute point, (μ_c, T_c). The three coordinates of the critical point are obtained by simultaneously solving the following three equations:

$$\left(\frac{\partial \mu}{\partial c}\right)_{T=T_c} = 0, \quad \left(\frac{\partial^2 \mu}{\partial c^2}\right)_{T=T_c} = 0, \quad \mu_c = \mu(c_c, T_c). \tag{2.4}$$

The projection on the (μ, c) plane reveals what the (μ, T) projection hides, and it is useful to consider the behavior of isotherms in this plane. To gain an understanding of the qualitative structure of the chemical potential $\mu(c, T)$, consider the free energy functions sketched in Fig. 2.2. Differentiation of these functions with respect to c will result in two isotherms, one for the one-phase region and the other for the two-phase region. The isotherm for the one-phase region is monotonic and starts with a negative value of μ for small c. It becomes zero at the free energy minimum and increases monotonically to positive values of μ beyond the minimum. In contrast, the two-phase isotherm has a so-called van der Waals loop, since the free energy has three extrema at which μ vanishes.

To examine these features quantitatively, consider an analytic free energy function obtained from a mean field theory of a binary mixture (Bragg and Williams, 1934; Bethe, 1935; Huang, 1987),

$$\frac{f(c, T)}{k_B T} = c \ln(c) + (1 - c) \ln(1 - c) + \chi c(1 - c). \tag{2.5}$$

The first two terms in this function arise from the increase in the translational entropy due to mixing. In the last term, χ is a parameter describing the enthalpic interaction between the two species. For small molecules it is

$$\chi = \frac{z}{k_B T}\left[\epsilon_{AB} - \frac{1}{2}(\epsilon_{AA} + \epsilon_{BB})\right], \tag{2.6}$$

where z is the effective coordination number and ϵ_{ij} is the interaction energy between monomers of species i and j. It is straightforward to obtain the corresponding chemical potential

$$\frac{\mu(c, T)}{k_B T} = \ln(c) - \ln(1 - c) - \chi(2c - 1),$$
$$= 2 \tanh^{-1}(2c - 1) - \chi(2c - 1). \tag{2.7}$$

By equating each of the first two derivatives of μ with respect to c to zero, one finds $c_c = 1/2$, $\chi_c = 2$ and $\mu_c = 0$. Since $\chi = \chi(T)$, one can obtain T_c from χ_c. If χ is large and positive, phase segregation is favorable. The $\tanh^{-1}(x)$ function increases monotonically from $-\infty$ to $+\infty$ as x goes from -1 to $+1$. For small χ (high T), the linear term in Eq. (2.7) does not change the monotonic nature of μ; however, for $\chi > \chi_c$, one obtains a van der Waals loop.

In the region around the critical point, it is appropriate to expand the free energy in powers of $c^* = (c - c_c)$. At the critical point $f_c/(k_B T_c) = \frac{1}{2} + \ln(\frac{1}{2})$, and the Taylor series expansion leads to the result

$$\frac{f(c^*, T)}{k_B T} - \frac{f_c}{k_B T_c} = \frac{a_2}{2} c^{*2} + \frac{a_4}{4} c^{*4} + \dots \tag{2.8}$$

where $a_2 = 2(\chi_c - \chi)$, which is proportional to $T - T_c$ for $\chi \sim T^{-1}$, and $a_4 = \frac{16}{3}$. This expansion is an example of what is generically called a Landau expansion of the free energy around the critical point.

An A–B polymer blend is a binary mixture of long-chain polymer molecules, and Flory–Huggins (FH) theory is a mean field theory for such a polymer mixture. In FH theory, polymer chains are placed on a lattice in such a way that each monomer unit occupies a lattice site, and connected polymer chains are placed so that they are locally self-avoiding. For an incompressible blend, all lattice sites are occupied by either A or B monomers. Let N_A (N_B) be the number of monomers (degree of polymerization) in an $A(B)$ polymer chain. If c is the concentration of A in the polymer blend, the free energy of mixing per site for an incompressible blend is

$$\frac{f(c, T)}{k_B T} = \frac{c}{N_A} \ln(c) + \frac{(1 - c)}{N_B} \ln(1 - c) + \chi \, c(1 - c), \tag{2.9}$$

where the Flory interaction parameter χ depends on c and T in a more complicated way than that for small molecule mixtures. It is often empirically fitted to a form $\chi = a + (b/T)$. The equation of state now takes the form

$$\frac{\mu(c, T)}{k_B T} = (N_A^{-1} - N_B^{-1}) + N_A^{-1} \ln(c) - N_B^{-1} \ln(1 - c) + \chi(1 - 2c). \tag{2.10}$$

For a symmetric polymer blend ($N_A = N_B = N$) the critical point coordinates are $\mu_c = 0$, $c_c = 1/2$, $\chi_c = 2/N$. More generally, for non-symmetric polymer blends, $c_c = N_B^{1/2}/(N_A^{1/2} + N_B^{1/2})$, $\chi_c = (N_A^{1/2} + N_B^{1/2})^2/(2N_A N_B)$. The Landau expansion for a symmetric blend has the same form as that for the small molecules with $a_2 = 2(\chi_c - \chi)$ and $a_4 = 16/3N$. Many blends have an upper critical point where the blend is miscible for $\chi < \chi_c$ ($T > T_c$) and immiscible for $\chi > \chi_c$ ($T < T_c$) at the critical concentration c_c. Two limiting cases of a phase-separated blend are often considered: a weak segregation limit where $\chi N \approx 2$ (close to the mean field critical point), and a strong segregation limit where $\chi N \gg 2$ and the blend is strongly immiscible.

2.1.2 Para-ferromagnetic transition

The Curie–Weiss theory is a mean field theory for the para-ferromagnetic phase transition in magnets. The free energy per spin for the Curie–Weiss model is expressed in terms of the magnetization per spin m and the temperature T, and is typically derived from an Ising model for a system of spin $\frac{1}{2}$ particles. Fluctuations in the magnetization are an important element for the para-ferromagnetic transition. The magnetization M has the magnetic field H as its thermodynamically conjugate variable, and the triplet (M, T, H) forms the relevant thermodynamic space. The functional relation $H = H(T, M)$ is the equation of state. The mean field critical temperature T_c for the para-ferromagnetic transition is $T_c = zJ/k_B$, where J is the spin–spin interaction constant in the Ising model on a lattice with coordination number z. The critical point of the system occurs at $(M = 0, T = T_c, H = 0)$. It is useful to introduce reduced variables: $m^* = M/M_o$, $T^* = T/T_c$, $H^* = (HM_o/k_B T_c)$, where $M_o = M(T = 0, H = 0)$. The Helmholtz free energy per spin is then given by

$$\frac{f}{k_B T} = f^*(T^*, m^*)$$

$$= -\ln 2 + \frac{1}{2}(1 - m^*)\ln(1 - m^*)$$

$$+ \frac{1}{2}(1 + m^*)\ln(1 + m^*) - \frac{m^{*2}}{2T^*}. \tag{2.11}$$

The equation of state for the system is obtained by using the relation $H^*/T^* = (\partial f^*/\partial m^*)_{T^*}$. The result is

$$m^* = \tanh[(H^* + m^*)/T^*]. \tag{2.12}$$

The scaled equation of state may also be written as

$$\tanh(H^*/T^*) = \frac{m^* - \tanh(m^*/T^*)}{1 - m^*\tanh(m^*/T^*)}. \tag{2.13}$$

For $T^* < 1$ this equation of state displays a van der Waals loop. Near the critical point both H^* and m^* are small, and one can use the small x expansions for $\tanh(x)$ and $\ln(1 + x)$ to construct a Landau expansion in order to study the behavior of the thermodynamic properties of the mean field model near the critical point.

2.1.3 Liquid–vapor transition

For the liquid–vapor transition the number density ρ is the order parameter and also a relevant thermodynamic variable. The pressure P is thermodynamically conjugate to $v = V/N = 1/\rho$. The appropriate thermodynamic space is (ρ, T, P), and the associated mean field theory is the well-known van der Waals theory (Kittel and Kroemer, 1980). In the van der Waals model, the Helmholtz free energy per particle $f(v, T)$ is given by

$$f(v, T) = -k_B T \left(1 + \ln[n_Q(v - b)] \right) - \frac{a}{v}. \tag{2.14}$$

The parameter $n_Q = \left((mk_B T)/(2\pi\hbar^2) \right)^{3/2}$ is called the quantum concentration for a system of N particles (each of mass m) in the volume V: it is roughly the concentration associated with one atom in a cube with linear dimension equal to the thermal average de Broglie wavelength. Using the thermodynamic relation, $P = -(\partial f/\partial v)_T$, one obtains the van der Waals equation of state:

$$P = \frac{k_B T}{v - b} - \frac{a}{v^2}. \tag{2.15}$$

Solution of the equations $(\partial P/\partial v)_T = 0$, $(\partial^2 P/\partial v^2)_T = 0$, and $P = P(v, T)$ yields the critical-point coordinates $P_c = a/(27b^2)$, $v_c = 3b$, and $T_c = (8a)/(27bk_B)$. The equation of state in scaled variables $P^* = P/P_c$, $v^* = v/v_c$, $T^* = T/T_c$ is

$$P^* = \frac{8}{3}\frac{T^*}{v^* - \frac{1}{3}} - \frac{3}{v^{*2}}. \tag{2.16}$$

For given values of P^* and T^*, the van der Waals equation is a cubic polynomial in v^*. Figure 2.3 shows sketches of three isotherms. The critical point is indicated by a dot on the critical isotherm ($T_c^* = 1$). For temperatures greater than T_c ($T^* > 1$), the isotherms are monotonic. For temperatures less than T_c ($T^* < 1$), the isotherms contain a van der Waals loop.

Near the liquid–vapor critical point, density fluctuations dominate the physics, and the intensive Gibbs free energy $g = f + Pv$ is the appropriate thermodynamic energy density. (For a one-component system, g is the chemical potential, but the symbol μ is not used, in order to avoid confusion with the binary mixture chemical potential.) The scaled equation of state can be obtained from the minimum of $g^* = g/P_c v_c$, $(\partial g^*/\partial v^*)_{T^*} = 0$. By integrating the scaled equation of state at constant T^* and P^*, one gets, apart from an additive function of T^*,

$$g^* = P^* v^* - \frac{3}{v^*} - \frac{8}{3} T^* \ln\left(v^* - \frac{1}{3} \right). \tag{2.17}$$

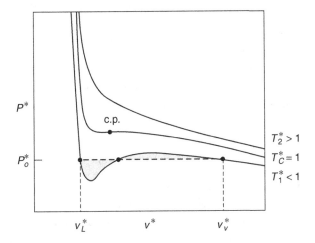

Fig. 2.3. Sketches of van der Waals isotherms and the Maxwell construction.

For $T < T_c$ ($T^* < 1$), in the region of a first-order transition, g^* must have two equal minima. Let these occur at v_V^* and v_L^*. Then at these minima, one has (partial derivatives taken at fixed P^* and T^*)

$$(\partial g^*/\partial v^*)|_{v_V^*} = (\partial g^*/\partial v^*)|_{v_L^*} = 0, \qquad (2.18)$$

and the equality of the minimum values implies

$$g^*(v_V^*) = g^*(v_L^*). \qquad (2.19)$$

The first of these two constraints arises from mechanical equilibrium and implies $P^*(v_V^*) = P^*(v_L^*)$. We denote this common value of the pressure by P_o^*. The second constraint, which arises from chemical equilibrium, can be rewritten as

$$g^*(v_V^*) - g^*(v_L^*) = P_o^*(v_V^* - v_L^*) - \int_{v_L^*}^{v_V^*} \left(\frac{8}{3}\frac{T^*}{v^* - \frac{1}{3}} - \frac{3}{v^{*2}}\right) dv^* = 0, \quad (2.20)$$

and is the prescription for the construction of Maxwell's equal area rule. Figure 2.3 shows this equal area construction. For a system in equilibrium, at a temperature less than T_c the isotherm is obtained by replacing the van der Waals loop with the constant-pressure line (horizontal dashed line in Fig. 2.3). The coexistence curve is the locus of the end points of such constant pressure lines obtained for temperatures less than T_c. For a system in thermodynamic equilibrium the isotherms in the (P^*, v^*) plane exhibit nonanalytic behavior at the two phase boundaries. Mean field equations of state such as the van der Waals equation are analytic through the two-phase region, since global minimization of the free energy is not incorporated in these models.

Phase separation is a nonequilibrium process. An appropriate mean field theory is often used as a mesoscopic model to study the dynamics of such processes. In phase segregation studies, the Maxwell construction is not necessary, and the nonequilibrium states within the coexistence curve are the central focus.

2.2 Order parameter

The work of Landau (Landau, 1937; Landau *et al.*, 1980) unified our understanding of phase transitions. The concept of an *order parameter* was introduced by Landau and, through its appearance in the pair of functions $(\phi, \partial f / \partial \phi)$, our understanding of the central role played by the free energy in the thermodynamics of phase transitions was enhanced.

The concepts illustrated by the examples discussed above can be extended to systems with a general order parameter. The three variables $(\phi, T, \partial f / \partial \phi)$ together form the relevant portion of the thermodynamic phase space of the system. The equation of state gives the functional interdependence among the three variables when the system is in equilibrium. The mean field form of the free energy $f(\phi, T)$ can be used to obtain an explicit expression for $(\partial f / \partial \phi)_T$, giving an approximate equation of state.

Figure 2.4 schematically shows the projections of the thermodynamic surface on the (T, ϕ) and $(T, \partial f / \partial \phi)$ planes. Figure 2.5 is a sketch of the $(\phi, \partial f / \partial \phi)$ projection. It has a van der Waals loop for $T < T_c$.

For equilibrium systems, thermodynamic stability requires that the thermodynamic force $\partial f / \partial \phi$ be a monotonically increasing function of the order parameter ϕ. We saw that mean field theories yield a van der Waals loop in the two-phase

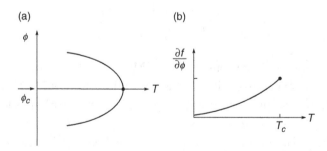

Fig. 2.4. Projections of the thermodynamic surface for a generic system with a scalar order parameter (a) on the (T, ϕ) plane and (b) on the $(T, \partial f / \partial \phi)$ plane. The heavy dots denote the critical point.

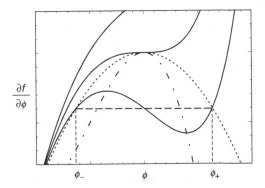

Fig. 2.5. Generic diagram showing thermodynamic force $\partial f / \partial \phi$ as a function of order parameter ϕ. The three solid lines are the mean field isotherms: the top curve is for $T > T_c$, the middle curve for $T = T_c$, and the bottom curve for $T < T_c$, which has a van der Waals loop. The dash-dot curve is the spinodal, which is the curve joining the extrema of the mean field isotherms for $T < T_c$. The Maxwell construction can be used to generate the equilibrium isotherm, which consists of two monotonic sections of the mean field isotherm connected by the horizontal dashed line. The curve joining the end points of the horizontal line is the coexistence curve (dotted line).

region of the phase diagram. The monotonicity condition is violated in the middle section of the van der Waals loop. Maxwell's construction replaces the loop in the mean field isotherm by a "constant thermodynamic force line" such that the area under the isotherm with the van der Waals loop is equal to the area under the isotherm with the horizontal constant thermodynamic force line (see Fig. 2.5). This construction yields the equilibrium isotherm. For $T < T_c$, the locus of the endpoints of the constant thermodynamic force line is the coexistence curve (dotted line in Fig. 2.5). The equilibrium system is inhomogeneous for $T < T_c$ and consists of two coexisting phases. The lever rule determines the volume fractions of the two coexisting phases: if ϕ_- and ϕ_+ are the order parameter values at the two end points of the constant thermodynamic force line, and the mean value of the order parameter is $\overline{\phi}$, then the equilibrium volume fraction of the $(+)$ phase is $(\overline{\phi} - \phi_-)/(\phi_+ - \phi_-)$, and that of the $(-)$ phase is $(\overline{\phi} - \phi_+)/(\phi_- - \phi_+)$.

For a homogeneous system in thermodynamic equilibrium, thermodynamic forces do not possess spatial gradients: thermal equilibrium implies a spatially uniform temperature, mechanical equilibrium requires a uniform pressure, and chemical equilibrium corresponds to a uniform chemical potential. A mean field isotherm augmented with the Maxwell construction leads to an isotherm that is appropriate to an inhomogeneous system in thermodynamic equilibrium: i.e. even though ϕ is an inhomogeneous function of space for a system with two coexisting phases, the thermodynamic force $\partial f / \partial \phi$ is constant over the entire system.

The locus of the extrema of the mean field isotherms for $T < T_c$ is the classical spinodal, which is shown in Fig. 2.5. All homogeneous states within the coexistence curve are *nonequilibrium* states. The states within the spinodal curve are thermodynamically unstable in a mean field phenomenological picture, and the states corresponding to points between the spinodal and coexistence curves are thermodynamically metastable.

In all of the examples discussed above, the system undergoes a continuous transition for a quench through the critical point, and a first-order transition when the system is quenched through any point on the coexistence curve other than the critical point. In both quenches, the system undergoes a reduction in some symmetry. The symmetry reduction is one of the features that is useful in identifying the order parameter, which is defined such that it vanishes in the higher-symmetry phase. For example, in the magnetic case the system is rotationally invariant in the paramagnetic phase above T_c, while below T_c spontaneous magnetization occurs and this rotational symmetry is lost. For the ferromagnetic phase the broken-symmetry variable is the magnetization, which is zero in the paramagnetic phase. A new excitation referred to as a Goldstone mode is associated with a spontaneously broken symmetry. For the ferromagnet, these excitations are spin waves. For the vapor to liquid transition, for which the order parameter can be identified as $(\rho_L - \rho_V)$, translational symmetry is broken in the direction of the gravitational field, and the associated Goldstone excitations are capillary waves at the liquid–vapor interface. For the first-order liquid to crystal transition, the continuous translational symmetry in the liquid gives way to a lower discrete lattice symmetry in the crystal. In this case the Goldstone modes are phonons (Forster, 1975; Mazenko, 2006). In each of the above examples, the order parameter was a real scalar variable. In general this need not be so, but for simplicity we continue the discussion of this simple case.

2.3 Order parameter and its spatial correlations

Near the critical point the order parameter ϕ is the important thermodynamic variable. The Landau expansion (Landau, 1937; Landau *et al.*, 1980) for the free energy (per unit volume), as described above for binary mixtures, takes the form

$$f(\phi^*, T) = f_c + \frac{a_2}{2}\phi^{*2} + \frac{a_4}{4}\phi^{*4} + \cdots, \tag{2.21}$$

where $a_2 = a_o(T - T_c)$, and a_o and a_4 are positive constants independent of temperature. This is an expansion of the free energy around the critical point, $\phi^* = \phi - \phi_c$. Often the zero level of energy is chosen so that $f_c = 0$. Since the order parameter is defined such that it is zero in the higher-symmetry phase, the expansion should be used in the lower-symmetry phase. The odd powers of ϕ^* are absent in Landau expansion on account of an assumed symmetry between ϕ^* and $-\phi^*$ as in a symmetric 50–50 binary mixture. This expansion yields

$$\frac{\partial f}{\partial \phi^*} = a_2 \phi^* + a_4 \phi^{*3} \tag{2.22}$$

for the thermodynamic force conjugate to the order parameter and

$$\frac{\partial^2 f}{\partial \phi^{*2}} = a_o (T - T_c) + 3a_4 \phi^{*2} \tag{2.23}$$

for the inverse susceptibility. At the critical point $T = T_c$, and the order parameter vanishes. As a consequence, due to the temperature dependence of a_2, the inverse susceptibility also vanishes. In the three examples considered earlier, the inverse susceptibilities that vanish are the following isothermal derivatives: $(\partial P/\partial \rho)_T$ for the vapor to liquid transition, $(\partial \mu/\partial c)_T$ for binary mixtures, and $(\partial H/\partial M)_T$ for the magnetic transition. At the liquid–vapor critical point, this leads to the divergence of the isothermal compressibility

$$\kappa_T = \frac{1}{\rho} \left(\frac{\partial \rho}{\partial P} \right)_T \tag{2.24}$$

and gives rise to the phenomenon of critical opalescence: the strong scattering of light that makes the fluid appear very turbid. The isothermal compressibility κ_T can also be related to density fluctuations in the system by using standard thermodynamic fluctuation theory (Pathria, 1972). If $n(\mathbf{r})$ is the microscopic local density at position $\mathbf{r}$ in a d-dimensional system with volume V, its average is the thermodynamic density,

$$\langle n \rangle = \rho, \tag{2.25}$$

where $\langle \ldots \rangle$ denotes a configurational average (Pathria, 1972), and

$$\kappa_T = \frac{V}{k_B T} \frac{\langle (\delta n)^2 \rangle}{\langle n \rangle^2} = \frac{V}{k_B T} \frac{\langle n^2 \rangle - \langle n \rangle^2}{\langle n \rangle^2}, \tag{2.26}$$

where $\delta n = n(\mathbf{r}) - \langle n \rangle$. The vanishing of the quadratic term in the Landau expansion at the critical point leads to the divergence of κ_T, which can now be seen to be connected to anomalous (diverging) density fluctuations in the system. In general, the divergence of the susceptibility is the consequence of the anomalous fluctuations of the order parameter at the critical point.

The quantity $n \, dV \equiv n \, d^d r$ is the number of particles present in $d^d r$ at a given instant. To represent the correlation between the densities of particles at two points in space, one can introduce

$$\langle \delta n_1 \, \delta n_2 \rangle = \langle n_1 \, n_2 \rangle - \langle n \rangle^2, \tag{2.27}$$

where the suffixes 1 and 2 denote values of local density at two points $\mathbf{r}_1$ and $\mathbf{r}_2$, and $\delta n = n - \langle n \rangle$. In a homogeneous isotropic medium, the correlation function depends only on the scalar distance $r_{12} = |\mathbf{r}_1 - \mathbf{r}_2|$. As this distance becomes large, the fluctuations at the two points become statistically independent and the correlation function tends to zero. We introduce the quantity $g(\mathbf{r}_{12})$ such that $(\langle n \rangle^2 g(\mathbf{r}_{12}) d^d r_1 d^d r_2)$ is the joint probability that there is a particle in $d^d r_1$ at $\mathbf{r}_1$ and a distinct particle in $d^d r_2$ at $\mathbf{r}_2$. This implies

$$\langle n_1 d^d r_1 \, n_2 d^d r_2 \rangle = \langle n \rangle^2 g(\mathbf{r}_{12}) \, d^d r_1 d^d r_2, \tag{2.28}$$

provided $\mathbf{r}_1 \neq \mathbf{r}_2$. In general,

$$\langle n_1 \, n_2 \rangle = \langle n \rangle^2 g(\mathbf{r}_{12}) + \langle n \rangle \delta(\mathbf{r}_1 - \mathbf{r}_2). \tag{2.29}$$

Since $g(\mathbf{r}_{12})$ approaches 1 as $\mathbf{r}_{12} \to \infty$, one has

$$\langle \delta n_1 \, \delta n_2 \rangle = \langle n \rangle^2 \, \Gamma(\mathbf{r}_{12}) + \langle n \rangle \delta(\mathbf{r}_1 - \mathbf{r}_2), \tag{2.30}$$

where

$$\Gamma(r) = g(\mathbf{r}) - 1. \tag{2.31}$$

The function $\Gamma(\mathbf{r})$ is the space-dependent density–density correlation function. Integrating Eq. (2.30) with respect to $d^d r_1$ and $d^d r_2$ over a finite volume V, one finds

$$V^2 \langle (\delta n)^2 \rangle = V \langle n \rangle \left(1 + \langle n \rangle \int d^d r \, \Gamma(\mathbf{r}) \right). \tag{2.32}$$

Comparing this result with Eq. (2.26), we obtain the connection between the isothermal compressibility and the density–density correlation function:

$$\kappa_T = \frac{1}{\rho k_B T} \left(1 + \langle n \rangle \int d^d r \, \Gamma(\mathbf{r}) \right). \tag{2.33}$$

It is clear from this relation that the divergence of κ_T can occur only if the correlation function $\Gamma(\mathbf{r})$ acquires an infinite correlation range at the critical point. The range of the correlation function is referred to as the correlation length ξ and is the length that dominates all other lengths in the critical region. Outside the critical region, for $r \gg \xi$, $\Gamma(\mathbf{r})$ behaves as

$$\Gamma(\mathbf{r}) \sim \frac{\exp(-|\mathbf{r}|/\xi)}{|\mathbf{r}|^{(d-1)/2} \, \xi^{(d-3)/2}}. \tag{2.34}$$

Near the critical point $\xi(T)$ diverges as $(T - T_c)^{-\nu}$, with $\nu = \frac{1}{2}$ in a mean field theory. A mean field theory leads to a power law behavior for the long-distance behavior of the correlation function near the critical point:

$$\Gamma(\mathbf{r}) \sim |\mathbf{r}|^{-(d-2+\eta)} \tag{2.35}$$

with $\eta = 0$. The exponents ν and η are examples of critical exponents. For additional information on critical phenomena see, for example, Goldenfeld (1992).

The relations in Eqs. (2.33)–(2.35) have analogs for the binary mixture phase transition and the magnetic (para-ferromagnetic) transition. More generally, one needs to consider fluctuations in the order parameter and introduce a space- and time-dependent order parameter density field $\phi(\mathbf{r}, t)$. The correlations of this field provide insight into the divergence of the generic susceptibility $(\partial^2 f/\partial \phi^2)_T^{-1}$ at the critical point.

From the above, we see that it is appropriate to introduce local quantities corresponding to the global thermodynamic and other variables. If the system is in equilibrium, local variables are useful for understanding the behavior of fluctuations in the system. If the system is inhomogeneous, the spatial dependence is manifest at the macroscopic level. If the system is in a nonequilibrium state, the global and local variables may also acquire time dependence. The free energy density $f(\phi, T)$ is a central quantity in the thermodynamic description of the system. Its extension to inhomogeneous systems and to nonequilibrium situations is considered in the next chapter.

3

Free energy functional

Self-organized and self-assembled structures are inhomogeneous, and many systems supporting such structures are in nonequilibrium states. In order to describe inhomogeneous and nonequilibrium systems, a generalization of the free energy function to a free energy functional is often useful. While this generalization is considered here, there are many nonequilibrium systems which cannot be described in terms of a free energy functional. A discussion of such more general situations is considered later in the book.

For inhomogeneous and nonequilibrium systems, it is natural to introduce the space- and time-dependent order parameter field, $\phi(\mathbf{r}, t)$. In a field theory description, the free energy $F(\phi, T)$ is generalized to a free energy functional $\mathcal{F}[\phi(\mathbf{r}, t)]$, where the dependence on other thermodynamic quantities which do not require a field description is suppressed. This formulation naturally leads to the extended thermodynamic force as the *functional* derivative of $\mathcal{F}$ with respect to ϕ.

The free energy functional $\mathcal{F}[\phi(\mathbf{r}, t)]$ has meaning only in terms of some coarse-grain or cellular approximation. The procedure for constructing the functional usually involves dividing the physical system into semi-macroscopic cells of fixed volume centered at positions $\mathbf{r}_i$. The order parameter takes on values ϕ_i in each of the cells. The partition function is written as a sum over the microscopic degrees of freedom, subject to constraints that keep the average order parameter in each cell i fixed at ϕ_i,

$$e^{-\mathcal{F}[\phi]/k_B T} = \sum{}' e^{-E/k_B T} \tag{3.1}$$

where E is the energy as a function of the microscopic variables, and the prime on the sum indicates that it is over the *constrained* microscopic variables. The order parameter field $\phi(\mathbf{r})$ has values ϕ_i at positions $\mathbf{r}_i$ and is allowed to vary smoothly only over distances comparable to the separation between cells. An equivalent way to obtain this result is to integrate out an appropriate set of short-wavelength Fourier components of the microscopic variables as in renormalization group methods (Wilson and Kogut, 1974; Wilson, 1975; Goldenfeld, 1992).

For the continuum description to make sense, the cells should be large compared with the volume per molecule (microscopic volume) and small enough that the material within each cell can rapidly reach local equilibrium. This implies that the linear dimension of each cell should be of the order of the correlation length.

In practice, such a coarse-graining procedure is rarely performed explicitly, and one simply assumes that a functional $\mathcal{F}[\phi]$ exists for a suitable choice of the coarse grain size and is physically plausible on the space of smooth functions $\phi(\mathbf{r})$.

3.1 Ginzburg–Landau–Wilson free energy functional

The Ginzburg–Landau–Wilson free energy functional is

$$\mathcal{F}_{GLW}[\phi(\mathbf{r}, t)] = \int d^d r \left[f(\phi) + \frac{\kappa}{2} (\nabla \phi)^2 \right], \tag{3.2}$$

where $f(\phi)$ is the local free energy density, which reduces to the thermodynamic free energy density for a uniform equilibrium system. The typical functional form of $f(\phi)$ was shown in Fig. 2.2. The gradient energy $\frac{\kappa}{2}(\nabla \phi)^2$ arises from the inhomogeneity of ϕ and is associated with short-range interactions. It accounts for the presence of interfaces within an equilibrium inhomogeneous system.

The dimension of $\mathcal{F}_{GLW}[\phi(\mathbf{r}, t)]$ is the same as that of the system's Helmholtz free energy F introduced in Section 2.1.1. For a homogeneous system, $\int d^d r \; f(\phi)$ integrated over the system volume V is the same as F. Thus the local free energy density $f(\phi)$ has the same dimensions as that of F/V.

The form of the free energy functional given in Eq. (3.2) is motivated as follows. In regions where the order parameter is nonuniform, the free energy density is expected to depend on its local value at a point in the system and on the values it takes in the immediate neighborhood of this point. Thus, it is appropriate to take ϕ and its spatial derivatives $\nabla \phi, \nabla^2 \phi, \ldots$ as independent variables. The free energy density f^* is assumed to be a continuous function of these variables. A Taylor expansion around the uniform state yields

$$\mathcal{F}[\phi(\mathbf{r}, t)] = \int d^d r \; f^*(\phi, \nabla \phi, \nabla^2 \phi, \ldots), \tag{3.3}$$

with

$$f^*(\phi, \nabla \phi, \nabla^2 \phi, \ldots) = f(\phi) + \sum_i L_i \frac{\partial \phi}{\partial r_i} \tag{3.4}$$

$$+ \frac{1}{2} \sum_{ij} \left[2\kappa_{ij}^{(1)} \frac{\partial^2 \phi}{\partial r_i \partial r_j} + \kappa_{ij}^{(2)} \frac{\partial \phi}{\partial r_i} \frac{\partial \phi}{\partial r_j} \right] + \cdots,$$

where $L_i = [\partial f^*/\partial(\partial\phi/\partial r_i)]_o, \kappa_{ij}^{(1)} = [\partial f^*/\partial(\partial^2\phi/\partial r_i \partial r_j)]_o$, and $\kappa_{ij}^{(2)} = [\partial^2 f^*/\partial(\partial\phi/\partial r_i)\partial(\partial\phi/\partial r_j)]_o$. For a d-dimensional system, subscripts i, j take values $1, 2, \ldots, d$.

In general, $\kappa_{ij}^{(1)}$ and $\kappa_{ij}^{(2)}$ are tensors reflecting the symmetry of the system. The L_i are polarization vector components if the system has a special symmetry direction, such as an electric polarization or a magnetization vector. For a cubic crystal or an isotropic medium (we restrict ourselves to these simple cases), the free energy is invariant under reflections ($r_i \to -r_i$) and under rotation about a fourfold axis ($r_i \to r_j$). These symmetry-related restrictions imply that L_i vanishes, $\kappa_{ij}^{(1)} = \kappa_1 \delta_{ij}$, and $\kappa_{ij}^{(2)} = \kappa_2 \delta_{ij}$, with $\kappa_1 = [\partial f^*/\partial\nabla^2\phi]_o$ and $\kappa_2 = [\partial^2 f^*/(\partial|\nabla\phi|)^2]_o$. Hence, for an isotropic medium or for a cubic lattice, one has

$$f^*(\phi, \nabla\phi, \nabla^2\phi, \ldots) = f(\phi) + \frac{1}{2}\kappa_2(\nabla\phi)^2 + \kappa_1\nabla^2\phi + \ldots. \quad (3.5)$$

The coefficients $\kappa_1, \kappa_2, \ldots$ can in general be ϕ-dependent. The term proportional to κ_1 can be further reduced by applying the divergence theorem over the entire system volume and choosing the external boundary such that the normal component of $\nabla\phi$ vanishes at every point on this boundary surface. This yields the Ginzburg–Landau–Wilson free energy functional given in Eq. (3.2) with $\kappa = \kappa_2 - 2\frac{d\kappa_1}{d\phi}$. The expression for κ is appropriate for small gradients. More generally, one can include the effects of all higher-order gradients of f^* by defining $\frac{\kappa}{2}(\nabla\phi)^2$ to be the difference $f^* - f$ (Cahn and Hilliard, 1958, 1959). Thus, the phenomenological form of the free energy functional given in Eq. (3.2) has a broad applicability to systems with short-range interactions.

3.2 Interfacial tension and the coefficient κ

The coefficient κ of the square gradient term in the Ginzburg–Landau–Wilson free energy functional is related to the interfacial tension. From a thermodynamic point of view, the presence of an interface introduces an additional surface contribution to the free energy. Consider an equilibrium system composed of two coexisting phases (denoted by subscripts $+$ and $-$) separated by a planar interface, whose normal is in the z direction. The free energy density of such a system is

$$f^* = f(\phi) + \frac{\kappa}{2}\left(\frac{d\phi(z)}{dz}\right)^2. \quad (3.6)$$

The interfacial tension $\tilde{\sigma}$ is the surface excess free energy per unit area when the reference surface (at $z = 0$) is chosen to make the surface excess 'mass' vanish: i.e.

$$\int_{-\infty}^{0} dz(\phi(z) - \phi_+) + \int_{0}^{\infty} dz(\phi(z) - \phi_-) = 0. \quad (3.7)$$

Then

$$\tilde{\sigma} = \int_{-\infty}^{\infty} dz \left[\Delta f + \frac{\kappa}{2} \left(\frac{d\phi(z)}{dz} \right)^2 \right], \tag{3.8}$$

where

$$\Delta f = \begin{cases} (f(\phi) - f_+), & \text{for } -\infty < z < 0, \\ (f(\phi) - f_-), & \text{for } 0 < z < \infty. \end{cases} \tag{3.9}$$

The equilibrium order parameter profile minimizes the excess surface free energy in Eq. (3.8). This leads to the condition that

$$\kappa \frac{d^2\phi(z)}{dz^2} - \frac{d\Delta f}{d\phi} = 0. \tag{3.10}$$

After a multiplication by $(d\phi/dz)$, this equation can be integrated to yield

$$\Delta f(\phi) = \frac{\kappa}{2} \left(\frac{d\phi(z)}{dz} \right)^2. \tag{3.11}$$

Using this result we may write the surface tension as

$$\tilde{\sigma} = \kappa \int_{-\infty}^{\infty} dz \left(\frac{d\phi(z)}{dz} \right)^2. \tag{3.12}$$

Since the creation of a stable surface requires energy, both $\tilde{\sigma}$ and κ must be positive for stability. Equation (3.12) is the mean field approximation for the surface tension. A detailed and exact treatment is given in Rowlinson and Widom (1982).

3.3 Landau expansion of the local free energy density

The local free energy density $f(\phi)$ can be thought of as a straightforward generaliza-tion of the thermodynamic free energy function in which the global thermodynamic variable ϕ is replaced by its local field value $\phi(\mathbf{r}, t)$. Many universal features of kinetics are insensitive to the detailed shape of $f(\phi)$. Following Landau (Landau, 1937; Landau *et al.*, 1980), one often uses a form for the free energy that is obtained by expanding $f(\phi)$ around the value of ϕ at the critical point, ϕ_c. If the mean value of ϕ is $\overline{\phi}$, then the order parameter fluctuation is

$$\delta\phi \equiv (\phi - \overline{\phi}) = (\phi - \phi_c) - (\overline{\phi} - \phi_c) \equiv \phi^* - \phi_o, \tag{3.13}$$

with $\phi^* = (\phi - \phi_c)$ and $\phi_o = (\overline{\phi} - \phi_c)$. The Landau expansion is written in terms of ϕ^* as

$$f(\phi^*) = f_c + \frac{1}{2} a_2 \phi^{*2} + \frac{1}{4} a_4 \phi^{*4} - \mathcal{H}\phi^*, \tag{3.14}$$

where $a_2 = a_o(T - T_c)$ and a_4 is a temperature-independent constant. The last term in Eq. (3.14) accounts for an external field $\mathcal{H}$ that couples linearly to ϕ^*. The coefficient a_2 is positive for temperatures above T_c, and negative below. The parameters a_o and a_4 are assumed to be positive constants. When appropriate, we shall set the external field to zero and choose the zero of energy to be f_c. This corresponds to a symmetric expression for Landau free energy density,

$$f(\phi^*) = \frac{1}{2}a_2\phi^{*2} + \frac{1}{4}a_4\phi^{*4}, \tag{3.15}$$

which leads to a ϕ-dependence analogous to that sketched in Fig. 2.2. At the critical point $f = f_c$, $T = T_c$, $\phi = \phi_c$, and $\mathcal{H} = 0$. For a symmetric system, $\overline{\phi} = \phi_c$ and $\delta\phi = \phi^*$. In the absence of $\mathcal{H}$, $f(\phi)$ has a single minimum for temperatures above T_c at $\phi^* = 0$, and two minima for temperatures below T_c at

$$\phi^* = \pm\phi^*_{min} = \pm[-a_2/a_4]^{1/2} \equiv \pm[a_o(T_c - T)/a_4]^{1/2}, \tag{3.16}$$

corresponding to the two coexisting ordered phases in equilibrium (see Fig. 2.2). At the two minima, the Landau free energy density is $f(\pm\phi^*_{min}) = -a_2^2/(4a_4)$.

In later chapters we shall often find it convenient to use a dimensionless Landau free energy density. Let us denote dimensional free energy density by $\tilde{f}(\phi^*)$. Using the dimensionless quantities $\psi = \phi^*/\phi^*_{min}$ and $f(\psi) = a_4\,\tilde{f}(\phi^*)/a_2^2$, the dimensionless Landau free energy density acquires the form $f(\psi) = \psi^4/4 - \psi^2/2$.

The quantities denoted as ϕ_- and ϕ_+ in Section 2.2 are related to $\pm\phi^*_{min}$ by $-\phi^*_{min} = \phi_- - \phi_c$ and $+\phi^*_{min} = \phi_+ - \phi_c$. Since $\phi_o = (\overline{\phi} - \phi_c)$ and $\phi_c = (\phi_- + \phi_+)/2$, the volume fraction of the $(-)$ phase is $(\phi^*_{min} - \phi_o)/(2\phi^*_{min})$ and that of the $(+)$ phase is $(\phi^*_{min} + \phi_o)/(2\phi^*_{min})$.

Comparison of the quadratic term $\frac{1}{2}a_2\phi^{*2}$ with the square gradient term $\frac{\kappa}{2}(\nabla\phi)^2$ allows one to identify an important characteristic length inherent in the model free energy functional: the correlation length

$$\xi = \sqrt{\kappa/|a_2|}. \tag{3.17}$$

At the critical point a_2 vanishes and ξ diverges. The correlation length is formally defined as the correlation range of the order parameter auto-correlation function $\Gamma(\mathbf{r})$, which was explicitly discussed earlier for the liquid–vapor critical point. The divergence of ξ in the neighborhood of the critical point leads to the divergence of susceptibility as well as singularities in other physical quantities.

4

Phase separation kinetics

Growth of order from disorder is a natural phenomenon which is seen in a variety of systems. An important class of such phenomena involves the kinetics of phase ordering and phase separation. The examples of such growth processes that were described in Chapter 2 had common characteristics. Now, we discuss the experimental results that point to common features of the kinetics of phase separation processes. A combination of techniques from nonequilibrium statistical mechanics and nonlinear dynamics is used to study the formation and evolution of spatial structures. Substantial progress in our understanding of the kinetics of domain growth during a first-order phase transition has been made over the past few decades. The knowledge gained in these studies forms the underpinning of the descriptions of many such processes which create order from disorder.

4.1 Kinetics of phase ordering and phase separation

Phase separation is usually initiated by a rapid change or quench in a thermodynamic variable (often temperature and sometimes pressure), which places a disordered system in a post-quench initial nonequilibrium state. The system then evolves towards an inhomogeneous ordered state of coexisting phases, which is its final equilibrium state. Depending on the nature of the quench, the post-quench state may be either thermodynamically unstable or metastable (see Fig. 2.1). In the former case, the onset of separation is spontaneous, and the kinetics that follows is known as *spinodal decomposition*. For the metastable case, nonlinear fluctuations are needed to initiate the separation process. The system is said to undergo phase separation through *homogeneous nucleation* if the system is pure. Phase separation occurs by *heterogeneous nucleation* if the system has impurities or surfaces which initiate nucleation events.

In both cases the late stages of the kinetics show power law domain growth. The power law depends on the nature of the fluctuating variables which drive the phase separation process: i.e. on the characteristics of the order parameter. For phase separation in a binary mixture, the order parameter $\phi(\mathbf{r}, t)$ is the relative concentration of one of the two species, and its fluctuation around the mean value c_o is $\delta c(\mathbf{r}, t) = c(\mathbf{r}, t) - c_o$. In the disordered phase the concentration is homogeneous, and the order parameter fluctuations are microscopic. In the ordered phase, the inhomogeneity created by two coexisting phases leads to a macroscopic spatial variation in the order parameter field near the interfacial region.

Depending on the system and the nature of the phase transition, the order parameter may be scalar, vector, or complex, and may be conserved or nonconserved. Here we consider nonconserved and conserved scalar order parameters. Binary mixture phase separation is an example of the latter case, while the former occurs in a variety of systems including some order–disorder transitions and antiferromagnets. For the para-ferromagnetic transition in magnets the magnetization is a conserved quantity in the absence of an external magnetic field but is nonconserved when such a field is present.

For a one-component fluid, the vapor–liquid transition is characterized by density fluctuations. For this transition the number density ρ is the order parameter, and it is conserved. The density–density correlation function $\Gamma(\mathbf{r})$ was introduced in Eq. (2.30). Its Fourier transform is the equilibrium structure factor,

$$S(\mathbf{k}) = 1 + \rho \int d^d r \, e^{-i\mathbf{k}\cdot\mathbf{r}} \Gamma(\mathbf{r}) \equiv \rho \int d^d r \, e^{-i\mathbf{k}\cdot\mathbf{r}} G(\mathbf{r}), \qquad (4.1)$$

where

$$G(\mathbf{r}) = \rho^{-2}\langle \delta n(\mathbf{r})\delta n(0)\rangle = \Gamma(\mathbf{r}) + \rho^{-1}\delta(\mathbf{r}). \qquad (4.2)$$

The functions $G(\mathbf{r})$ and $S(\mathbf{k})$ can also be defined for binary mixtures and magnetic systems. In many instances, $S(\mathbf{k})$ may be measured in appropriate elastic scattering experiments. Figure 4.1 reiterates the thermodynamic aspects of phase separation discussed in Chapter 2, and elucidates the role of density fluctuations in the vapor $\rightarrow$ liquid phase transition by either nucleation or spinodal decomposition.

In a quench experiment which monitors the kinetics of the phase transition, the spatial structure of the system evolves in time. As a result, the equal-time correlation function of the order parameter fluctuations,

$$G(\mathbf{r}, t) \equiv \langle \delta\phi(\mathbf{r}, t)\delta\phi(0, t)\rangle_{ne}, \qquad (4.3)$$

which would be time independent in equilibrium, acquires time dependence associated with the growth of order in the nonequilibrium system. The spatial Fourier

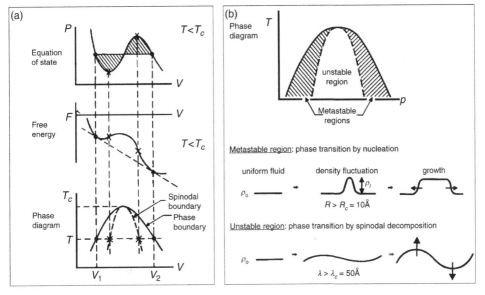

Fig. 4.1. Vapor–liquid phase transition. Panel (a) shows how one can obtain the free energy and the coexistence and spinodal curves of the phase diagram for $T < T_c$ from an equation of state with a van der Waals loop. The homogeneous states within the coexistence curves are nonequilibrium states, and the spinodal curve is the result of a mean field approximation. Panel (b) depicts the unstable and metastable regions for a homogeneous density state. The lower part of this panel shows the types of density fluctuation that lead to nucleation in the metastable region and to spinodal decomposition in the unstable phase. The spinodal boundary between the two regions is sharp only in the mean field approximation. Reprinted from Abraham (1979). Copyright 1979, with permission from Elsevier.

transform of $G(\mathbf{r}, t)$,

$$S(\mathbf{k}, t) = \int d^d r \, e^{-i\mathbf{k}\cdot\mathbf{r}} \, G(\mathbf{r}, t), \qquad (4.4)$$

is called the time-dependent structure factor.

In some experiments, the system morphology can be observed directly. An early example is the work by Oki et al. (1977) and Sagane et al. (1977) on iron-rich iron–aluminum alloys, where electron microscopy was used to unfold the time evolution of the morphology of ordered and disordered domains in Fe_3Al. The results are reproduced in Fig. 4.2. In this figure time increases from top to bottom. The central set of panels corresponds to a near-critical quench and the two side panels to off-critical quenches. For the near-critical quench, after an initial transient period, well-defined domains with a "stringy" complex morphology form. As time evolves this structure coarsens. In contrast, in the off-critical quenches the phase segregation

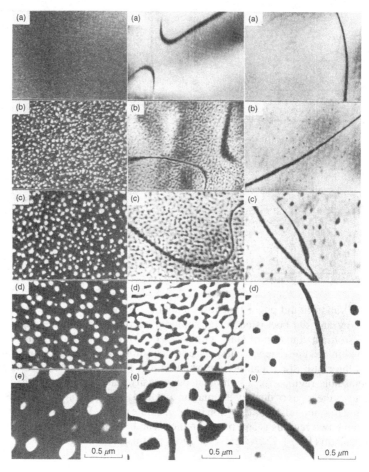

Fig. 4.2. Coarsening of domain structure following quenches from the disordered phase region at $T = 630°C$ to the region of two coexisting phases at $T = 570°C$. Two concentrations corresponding to off-critical paths (left and right panels) and a concentration corresponding to a near-critical path (center set of panels) are shown. Reprinted photo 1 with permission Oki *et al.* (1977).

patterns take the form of droplets. As time evolves, the coarsening consists of the growth of large droplets at the expense of small droplets.

The evolution of a system following the quench comprises different stages. The early stage involves the emergence of macroscopic domains from the initial post-quench state. It is characterized by the formation of interfaces (domain walls) separating regions of space within which the system approaches one of its final coexisting states (domains). Late stages are dominated by the motion of these interfaces as the system acts to minimize its surface free energy. During this stage the mean size of the domains grows with time while the total amount of interface decreases.

4.2 Dynamical scaling

The late-stage domain growth kinetics exhibits *dynamical scaling*, which arises when a single characteristic length dominates the time evolution. The structure of the space-dependent order parameter at a given time instant is referred to as the system's morphology at that time. Various measures of the morphology depend on time only through the characteristic length. The evolution of the system then acquires self-similarity in the sense that the spatial patterns formed by the domains at two different times are statistically identical apart from a global change of length scale. Dynamical scaling has been observed in binary alloys (Gaulin *et al.*, 1987), binary fluids (Wong and Knobler, 1978; Chou and Goldburg, 1981), glasses (Craievich and Sanchez, 1981), and polymer blends (Hashimoto *et al.*, 1986a, 1986b).

The time-dependent structure factor $S(\mathbf{k}, t)$, which is proportional to the intensity $I(\mathbf{k}, t)$ measured in an elastic scattering experiment (Wong and Knobler, 1978; Chou and Goldburg, 1979), is a measure of the strength of spatial correlations in a phase-ordering system with wavenumber k at time t. It exhibits a peak whose position is inversely proportional to the average domain size. As the system phase separates (orders), the peak moves towards increasingly smaller wavenumbers (see Fig. 4.3).

A signature of dynamical scaling is the collapse of the experimental data to a scaled form. For a d-dimensional system we have

$$S(k, t) = \left(R(t)\right)^d S_o\left(kR(t)\right), \tag{4.5}$$

where S_o is a universal function and $R(t)$ is a characteristic length, such as the average domain size (see Fig. 4.4). Other lengths in the system, such as the interfacial width, may play an important role in the kinetics; dynamical scaling may be valid only asymptotically at very late times. The power law growth, $R(t) \sim t^n$, is also an asymptotic relation, and extraction of the exponent n from experimental data requires care. For systems which can be described using a scalar nonconserved order parameter the growth exponent is $n = \frac{1}{2}$, while for systems with a scalar

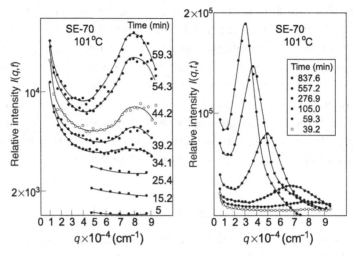

Fig. 4.3. Time-dependent structure factor as measured in light-scattering experiments on a phase-separating mixture of polystyrene ($M = 1.5 \times 10^5$) and poly(vinylmethylether) ($M = 4.6 \times 10^4$), following a fast quench from a homogeneous state to $T = 101°C$ located in the two-phase region. The time in minutes following the quench is indicated for each structure factor curve. Reprinted with permission from Hashimoto *et al.* (1983). Copyright 1983, American Chemical Society.

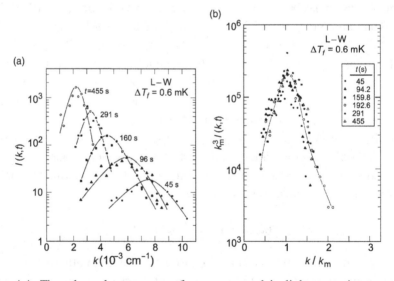

Fig. 4.4. Time-dependent structure factor measured in light-scattering experiments on a phase-separating mixture of 2,6-lutidine and water. The results of a fast quench from a homogeneous state through the critical point to a temperature 0.6 mK below the critical temperature are shown. The time (in seconds) following the quench is indicated for each structure factor curve. The collapse of data in panel (b) indicates dynamic scaling. Reprinted Figures 2 and 4 with permission from Chou and Goldburg (1981). Copyright 1981 by the American Physical Society.

conserved order parameter it is $\frac{1}{3}$. The growth exponent can also be obtained from the inverse perimeter density $P(t)$, defined as

$$P(t) = (1 - \psi_o)^{1/2} \frac{V}{V_p},\tag{4.6}$$

where V_p is the volume covered by interfaces in a system with volume V and $\psi_o = \phi_o/\phi^*_{min}$ (see Section 3.3). One expects $P \sim P_o t^n$. Dynamic scaling and power law growth together also imply that

$$G(r,t) = (1 - \phi_o^2) G_o(rt^{-n}).\tag{4.7}$$

During domain coarsening following a quench, the pattern formation process can be experimentally observed using phase-contrast microscopy and quantified through a digital image analysis (Tanaka *et al.*, 1990; Tanaka, 1994). An important characteristic of the late stages of phase separation kinetics for asymmetric mixtures is the cluster size distribution function of the minority phase clusters. The number of clusters of the minority phase per unit volume with radii between R and $R + dR$ is $n(R, \tau)dR$. The zeroth moment of this distribution gives the mean number of clusters at time τ, and the first moment is proportional to the mean cluster size.

If the shape of the cluster evolves with time, one can analyze the dynamics of the deviations from a symmetric shape. Domain growth is significantly altered if the system also has long-range interactions and supercrystal periodic equilibrium states. Systems with such long-range interactions are discussed in detail in Chapters 13, 14, and 15.

5

Langevin model for nonconserved order parameter systems

Although the microscopic Hamiltonian contains all of the information needed to describe phase separation kinetics, in practice the large number of degrees of freedom in the system makes it necessary to construct reduced descriptions. Generally a subset of slowly varying macrovariables, such as the hydrodynamic modes, is a useful starting point for theoretical models. The equations of motion of the macrovariables can be derived from the microscopic Hamiltonian, but in practice one often begins with a phenomenological description. The set of macrovariables is chosen to include the order parameter and all other slow variables to which it is coupled. Such slow variables are typically obtained from consideration of the conservation laws and broken symmetries of the system. The remaining degrees of freedom are assumed to vary on a much faster time scale and enter the phenomenological description as random thermal noise. The resulting coupled nonlinear stochastic differential equations for such a chosen "relevant" set of macrovariables are collectively referred to as the Langevin field theory description. In two of the simplest Langevin models, the order parameter ϕ is the only relevant macrovariable; in model A (introduced in this chapter) it is nonconserved and in model B (described in the next chapter) it is conserved. The labels A, B, etc. have an historical origin from the Langevin models of critical dynamics. The scheme is often referred to as the Hohenberg–Halperin classification scheme (Hohenberg and Halperin, 1977).

5.1 Langevin model A

For model A the Langevin description assumes that, on average, the time rate of change of the order parameter is proportional to the thermodynamic force that drives the phase transition. For this single-variable case, the thermodynamic force is canonically conjugate to the order parameter; if ϕ is a state variable, then its canonically conjugate force is $\partial f / \partial \phi$ (see Fig. 2.5), where f is the local free energy density.

In a field theory description, the thermodynamic free energy F is generalized to a free energy functional $\mathcal{F}[\phi(\mathbf{r}, t)]$, leading to the thermodynamic force as the analogous functional derivative. The Langevin equation for model A is then

$$\frac{\partial \phi}{\partial t} = -M \frac{\delta \mathcal{F}}{\delta \phi} + \eta^*(\mathbf{r}, t), \tag{5.1}$$

where the proportionality coefficient M is the mobility coefficient, which is related to the random thermal noise η^* through the fluctuation–dissipation relation,

$$\langle \eta^*(\mathbf{r}, t) \eta^*(\mathbf{r}', t') \rangle = 2k_B T M \delta(\mathbf{r} - \mathbf{r}') \delta(t - t'). \tag{5.2}$$

The Langevin equation for model A is often referred to as the time-dependent Ginzburg–Landau (TDGL) equation. Its name derives from its use in the analysis of superconductivity and superfluidity, for which the order parameter is complex.

The free energy functional $\mathcal{F}$ plays a crucial role in the kinetics of model A. Its functional form depends on the nature of the system under investigation. We first consider systems with only short-range interactions, where $\mathcal{F} = \mathcal{F}_{GLW}$, and use the form given in Eqs. (3.2) and (3.15) with the external field $\mathcal{H}$ set to zero. Systems with nonzero $\mathcal{H}$ are treated in Chapter 13. The Landau expansion of $f(\phi)$ given in Eqs. (3.14) and (3.15) is an expansion around the critical point: i.e. in $\phi^* \equiv (\phi - \phi_c)$. For critical quenches, the average order parameter is ϕ_c, and ϕ^* is also the order parameter fluctuation $\delta\phi$. For off-critical quenches, the order parameter fluctuation is $\delta\phi = (\phi - \overline{\phi}) = (\phi^* - \phi_o)$, where $\phi_o = (\overline{\phi} - \phi_c)$. Thus, for off-critical quenches, one needs to replace ϕ^* by $(\delta\phi + \phi_o)$ in the Landau expansion. The Langevin equations describe the dynamics of the order parameter fluctuation $\delta\phi$. Using the GLW form of free energy functional, Eq. (5.1) becomes

$$\frac{\partial \delta\phi}{\partial t} = -M[(a_2 - \kappa \nabla^2)\phi^* + a_4 \phi^{*3}] + \eta^*(\mathbf{r}, t), \tag{5.3}$$

which defines the kinetics of model A for time-independent $\overline{\phi}$. (See Chapter 13 for an example where $\overline{\phi}$ is time-dependent.) The quadratic coefficient a_2 in the Landau expansion of the local free energy density has a temperature dependence $a_2 = -a_o(T_c - T)$ such that a_2 is negative for $T < T_c$.

The Langevin equation for model A may be rewritten in terms of scaled (dimensionless) variables. For critical quenches $\phi^* = \delta\phi$. The various scaling quantities were introduced in Section 3.3 and may be expressed in terms of the square gradient coefficient κ that appears in the Ginzburg term and the quadratic and quartic coefficients a_2 and a_4 in the Landau expansion of the local free energy density. They are the correlation length $\xi = (-a_2/\kappa)^{-1/2}$, the magnitude of the order parameter $\phi^*_{min} = (-a_2/a_4)^{1/2}$ for which the double-well local free energy density is minimum, and the mobility coefficient M. The scaled variables are $\psi(\mathbf{x}, \tau) = \delta\phi/\phi^*_{min}$, $\mathbf{x} = \mathbf{r}/\xi$, and the dimensionless time τ, which is defined as $\tau = 2M|a_2|t$. In terms of these variables, the model A Langevin equation for critical quenches in absence of an external field reduces to a dimensionless form

$$\frac{\partial\psi(\mathbf{x}, \tau)}{\partial\tau} = \frac{1}{2}(\nabla^2_\mathbf{x}\psi + \psi - \psi^3) + \epsilon^{1/2}\eta(\mathbf{x}, \tau), \tag{5.4}$$

where the fluctuation–dissipation relation is

$$\langle\eta(\mathbf{x}, \tau)\eta(\mathbf{x}', \tau')\rangle = \delta(\mathbf{x} - \mathbf{x}')\delta(\tau - \tau'), \tag{5.5}$$

and the strength of the thermal noise ϵ is

$$\epsilon = \frac{k_BT\,a_4}{a_0^2(T_c - T)^2}\left(\frac{a_0|T_c - T|}{\kappa}\right)^{d/2} = \frac{k_BT}{\phi^{*2}_{min}a_0(T_c - T)}\left(\frac{a_0|T_c - T|}{\kappa}\right)^{d/2}. \tag{5.6}$$

The transformation used to obtain Eq. (5.4) is useful for kinetics following *deep* quenches in the unstable regions. The dimensionless noise strength ϵ and the correlation length ξ diverge at the critical point (for $d < 4$). Thus, the scaling transformation is not so useful for shallow quenches close to the critical point.

5.2 Model A reaction–diffusion system

Some chemical reactions carried out under far-from-equilibrium conditions can be described by the model A equation. An often-studied model of a chemical system exhibiting bistability is the Schlögl model (Schlögl, 1972). It is defined by the reaction scheme

$$A \underset{k_{-1}}{\overset{k_1}{\rightleftharpoons}} X, \quad 2X + B \underset{k_{-2}}{\overset{k_2}{\rightleftharpoons}} 3X. \tag{5.7}$$

In thermodynamic equilibrium,

$$k_1\,c_A^{eq} = k_{-1}\,c_X^{eq}, \\ k_2\,(c_X^{eq})^2\,c_B^{eq} = k_{-2}\,(c_X^{eq})^3, \tag{5.8}$$

where c_A^{eq}, c_B^{eq} and c_X^{eq} are the equilibrium densities of A, B, and X, respectively. The ratio of c_A^{eq} and c_B^{eq} is related to the ratio of the rate constants by detailed balance:

$$K_{eq} = c_A^{eq}/c_B^{eq} = (k_{-1}k_2)/(k_1k_{-2}). \tag{5.9}$$

If the concentrations of species A and B are held at constant values c_A^0 and c_B^0 by flows of reagents into and out of the system, the system is in a nonequilibrium state.

Only the concentration of species X varies with time and space. We let $c_X \equiv c$, $c_A^0 \equiv a$, and $c_B^0 \equiv b$. The time evolution of $c(\mathbf{r}, t)$ satisfies the reaction–diffusion equation

$$\frac{\partial c(\mathbf{r}, t)}{\partial t} = k_1 a - k_{-1} c + k_2 b c^2 - k_{-2} c^3 + D \nabla^2 c. \tag{5.10}$$

If we define the free energy functional

$$\mathcal{F}[c(\mathbf{r}, t)] = \int d^3 r \left\{ f[c(\mathbf{r}, t)] + \frac{1}{2} D |\nabla c|^2 \right\}, \tag{5.11}$$

where

$$f[c] = -k_1 a c + \frac{k_{-1}}{2} c^2 - \frac{k_2 b}{3} c^3 + \frac{k_{-2}}{4} c^4, \tag{5.12}$$

then the reaction–diffusion equation may be rewritten as

$$\frac{\partial c(\mathbf{r}, t)}{\partial t} = -\frac{\delta \mathcal{F}[c(\mathbf{r}, t)]}{\delta c(\mathbf{r}, t)}. \tag{5.13}$$

The local free energy functional $f[c]$ can be transformed to the form in Eq. (3.14) by a suitable choice of c_o and defining $c = c_o + \delta c$. The resulting one-variable reaction–diffusion equation has the same form as the time-dependent Ginzburg–Landau equation without the noise term. The thermal noise term can be added to account for fluctuations. (Compare Eq. (5.1) with Eq. (5.13) and Eq. (5.4) with Eq. (5.10).) Consequently, the domain-coarsening phenomena for a system with a nonconserved order parameter can be observed in this reacting system. In general it is not always possible to write a reaction–diffusion system in the gradient form of model A involving a free energy functional.

The homogeneous steady states of Eq. (5.10) occur when $k_1 a - k_{-1} c + k_2 b c^2 - k_{-2} c^3 = 0$. This cubic equation is similar to the van der Waals equation (see Section 2.1.3, Eqs. (2.15) and (2.16)). If the ratio a/b is held fixed, the three steady-state values of c, $(c_o, c_\pm)$, need not correspond to a state of thermodynamic equilibrium. As in the case of the van der Waals equation, it can be shown that the two states $c_\pm$ are stable and the state c_o is unstable. The Schlögl model has been studied often. Its stochastic dynamics has been modeled using reactive lattice-gas, master, Fokker–Planck and Langevin equations. In the presence of fluctuations the bistable states are metastable, since noise can induce transitions between them. The lifetimes of such metastable states can be extremely long (Langer, 1969; Gunton and Droz, 1983; Debenedetti, 1996). In simulations on finite systems fluctuation-induced transitions occur only rarely.

Figure 5.1 shows domain formation and growth for this system following a critical quench starting from an unstable state (see Gruner *et al.*, 1993). The long-time dynamics is driven by the curvature of the boundaries separating the stable

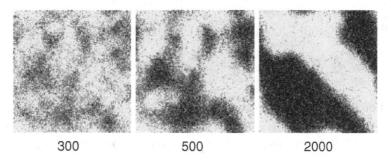

300   500   2000

Fig. 5.1. Phase separation dynamics of a critically quenched Schlögl system. The times of the configurations are indicated below the panels.

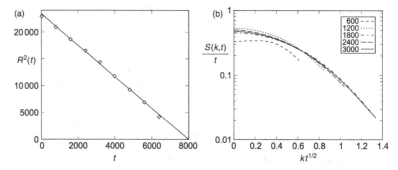

Fig. 5.2. Results for a system at equistability: (a) square of the domain size as a function of time. The system was prepared with a disk of radius $R(0) = 150$ of the high-density steady state in a sea of low-density steady state. (b) Dynamic structure factor at different times. The curves are the average of many realizations of a critically quenched system with an initial density corresponding to the deterministic unstable state, c_o. Reprinted with permission from Gruner *et al.* (1993). Copyright 1993, American Institute of Physics.

phases. In an infinite system, although the average order parameter $(c - c_o)$ is zero, domains of arbitrarily large size can exist. In finite systems, different realizations of the evolution process lead to pure c_+ or c_- phases, or mixtures of these phases separated by zero-curvature interfaces. The order parameter is zero when averaged over many realizations. At early times the boundaries are indistinct, while at later times sharp boundaries form and evolve slowly due to diffusive motion of the interfaces.

The evolution of domains may be monitored by the nonequilibrium correlation function $\langle \delta c(\mathbf{r},t)\delta c(0,t)\rangle_{ne} \equiv G(\mathbf{r},t)$ and its Fourier transform $S(\mathbf{k},t)$. If one assumes that the domain size $R(t)$ is the only characteristic length in the system, and its dynamics is governed by the interfacial curvature, then $R(t) \sim t^{1/2}$, and $S(\mathbf{k},t)$

is predicted to have a scaling form $S(k,t) \sim t S_o(kt^{1/2})$ for this two-dimensional system. Figure 5.2 shows the linear behavior of $R^2(t)$ along with the results of simulations of the universal scaling function $S_o(x)$ through a plot of $\log(S(k,t)/t)$ as a function of $kt^{1/2}$. Collapse of the data for sufficiently long times confirms the scaling structure.

6

Langevin model for conserved order parameter systems

In model B systems of the Hohenberg–Halperin classification scheme the order parameter ϕ is conserved and is the only relevant macrovariable. The model B Langevin equation can be used to describe binary alloys and spin-exchange kinetic Ising models. Phase separation in binary fluid mixtures is more complex than that in Ising models due to a possible coupling of the concentration order parameter to a transverse velocity field. The model B description can be used for binary fluid mixtures only if this fluid flow effect is neglected. The model B Langevin description has been put on a firm microscopic basis by Langer (1971), who examined the dynamics of the spin-exchange kinetic Ising model. Langer's analysis was carried out by coarse-graining the master equation, which describes the dynamics of the conserved Ising spins, in order to obtain a Fokker–Planck equation for the probability distribution of the magnetization order parameter. The Fokker–Planck equation is equivalent to the Langevin equation for model B. The discussion in this chapter is limited to the Langevin equation.

6.1 Langevin model B

The structure of the Langevin equation for model B can be deduced from the following considerations. Since ϕ is conserved, it obeys a conservation law (continuity equation),

$$\frac{\partial \phi}{\partial t} = -\nabla \cdot \mathbf{j}(\mathbf{r}, t),$$ (6.1)

where $\mathbf{j}$ is the current associated with ϕ. Provided $\mathbf{j}$ itself is not a conserved variable, one can write the transport law,

$$\mathbf{j}(\mathbf{r}, t) = -\left[M\nabla\mu(\mathbf{r}, t) + \zeta^* \right].$$ (6.2)

Here ζ^* is the order parameter current arising from thermal noise and $\mu(\mathbf{r}, t)$ is the local chemical potential, which is synonymous with the thermodynamic force introduced in Chapter 2. More specifically, the local chemical potential is related to the free energy functional by

$$\mu(\mathbf{r}, t) = \frac{\delta \mathcal{F}}{\delta \phi}. \tag{6.3}$$

Combining these results, one has the Langevin equation for model B,

$$\frac{\partial \phi}{\partial t} = M \nabla^2 \left(\frac{\delta \mathcal{F}}{\delta \phi} \right) + \zeta(\mathbf{r}, t), \tag{6.4}$$

where $\zeta = \nabla \cdot \boldsymbol{\zeta}^*$ is the random thermal noise, which satisfies the fluctuation dissipation theorem

$$\langle \zeta(\mathbf{r}, t) \zeta(\mathbf{r}', t') \rangle = -2k_B T M \nabla^2 \delta(\mathbf{r} - \mathbf{r}') \delta(t - t'). \tag{6.5}$$

The Langevin equation for model B is also known as the Cahn–Hilliard–Cook equation due to its applications in the kinetics of alloy ordering and segregation.

As in the case of model A, we set $\mathcal{F} = \mathcal{F}_{GLW}$ and use the form given in Eqs. (3.2) and (3.14). On account of the overall Laplacian on the right hand side of Eq. (6.4), the term arising from the external field vanishes, and the model B Langevin equation reduces to

$$\frac{\partial \delta \phi}{\partial t} = M \nabla^2 [(a_2 - \kappa \nabla^2) \phi^* + a_4 \phi^{*3}] + \zeta(\mathbf{r}, t), \tag{6.6}$$

where $\phi^* = (\delta \phi + \phi_o)$ as in model A.

6.2 Critical quench in a model B system

For critical quenches $\phi_o = 0$, and Eq. (6.6) can be cast in terms of the dimensionless variables $\psi(\mathbf{x}, \tau) = \delta\phi/\phi^*_{min}$, $\mathbf{x} = \mathbf{r}/\xi$, and the dimensionless time τ defined as $\tau = \left(2Ma_o^2(T_c - T)^2/\kappa\right) t$. In these dimensionless variables the model B Langevin equation that is appropriate for the description of critical quenches takes the form

$$\frac{\partial \psi(\mathbf{x}, \tau)}{\partial \tau} = -\frac{1}{2} \nabla^2 (\nabla^2 \psi + \psi - \psi^3) + \epsilon^{1/2} \eta(\mathbf{x}, \tau). \tag{6.7}$$

The random thermal noise η satisfies the fluctuation–dissipation relation

$$\langle \eta(\mathbf{x}, \tau) \eta(\mathbf{x}', \tau') \rangle = -\nabla^2 \delta(\mathbf{x} - \mathbf{x}') \delta(\tau - \tau'). \tag{6.8}$$

Figure 6.1 shows how spinodal decomposition proceeds for a symmetric mixture following a critical quench (see Rogers *et al.*, 1988; Rogers, 1989). The growth of the domain size $R(\tau)$ is evident from the morphology of the pattern. The effect of

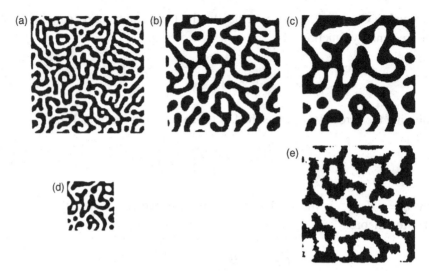

Fig. 6.1. Time evolution of the order parameter field during phase separation of a system with a conserved order parameter for $\psi_o = 0.0$. The shaded regions correspond to one phase ($\psi > 0$) and the white regions to the other phase. For panels (a), (b), (c), and (d), $\epsilon = 0$. Panels (a), (b), and (c) correspond to times $\tau = 150, 600$, and 2200, respectively. Panel (d) shows a uniform contraction of the configuration (c) to match the average domain size of (a). Self-similar growth implies that, on average, (d) and (a) have the same domain morphology. Panel (e) shows a snapshot of the configuration for $\epsilon = 0.5$ and $\tau = 3000$.

noise on the structure of the pattern is shown in panel (e). Panels (a) and (d) illustrate self-similarity and dynamic scaling: the spatial patterns formed by domains at two different times are statistically identical, apart from a global change of the length scale.

As discussed in Chapter 4, dynamic scaling arises when a single length dominates the time evolution. As a result, various measures of the morphology depend on time only through this length. Dynamical scaling is valid only at late times. At early and intermediate times, the interfacial width ξ is another important length that enters the description and leads to a breakdown of scaling. The ratio $\xi/R(\tau)$ is negligibly small only at late times, and dynamic scaling can be justified only when this ratio vanishes. Figure 6.1(e) shows that the interfacial structure is significantly more diffuse when thermal noise is present during the time evolution. The larger the value of the noise strength ϵ, the longer it takes for sharp interfacial structure to develop. Thus, longer times are required for the onset of dynamical scaling. However, the spatial distribution of domains in late-stage coarsening is qualitatively similar for zero and nonzero ϵ.

Using renormalization group techniques, Bray (1994) showed that thermal noise is irrelevant for deep-quench kinetics. This is so because the free energy has two stable fixed points to which the system can flow: the infinite-temperature fixed point for $T > T_c$, and the zero-temperature strong-coupling fixed point for $T < T_c$. Since the strength of the noise ϵ vanishes at $T = 0$, the thermal noise term η can be neglected in model B phase separation kinetics for $T < T_c$. While thermal noise may often be neglected, there are many examples of kinetics where it can play an important role.

6.3 Off-critical quench in a model B system

If the thermal noise term is ignored in the model B Langevin equation, the resulting deterministic equation is known as the Cahn–Hilliard (CH) equation (see Eq. (6.6)),

$$\frac{\partial \delta \phi}{\partial t} = M \nabla^2 [(a_2 - \kappa \nabla^2) \phi^* + a_4 \phi^{*3}], \tag{6.9}$$

where $\phi^* = (\delta \phi + \phi_o)$. In critical quench experiments on symmetric systems the right-hand side of the CH equation has only linear and cubic terms. For asymmetric mixtures and off-critical quenches ϕ_o is nonzero, and for $T < T_c$ the CH equation has an additional quadratic nonlinear term. Its explicit form can be deduced from Eq. (6.9), and is given by

$$\frac{\partial \delta \phi}{\partial t} = -M \nabla^2 [(|a_2| - 3a_4 \phi_o^2 + \kappa \nabla^2) \delta \phi - 3a_4 \phi_o (\delta \phi)^2 - a_4 (\delta \phi)^3]. \tag{6.10}$$

In terms of dimensionless variables, $\psi(\mathbf{x}, \tau) = \delta \phi / \phi_{min}^*$, $\mathbf{x} = \mathbf{r}/\xi$, $\tau = (2Ma_2^2/\kappa) t$, and $\psi_o = \phi_o / \phi_{min}^*$, which are defined for $T < T_c$, the resulting dimensionless CH equation is

$$\frac{\partial \psi}{\partial \tau} = -\frac{1}{2} \nabla^2 ([\nabla^2 + q_c^2] \psi - 3 \psi_o \psi^2 - \psi^3), \tag{6.11}$$

where $q_c^2 = (1 - 3\psi_o^2)$. This is the scaled Cahn–Hilliard equation for an off-critical quench where the field variable ψ is the scaled order parameter fluctuation around the average value $\psi_o = \phi_o / \phi_{min}^*$.

For a critical quench where $\psi_o = 0$, the bilinear term vanishes, $q_c^2 = 1$, and the CH equation reduces to its symmetric parameter-free form (see Eq. (6.7)). For an off-critical quench, ψ_o is nonzero and is the only parameter in Eq. (6.11). Thus, it is useful to summarize its attributes. It is a measure of how far the system is from a critical quench. The volume fraction of the $(-)$ phase is $(1 - \psi_o)/2$ and that of the $(+)$ phase is $(1 + \psi_o)/2$. For $\psi_o = 0$, one has a symmetric system that is quenched through the critical point, while for $\psi_o = \pm 1$ the system is quenched to the coexistence curve. The conservation law for model B dictates that the average value of the order parameter remain time-invariant: as a consequence ψ_o remains constant throughout the time evolution. Since the final equilibrium phase corresponds to two coexisting phases with $\phi = \pm\phi^*_{min}$, a nonzero value of ψ_o reflects an asymmetry in the spatial extent of these two phases. The degree of asymmetry is given by the lever rule. Finally, it can be shown that the equation of the classical spinodal, shown in Figs. 2.1 and 2.5, is $q_c^2 = 0$ or $|\psi_o| = \psi_{sp} \equiv 1/\sqrt{3}$. For unstable states within the classical mean field spinodal, $q_c^2 \equiv (1 - (\psi_o/\psi_{sp})^2) > 0$.

Equation (6.11) must be supplied with appropriate initial conditions describing the system prior to the onset of phase separation. The initial post-quench state is characterized by the pre-quench order parameter fluctuations at the initial temperature T_o (Elder *et al.*, 1988). From renormalization group arguments, any initial short-range correlations should be irrelevant, and one can take the initial conditions to represent a completely disordered state at $T = \infty$. For example, one can choose the white-noise form $\langle \psi(\mathbf{x}, 0)\psi(\mathbf{x}', 0)\rangle = \epsilon_o\delta(\mathbf{x} - \mathbf{x}')$, where the angular brackets represent an average over an ensemble of initial conditions, and ϵ_o ($\epsilon_o \ll 1$) controls the size of the initial fluctuations in ψ. In this context, the fundamental problem in phase separation kinetics is the construction of the late-time solutions of deterministic equations, such as Eq. (6.11), subject to random initial conditions. Even though Eq. (6.11) is a mathematically intractable nonlinear partial differential equation, some qualitative observations concerning its structure and solution can be made.

The cubic term in this equation treats both phases in a symmetric manner. This is the only nonlinear term for a symmetric binary mixture, and it leads to a labyrinthine morphology in which both the phases have an equal share of the system volume. Labyrinthine patterns are a characteristic of systems undergoing spinodal decomposition. The partial differential equation is parameter-free for the symmetric case, and there is no convenient small expansion parameter, especially for early times. The linear approximation loses its validity around $\tau \sim 10$. For late times, the ratio of the interfacial width to the time-dependent domain size $\xi/R(\tau)$ has been used as a small parameter by Pego (1989) in a matched asymptotic expansion analysis. It provides a connection between this nonlinear problem and the Mullins–Sekerka instability on long time scales and to the classic Stefan problem on short time scales.

The quadratic term in Eq. (6.11) treats the two phases in an asymmetric manner and is the source of droplet-like morphology. As the off-criticality ψ_o increases, the quadratic nonlinearity gradually assumes a greater role compared with the cubic nonlinear term and leads to a transition from the labyrinthine to the droplet-like morphology.

The different morphologies that the order parameter field adopts are shown in Fig. 6.2 (see Rogers and Desai, 1989). For the critical quench shown in panels (a) and (b), the symmetry of ψ between the two phases is apparent and expected, since neither phase can be a minority phase for a 50–50 mixture. The qualitative

Fig. 6.2. The order parameter field morphology from two-dimensional simulations of Eq. (6.11). A critical quench ($\psi_o = 0.0$) at (a) $\tau = 500$ and (b) $\tau = 5000$; and an off-critical quench ($\psi_o = 0.4$) at (c) $\tau = 500$ and (d) $\tau = 5000$. The dark regions have $\psi < 0$.

differences in the morphology for critical and off-critical quench evolutions at late times are also clear: bicontinuous for a critical quench, and isolated closed-cluster morphology for an asymmetric off-critical quench. Domain coarsening is also evident in each of the two cases.

The spatial correlation function

$$G(\mathbf{x}, \tau) \equiv \langle \psi(\mathbf{x}, \tau) \psi(\mathbf{0}, \tau) \rangle \tag{6.12}$$

and the time-dependent structure factor $S(\mathbf{k}, \tau)$

$$S(\mathbf{k}, \tau) = \int d^d x \, e^{-i\mathbf{k}\cdot\mathbf{x}} \, G(\mathbf{x}, \tau) \tag{6.13}$$

can be computed from simulations of the model B order parameter equations. There are a number of ways to obtain the time-dependent domain size, $R(\tau)$, from such results. The first zero of $G(x, \tau)$ provides a length which is a measure of $R(\tau)$. One may also extract a characteristic wavelength from either the first moment of $S(k, \tau)$ or the value k_m where $S(k, \tau)$ is maximum and identify such a wavelength with

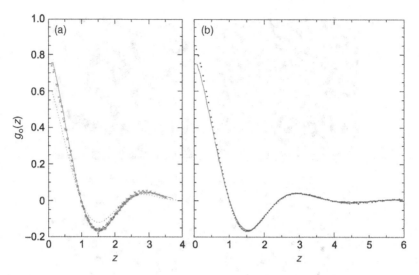

Fig. 6.3. Late-stage scaling function $g_o(z)$. (a) Solid line, $\epsilon = 0$; open squares, $\epsilon = 0.05$; crosses, $\epsilon = 0.2$. The dashed line corresponds to the longest simulation time for $\epsilon = 0.5$ and indicates that the scaling regime has not yet been reached. (b) $g_o(z)$ at $\tau = 5000$: solid line, $\psi_o = 0.4$; circles, $\psi_o = 0$. Reprinted Figure 8 with permission from Rogers *et al.* (1988) and Figure 9 from Rogers and Desai (1989). Copyright 1988 and 1989 by the American Physical Society.

$R(\tau)$. The result that is now firmly established from experiments and simulations is that

$$R(\tau) \sim \tau^{1/3}, \tag{6.14}$$

independent of the system dimensionality d.

Simulation results confirm the existence of dynamic scaling of the time-dependent structure factor, $S(k, \tau) = (R(\tau))^d S_o(kR(\tau))$, at late times. Dynamic scaling can also be discussed in terms of $G(x, \tau)$, which may be written in the form

$$G(x, \tau) = G(x/R(\tau)) \equiv (1 - \psi_o^2)^{-1} g_o(z), \tag{6.15}$$

where $z = x/R(\tau)$. The factor $(1 - \psi_o^2)$ is included to ensure that $g_o(z) \to 1$ as $z \to 0$ for an infinite system. This normalization factor is unity for a critical quench. Figure 6.3 shows $g_o(z)$ obtained from the simulation of the model B Langevin equation for a $2d$ system. The scaling function $g_o(z)$ is independent of thermal noise strength and the off-criticality parameter ψ_o. Note that the larger the value of ϵ, the longer it takes a simulation to reach the asymptotic scaling regime. The dependence of $G(x, \tau)$ on ψ_o is removed through the scaling by $R(\tau)$ and the factor $(1 - \psi_o^2)^{-1}$ in Eq. (6.15).

In Fig. 6.3, $g_o(0)$ has not yet attained its asymptotic value of unity, because the ratio of the interfacial width to the domain size is not small enough at the longest

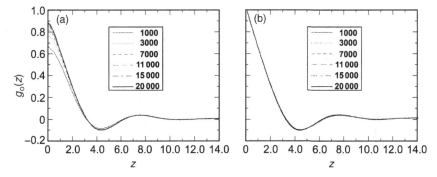

Fig. 6.4. Late-stage scaling function $g_o(z)$ for a $3d$ cell dynamical model of a binary alloy. (a) Simulation results for various values of τ. (b) Hardened data corresponding to (a). Note that the curves from each time step are nearly indistinguishable. Reprinted Figures 18 and 20 with permission from Shinozaki and Oono (1993). Copyright 1993 by the American Physical Society.

times probed in the simulation. One way to obtain the asymptotic scaled correlation function is to "harden" the morphology data by applying a transformation $\psi \rightarrow$ sign(ψ), so that $\psi(\mathbf{x}, \tau)$ becomes a field with only two possible values, ± 1. The implicit value of the ratio of the interfacial width to the domain size is zero for such hardened data, and the resulting scaled correlation function $g_o(z)$ has the correct limit $g_o(0) = 1$. This is illustrated in Fig. 6.4, where a $3d$ symmetric binary alloy was simulated using a discrete cell-dynamical system to probe the phase separation process. The original simulation data on the left may be contrasted with the hardened data on the right. A comparison of Figs. 6.3 and 6.4 shows that the dynamically scaled correlation function is also independent of the system dimensionality.

6.4 Model B interfacial structure

Interfaces play a central role in the phase transition kinetics of both models A and B. Figure 6.5 shows the interfacial structure corresponding to Fig. 6.2(b) (see Rogers, 1989). The relative magnitudes of the interfacial width and the domain size for a late-stage configuration are clearly seen in this figure. In Fig. 6.6, the interfacial structure at early times is shown for an off-critical quench (see Elder and Desai, 1989). The interfacial evolution at early times is apparent. Large interconnected clusters begin to break apart into small circular droplets because the quadratic nonlinearity in the free energy functional eventually outpaces the cubic nonlinearity when off-criticality ψ_o is large.

Interface morphologies, such as those shown in Figs. 6.5 and 6.6, are highly complex structures. While the correlation function and the structure factor provide

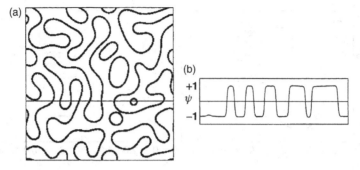

Fig. 6.5. Interface structure for $\tau = 5000$, $\psi_o = 0$. In (a) the shaded regions correspond to interfaces separating the domains, defined as regions where $0.75\psi_- < \psi < 0.75\psi_+$. In (b) a cross-sectional view of the order parameter ψ along the horizontal line in (a) is given.

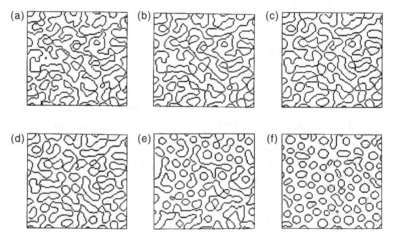

Fig. 6.6. Time dependence of the domain boundary morphology for an off-critical quench with $\psi_o = 0.48$. The classical spinodal corresponds to $\psi_o \sim 0.58$. The interface separating two domains is defined as the locus of points where $\psi = 0$. The evolution is shown for early times τ of (a) 50, (b) 100, (c) 150, (d) 200, (e) 250, and (f) 300.

information on the time evolution of the characteristic length scale, averaging removes morphological details. Topological methods have been used to obtain quantitative information on the connectivity within such structures (Jinnai *et al.*, 2000; Gameiro *et al.*, 2005; Kwon *et al.*, 2007). Two structures are topologically equivalent if they can be deformed into each other without cutting or gluing. Thus, a pyramid, a sphere, and a cube are topologically equivalent. Similarly, a doughnut and a cube with a handle are equivalent.

The Betti numbers β_i, which are topological invariants, provide a measure of the similarity among microstructures. They have been used to quantify the microstructure geometry, including the boundary effects that exist in finite systems. They can also be used to monitor basic topological changes, such as the transition from a bicontinuous structure to an isolated cluster morphology. Within a microstructure, the zeroth Betti number β_0 is the number of different connected components, the first Betti number β_1 is the number of tunnels (in $2d$ tunnels reduce to loops), and the second Betti number β_2 is the number of enclosed volumes (cavities). In systems that can be described by models A and B, the time variation of the Betti numbers indicates how the structural connectivity evolves as the systems coarsen. The Betti number β_1 is equivalent to the genus g of a structure, which is defined as the number of cuts that can be made on a closed surface without separating it into two disconnected bodies. Thus, a pyramid, a sphere, and a cube have $\beta_1 = g = 0$ while a doughnut and a cube with a handle have $\beta_1 = g = 1$. The pinching of a tube that can occur during the evolution of a bicontinuous structure is a topologically singular event and decreases g.

For coarsening in systems governed by the Allen–Cahn (model A) and Cahn–Hilliard (model B) equations, the genus per unit volume $g_V \equiv g/V$ decays in time as $g_V \sim t^{-n_g}$, with n_g approximately $\frac{3}{2}$ for model A and 1 for model B (Kwon *et al.*, 2007). If we identify the interface area per unit volume S_V as an inverse characteristic length of the system, then $S_V \sim t^{-n}$, with $n = \frac{1}{2}$ for model A and $n = \frac{1}{3}$ for model B and $n_g = 3n$. As a consequence, a scaled genus, defined as $g_V S_V^{-3}$, is dimensionless. It becomes time-independent at late times in the scaling region. Its value is approximately 0.14.

Another quantity that is used to characterize an interface is its curvature, which is associated with the change in the normal as one moves along the surface. Since the normal direction and the directions along the surface are vectors, the curvature is a tensor quantity. It is useful to describe the curvature tensor by its invariants, since these quantities do not change if the coordinate system used to describe the surface is rotated. Invariants are intrinsic properties of the surface. In a $2d$ system, the surface is a $1d$ curve and the curvature tensor can be expressed as a 2×2 matrix, which has two invariants: the curvature K and its determinant, which is zero. In a $3d$ system, the surface is two-dimensional, and the curvature tensor, expressed as a 3×3 matrix, has three invariants: its determinant, its trace, and the sum of three principal minors (minors of the three diagonal elements). Again the determinant vanishes. It is useful to diagonalize the curvature tensor. Out of the three eigenvalues, one is zero and the other two are the principal curvatures, denoted as κ_1 and κ_2. Eigenvectors corresponding to the two nonzero eigenvalues are known as principal directions of the surface, and the surface curvatures κ_1 and κ_2 along these directions are also extrema (κ_1 is the minimum and κ_2 the maximum). It is then easy to see that the invariant trace is $(\kappa_1 + \kappa_2)$. The mean curvature $H \equiv (\kappa_1 + \kappa_2)/2$ is the more commonly used invariant. From the diagonalized curvature tensor,

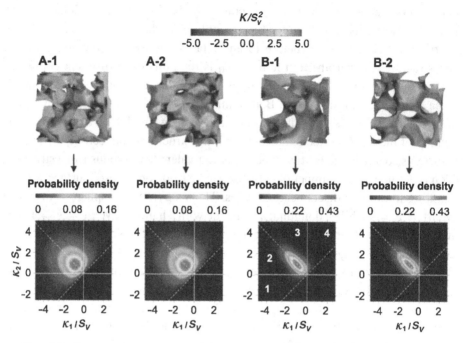

Fig. 6.7. Top panels: Interface structures color-coded by the scaled Gaussian curvature. Bottom panels: Scaled interfacial shape distribution (ISD), normalized to unity. A-1: model A, $\tau = 480$, $S_V^{-1} = 56.79$; A-2: model A, $\tau = 4000$, $S_V^{-1} = 158.93$; B-1: model B, $\tau = 64000$, $S_V^{-1} = 35.61$; B-2: model B, $\tau = 179200$, $S_V^{-1} = 77.81$. Four regions in the ISDs (as marked in B-1: bottom panel) are divided by lines where $\kappa_1 = \kappa_2$ on which spherical patches lie, $\kappa_1 = -\kappa_2$ on which patches with $H = 0$ lie, and $\kappa_i = 0$ ($i = 1, 2$: two axes) on which cylindrical patches with $K = 0$ lie. The planar patches ($\kappa_1 = \kappa_2 = 0$) lie at the origin. Physically attainable surfaces are to the left of the $\kappa_1 = \kappa_2$ line, since $\kappa_2 \geq \kappa_1$ by definition. Much of the distribution lies in regions 2 and 3 where $K < 0$, indicating a predominance of saddle-shaped patches. The scaled ISDs are symmetric with respect to the $H = 0$ line; the distribution width $\sqrt{\langle H^2 \rangle}/S_V$ is much larger for model A than for model B. Reprinted Figure 2 with permission from Kwon *et al.* (2007). Copyright 2007 by the American Physical Society.

whose diagonal elements are $(0, \kappa_1, \kappa_2)$, one can obtain the three principal minors as $(\kappa_1\kappa_2, 0, 0)$, which gives the invariant sum as $\kappa_1\kappa_2$. This invariant sum is referred to as the Gaussian curvature K. An essential measure of the morphology is provided by K: for a saddle-shaped or hyperbolic surface it is negative, and for an elliptic or parabolic surface it is positive. These two- and three-dimensional results can be generalized to higher dimensions.

A detailed characterization of the interfacial structure can be made through a study of the interfacial shape distribution (ISD), $P(\kappa_1, \kappa_2)$, which is the probability of finding a patch of interface with (κ_1, κ_2). Figure 6.7 displays the $3d$ interface structures and $P(\kappa_1/S_V, \kappa_2/S_V)$ for critical quenches corresponding to a symmetric two-phase system with equal volume fractions. Results are shown for both nonconserved and conserved dynamics at two different late-stage times. Time-independence of $P(\kappa_1/S_V, \kappa_2/S_V)$ indicates self-similarity at late times in the coarsening process.

Jinnai *et al.* (2000) obtained the joint probability density $P(H, K)$ for the gyroid structure from an experimental measurement of the microstructure of a triblock copolymer. This unusual structure also occurs in various lipid–water systems. It consists of two interpenetrating networks, one of which is right-handed and the other is its mirror image (space group $I a\overline{3}d$); it is a three-dimensional analog of the honeycomb lattice. A gyroid surface is a periodic surface with a constant (nonzero) mean curvature H, and has smooth internal interfaces which form three tubes at a junction (Schick, 1998).

While the model B Langevin equation captures the essence of the phase segregation kinetics, real binary mixtures have additional physical effects. Hydrodynamic interactions contribute at late times for binary polymer melts (Glotzer, 1995), hydrodynamic flow effects become important at late times for small molecule binary fluid mixtures (Siggia, 1979), and elastic effects play a subsidiary, but important, role for binary alloys (Cahn, 1961, 1966, 1968). For small molecule binary fluid mixtures the concentration order parameter can also couple to the total density, especially within the interface region if the interaction between the unlike molecules is strongly repulsive (Denniston and Robbins, 2004). The kinetics of phase ordering and phase separating systems is complex. Laboratory experiments, model simulations, and analytical theories have contributed to our current understanding.

For systems whose kinetics can be modeled through a free energy functional, our discussion was limited to the two simplest models, A and B. In many instances more complex models are appropriate. For example, the coarsening of the domain structure shown in Fig. 4.2 is for the disorder–order transition in Fe_3Al, which has a tricritical point. The kinetics of such first-order transitions requires the consideration of elastic fields (Sagui *et al.*, 1994) where a nonconserved order parameter is coupled to a conserved variable (model C).

7

Interface dynamics at late times

The dynamics of interfaces plays an important role in the evolution of the morphology of phase-ordering and phase-separating systems. After a short transient period, well-defined domains of the coexisting phases are separated by sharp interfaces, often with a complex structure. Curvature-driven interface dynamics governs the manner in which the system evolves in the late stages of the domain-coarsening process. The focus in this chapter is on the quantitative description of interface structure and dynamics for models A and B; however, the results have more general applicability and will be used in applications to diverse systems in later chapters.

7.1 Model B interface

In order to construct the interfacial profile for model B, we consider a d-dimensional system and start with Eq. (6.7), which can be rewritten as

$$\frac{\partial \psi}{\partial \tau} = -\nabla^2 \left(\nabla^2 \psi - \frac{\delta f}{\delta \psi} \right). \tag{7.1}$$

We have set $\epsilon = 0$ in order to neglect thermal noise and have absorbed the factor of $\frac{1}{2}$ in the unit of dimensionless time τ, which for model B is now $\tau = \left(Ma_o^2 (T_c - T)^2/\kappa \right) t$. For a symmetric system, we use the dimensionless form of the bulk free energy density $f(\psi) = \psi^4/4 - \psi^2/2$ (see discussion following Eq. (3.15)). Equation (7.1) is often called the Cahn–Hilliard equation (Cahn, 1961; Bray, 1994). In the arguments given below, only the double-well nature of $f(\psi)$, and not its explicit analytic form, is needed. Since the correlation length ξ plays a central role in the analysis, we rewrite Eq. (7.1) as

$$\frac{\partial \psi}{\partial \tau} = -\nabla^2 \left(\xi^2 \nabla^2 \psi - \frac{\delta f}{\delta \psi} \right), \tag{7.2}$$

even though $\xi = 1$ in our dimensionless units. The domain size $R(\tau)$ is much larger than ξ at late times, and it is appropriate to consider first a locally planar interface with its normal in the x direction. Let $x = x_o$ be the interface position. Then $\psi \sim \psi(x, \tau)$, and the stationary solution of Eq. (7.2), $\psi_o(x)$, satisfies

$$\xi^2 \frac{d^2 \psi_o}{dx^2} = \frac{\delta f}{\delta \psi_o}, \tag{7.3}$$

which has a kink profile solution

$$\psi_o(x) = \pm \tanh((x - x_o)/\sqrt{2}\xi), \tag{7.4}$$

for a free energy with the form $f(\psi) = \psi^4/4 - \psi^2/2$. This can be verified either by integrating Eq. (7.3) with the boundary conditions $\psi_o(\pm\infty) = \pm 1$ and $\psi_o(x_o) = 0$, or by direct substitution. It is instructive to expand ψ around ψ_o and convert Eq. (7.2) to a linear eigenvalue problem. With $\psi(x, \tau) = \psi_o(x) + \zeta(x, \tau)$, one gets

$$\frac{\partial \zeta}{\partial \tau} = \hat{\Omega}[\psi_o(x)]\zeta \tag{7.5}$$

where the linear operator $\hat{\Omega}$ is

$$\hat{\Omega} = \frac{d^2}{dx^2} \left[\frac{\delta^2 f}{\delta \psi_o^2} - \xi^2 \frac{d^2}{dx^2} \right]. \tag{7.6}$$

Let $\zeta_n(x)$ be an eigenfunction of $\hat{\Omega}$ with eigenvalue $-\omega_n$ so that $\hat{\Omega}\zeta_n(x) = -\omega_n \zeta_n(x)$. We may expand $\zeta(x, \tau)$ in a complete set of eigenfunctions, $\{\zeta_n(x)\}$,

$$\zeta(x, \tau) = \sum_n a_n \zeta_n(x) e^{-\omega_n \tau}. \tag{7.7}$$

Differentiation of Eq. (7.3) with respect to x shows that $\zeta_o(x) \equiv d\psi_o(x)/dx$ is an eigenfunction of $\hat{\Omega}$ corresponding to an eigenvalue zero. This eigenfunction goes to $|\psi_+ - \psi_-|\delta(x - x_o)$ as $\xi \to 0$ (in comparison with the domain size $R(\tau)$), where $\psi_\pm$ are the values of $\psi_o(x)$ at $\pm\infty$. The eigenfunction $\zeta_o(x)$ is the Goldstone mode arising from the broken translational invariance symmetry (see Fig. 7.1). Since $\psi_o(x)$ is a stationary solution, $\zeta(x, \tau)$ must be orthogonal to $\zeta_o(x)$, which implies that $a_o = 0$. This is a consequence of the broken translational symmetry.

Now consider the more general case of a curved interface which is far from other interfaces. This assumption is valid at late times. For a one-dimensional system the interface is a point: $x = x_o$. In d dimensions, the interface is a $(d - 1)$-dimensional surface. The interfacial profile is the variation of ψ along the coordinate u orthogonal to the $(d - 1)$-dimensional interface, which is described by the coordinates $(s_1, s_2, \ldots, s_{(d-1)})$. In two dimensions (compared with $d = 1$), the additional

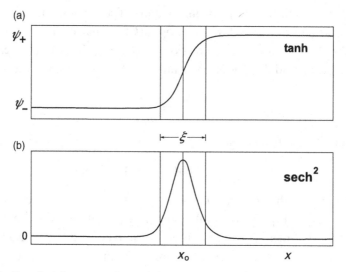

Fig. 7.1. Interfacial structure in model B. (a) The hyperbolic tangent profile $\psi_o(x)$ of the one-dimensional equilibrium solution corresponding to two coexisting phases. (b) The Goldstone mode $d\psi_o(x)/dx$ for this model.

qualitative feature that must be accounted for is the curvature K of the $1d$ interface. Thus, we have two lengths, ξ and K^{-1}, in the problem. The length K^{-1} is related to the domain size $R(\tau)$. In the late-stage regime K^{-1} is large, and the parameter $\lambda \equiv \xi K$ is a natural small parameter for expansion.

It is useful to introduce normal coordinates that move with the interface (Zia, 1985). To be explicit, we consider $d = 2$, and write (see Fig. 7.2)

$$\mathbf{x}(s,u) = \mathbf{R}(s) + u\hat{\mathbf{n}}(s), \qquad (7.8)$$

where $\mathbf{R}(s)$ is the position of the interface as a function of the arc length s, a curvilinear coordinate parallel to the interface, and $\hat{\mathbf{n}}(s)$ is the unit vector normal to the interface at the arc length value s. For interface dynamics, only points in the system with u comparable to ξ and much smaller than K^{-1} are relevant. Typically $-\xi < u < +\xi$ is often an adequate range for u. In Fig. 7.2 the edges of the diffuse interface correspond to $u = \pm\xi$. Clearly $\mathbf{x}(s,0) = \mathbf{R}(s)$ is the center of the interface, and the vector

$$\hat{\mathbf{t}} \equiv \frac{d\mathbf{R}}{ds} \qquad (7.9)$$

is the unit vector tangent to the interface at s, so that $\hat{\mathbf{n}} \cdot \hat{\mathbf{t}} = 0$. It is convenient to transform the Cartesian components (x, y) of the vector $\mathbf{x}$ to the curvilinear

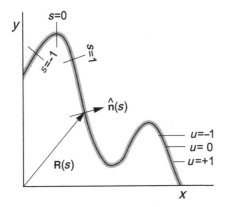

Fig. 7.2. Curvilinear coordinate system for the moving interface. The central darker solid line ($u = 0$) denotes the position of the interface. The figure also shows a portion of an interface that separates two phases in the late stage of the evolution.

interfacial coordinates $(u, s) \equiv (u_1, u_2)$. The unit vectors $(\hat{\mathbf{x}}, \hat{\mathbf{y}})$ may also be transformed to the unit vectors $(\hat{\mathbf{n}}, \hat{\mathbf{t}})$.

Let $\theta(s)$ be the angle between $\hat{\mathbf{n}}$ and $\hat{\mathbf{x}}$. In terms of $\theta(s)$,

$$\hat{\mathbf{t}}(s) = \hat{\mathbf{x}} \sin \theta(s) - \hat{\mathbf{y}} \cos \theta(s)$$
$$\hat{\mathbf{n}}(s) = \hat{\mathbf{x}} \cos \theta(s) + \hat{\mathbf{y}} \sin \theta(s). \tag{7.10}$$

As one moves along the interface, $\hat{\mathbf{t}}$ and $\hat{\mathbf{n}}$ change their orientation. The relationship of these changes to the curvature K is given by the geometrical relationships embodied in the Frenet equations:

$$\frac{d\hat{\mathbf{t}}}{ds} = -K\hat{\mathbf{n}}, \qquad \frac{d\hat{\mathbf{n}}}{ds} = K\hat{\mathbf{t}}. \tag{7.11}$$

By convention, a protrusion of the ψ_- phase into the ψ_+ phase is chosen to have a positive curvature. In Fig. 7.2, the region with positive (negative) u is the ψ_+ (ψ_-) phase. Substituting Eq. (7.10) into Eq. (7.11), one can obtain the relationship between the curvature K and the angle θ,

$$K(s) = -\frac{d\theta(s)}{ds}. \tag{7.12}$$

Another important relationship connects the curvature K and the normal to the interface $\hat{\mathbf{n}}$:

$$\nabla \cdot \hat{\mathbf{n}} = \left(\hat{\mathbf{n}} \frac{\partial}{\partial u} + \hat{\mathbf{t}} \frac{\partial}{\partial s}\right) \cdot \hat{\mathbf{n}}$$
$$= \hat{\mathbf{t}} \cdot \frac{\partial \hat{\mathbf{n}}}{\partial s} = \hat{\mathbf{t}} \cdot K(s)\hat{\mathbf{t}} = K(s), \tag{7.13}$$

where Eq. (7.11) has been used in writing the second line. The orientation of $\hat{t}$ and $\hat{n}$ with respect to the fixed Cartesian system (x, y) varies from point to point, and Eq. (7.13) explicitly describes the relation of this change along the interface with the curvature. With this definition the curvature of a straight line is zero, while the curvature of a circle is the inverse of its radius. The generalization of Eq. (7.13) to a d-dimensional system is $\nabla \cdot \hat{n} = (d - 1)H$, where H is the mean curvature.

As the interface moves, its velocity is given by

$$v(s, \tau) = -\frac{du(\mathbf{X}, \tau)}{d\tau}, \tag{7.14}$$

where $\mathbf{X} \sim (s, u = 0)$ is a point on the interface. The need for the negative sign is clear since, if the interface moves in positive u direction by an amount Δ in time $d\tau$, the coordinate value u at point $\mathbf{x}$ decreases to $(u - \Delta)$. We now write $\psi(\mathbf{x}, \tau)$ in the form

$$\psi(\mathbf{x}, \tau) = \psi_o[u(\mathbf{x}, \tau)] + \delta\psi, \tag{7.15}$$

where ψ_o is given by (see Eq. (7.4))

$$\psi_o(u) = \pm \tanh(u/\sqrt{2}\xi), \tag{7.16}$$

and assume that $\delta\psi$ is small for small values of u. The function $\delta\psi$ arises from degrees of freedom associated with the bulk. Substituting this form into Eq. (7.2) and keeping leading-order terms yields

$$-v\frac{d\psi_o}{du} = \nabla^2\left[\hat{\Omega}_2\delta\psi - \xi^2 K\frac{d\psi_o}{du}\right], \tag{7.17}$$

where, to lowest order in ξK, the expressions for ∇^2 and $\hat{\Omega}_2$ are

$$\nabla^2 \approx \frac{\partial^2}{\partial u^2} + \frac{\partial^2}{\partial s^2} + K\frac{\partial}{\partial u}, \tag{7.18}$$

and

$$\hat{\Omega}_2 \approx \frac{\delta^2 f\{\psi_o\}}{\delta\psi_o^2} - \xi^2\left(\frac{\partial^2}{\partial u^2} + \frac{\partial^2}{\partial s^2}\right). \tag{7.19}$$

The details of the derivation of these results are given in the Appendix. In Eq. (7.17) the outer Laplacian can be inverted by introducing the diffusion Green function

$$\nabla^2 G(\mathbf{x}, \mathbf{x}') = \delta(\mathbf{x}, \mathbf{x}'), \tag{7.20}$$

which converts Eq. (7.17) to

$$\hat{\Omega}_2 \delta\psi - \xi^2 K(s)\frac{d\psi_o}{du} = -\int d^2x'\, G(\mathbf{x}, \mathbf{x}')\frac{d\psi_o(u')}{du'}v(\mathbf{x}'). \qquad (7.21)$$

The operator $\hat{\Omega}_2$ has the form

$$\hat{\Omega}_2(u, s) = \hat{\Omega}_A(u) - \xi^2\frac{\partial^2}{\partial s^2}. \qquad (7.22)$$

The eigenvalue problem for this operator can be written as

$$\hat{\Omega}_2 \tilde{\zeta}_n(u)\tilde{\mathcal{X}}_m(s) = -(\tilde{\omega}_n + \tilde{\alpha}_m)\tilde{\zeta}_n(u)\tilde{\mathcal{X}}_m(s), \qquad (7.23)$$

where the eigenfunctions of $\hat{\Omega}_2$ have been expressed as products, $\tilde{\zeta}_n(u)\tilde{\mathcal{X}}_m(s)$, of eigenfunctions of $\hat{\Omega}_A$ and $\xi^2\partial^2/\partial s^2$,

$$\hat{\Omega}_A(u)\tilde{\zeta}_n(u) = -\tilde{\omega}_n\tilde{\zeta}_n(u), \qquad (7.24)$$

and

$$\xi^2\frac{\partial^2}{\partial s^2}\tilde{\mathcal{X}}_m(s) = \tilde{\alpha}_m\tilde{\mathcal{X}}_m(s). \qquad (7.25)$$

It is straightforward to show that $\tilde{\mathcal{X}}_m(s) = N\exp(-ims)$, where N is a normalization factor and $\tilde{\alpha}_m = -m^2\xi^2$. The operator $\hat{\Omega}_A(u)$ has a nontrivial null space associated with the Goldstone mode, $\hat{\Omega}_A(u)\,(d\psi_o(u)/du) = 0$. Thus, the interface dynamics can be projected out of the full system dynamics through this Goldstone mode as follows. We expand the contribution from the bulk degrees of freedom, $\delta\psi$, in terms of the eigenfunctions of $\hat{\Omega}_A$: $\delta\psi = \sum_{n=0}^{\infty} a_n(s)\tilde{\zeta}_n(u)$. The bulk degrees of freedom have to remain orthogonal to the interfacial degrees of freedom, which are represented by the Goldstone mode, $\tilde{\zeta}_o(u) \equiv (d\psi_o(u)/du)$, with $\tilde{\omega}_o = 0$, the right eigenmode of $\hat{\Omega}_A$ with zero eigenvalue. It is also a left eigenvector. The orthogonality condition reduces to $a_o = 0$. We can use this condition by left-multiplying Eq. (7.21) by $\tilde{\zeta}_o(u)$ and integrating over u. This eliminates the $\hat{\Omega}_2$ term. The second term on the left-hand side becomes $(-\sigma\xi^2 K)$, where we identify the surface tension (dimensionless) with

$$\sigma = \int du\left(\frac{d\psi_o(u)}{du}\right)^2. \qquad (7.26)$$

By comparing this result with Eq. (3.12) we see that $\tilde{\sigma} = \sigma \xi / (\kappa \phi_{min}^{*2})$. For the double-well free energy $f(\psi) = \psi^4/4 - \psi^2/2$, this reduces to $\tilde{\sigma} = \sigma a_4 / (a_2 \sqrt{a_2 \kappa})$ (see also Eq. (3.15)). The resulting u-integrated equation becomes

$$\sigma \xi^2 K(s) = \int du \, d\mathbf{x}' G(\mathbf{x}, \mathbf{x}') \frac{d\psi_o(u')}{du'} \frac{d\psi_o(u)}{du} v(\mathbf{x}'), \qquad (7.27)$$

which is correct to lowest order in ξK. To the extent that $d\psi_o(u')/du'$ has short range, this result connects the interfacial curvature $K(s)$ at a point s to the interfacial velocity at points in its neighborhood. The (mean field) surface tension is seen to be the driving force for the dynamics of the interfaces. The diffusion Green function inextricably couples the interface motion at two points s and s' on the interface ($u = u' = 0$) to the bulk dynamics. For a conserved order parameter the interface dynamics and domain growth at late times are governed by an evaporation–diffusion–condensation mechanism.

7.2 Model A interface

The interface analysis is considerably simpler for model A. There is no outer Laplacian in Eq. (7.2) or in Eq. (7.17). The Goldstone mode $\tilde{\zeta}_o$ is still the interface profile gradient, but the Green function is replaced by $G(\mathbf{x}, \mathbf{x}') = -\delta(\mathbf{x} - \mathbf{x}')$, which has the consequence that Eq. (7.27) simplifies to

$$\sigma \xi^2 K(s) = -\left[\int du \left(\frac{d\psi_o(u)}{du}\right)^2\right] v(s) = -\sigma v(s), \qquad (7.28)$$

after performing the integration over $\mathbf{x}'$. The surface tension σ cancels from both sides and we get the well-known Allen–Cahn result (Lifshitz, 1962; Allen and Cahn, 1979),

$$v(s) = -\xi^2 K(s). \qquad (7.29)$$

This equation for model A shows that the interface motion is driven entirely by the local curvature, and interface dynamics decouples from the bulk dynamics. Even though the surface tension plays an important implicit role, it is the local curvature that drives the interface velocity.

The quantities K, v, and σ are dimensionless. We denote the corresponding quantities with dimensions by $\tilde{K}$, $\tilde{v}$, and $\tilde{\sigma}$. A similar analysis starting from the model A Langevin equation (5.1), which contains the thermal noise and is not in dimensionless form, leads to the result that interface velocity $\tilde{v}$ is related to the curvature $\tilde{K}$ by

$$\tilde{v}(s) = -\tilde{\lambda} \, \tilde{\sigma} \tilde{K}(s) + \tilde{\eta}, \qquad (7.30)$$

where $\tilde{\sigma}$ is given by Eq. (3.12). The parameter $\tilde{\lambda}$ is

$$\tilde{\lambda} = M\left[\int_{-\infty}^{\infty} dz \left(\frac{d\phi_0(z)}{dz}\right)^2\right]^{-1}, \tag{7.31}$$

and the random noise $\tilde{\eta}$ satisfies the fluctuation–dissipation relation,

$$\langle \tilde{\eta}(\mathbf{R}(s), t)\tilde{\eta}(\mathbf{R}(s'), t')\rangle = 2\tilde{\lambda} k_B T \delta(\mathbf{r} - \mathbf{r}')\delta(t - t'), \tag{7.32}$$

where $\mathbf{r} \equiv \mathbf{R}(s)$, $\mathbf{r}' \equiv \mathbf{R}(s')$ are two points on the interface. From Eq. (3.12), the integral within the square brackets in Eq. (7.31) is $\tilde{\sigma}/\kappa$. Thus, the product $\tilde{\lambda}\tilde{\sigma}$ in Eq. (7.30) reduces to $M\kappa$. Even though the interface motion decouples from the bulk dynamics in both derivations of the Allen–Cahn result, the surface tension plays an important role.

The results in either Eq. (7.29) or (7.30) are also starting points for demonstrating the following interesting result for model A kinetics. Consider a circular domain with average radius $R(\tau)$ growing in time during the late-stage regime. Since $K \sim 1/R$ and $v \sim -dR/d\tau$, one has $R^2(\tau) \sim \tau$. Thus, for model A, $R(\tau) \sim \tau^n$ with $n = 1/2$. For a sphere in d dimensions the result is

$$\frac{dR}{d\tau} \sim \frac{(d-1)}{R}, \tag{7.33}$$

since

$$K \sim \frac{(d-1)}{R}. \tag{7.34}$$

This in turn leads to the result

$$R^2(\tau) \sim (d-1)\tau. \tag{7.35}$$

Starting from Eq. (7.27), in the next chapter we carry out an analysis which leads to the domain growth law for model B.

Appendix: Derivation of equation for interface velocity

Here we give details necessary for the derivation of Eqs. (7.17), (7.18), and (7.19). As one moves along the interface, $\hat{\mathbf{t}}$ and $\hat{\mathbf{n}}$ change their orientation. The relationship between these changes and the curvature K is given by the Frenet equations (7.11). For planar curves of length L with periodic boundary conditions, where the origin is chosen so that $-L/2 < s < L/2$, one can integrate Eq. (7.12) to get

$$\theta(s) = \theta(-\frac{L}{2}) - \int_{-\frac{L}{2}}^{s} K(s')ds'. \tag{7.36}$$

Examples of curves for which this result applies are the boundary of a closed planar domain and a percolating curve within a planar cluster with periodic boundary conditions. In Eq. (7.9), if we use the definition of $\hat{\mathbf{t}}$ given in Eq. (7.10) and then integrate the result, we obtain the interface position

$$\mathbf{R}(s) = \mathbf{R}(-\frac{L}{2}) + \hat{\mathbf{x}} \int_{-\frac{L}{2}}^{s} \sin\theta(s')ds' - \hat{\mathbf{y}} \int_{-\frac{L}{2}}^{s} \cos\theta(s')ds'. \qquad (7.37)$$

Thus, the curvature $K(s)$ defines the curve uniquely to within a rigid translation $\mathbf{R}(-\frac{L}{2})$ and a rotation $\theta(-\frac{L}{2})$. Therefore, it is useful to focus attention on the intrinsic function $K(s, \tau)$ to define the time evolution of the domain morphology.

The transformation $(x, y) \to (u, s)$ is governed by the covariant metric tensor $g_{\alpha\beta} = \frac{\partial \mathbf{x}}{\partial \omega^\alpha} \cdot \frac{\partial \mathbf{x}}{\partial \omega^\beta}$, $\alpha, \beta = 1, 2$ with $\omega^1 = u$ and $\omega^2 = s$. The determinant of $g_{\alpha\beta}$ is g, and it appears in the transformation of the area element,

$$dxdy = \sqrt{g}\,duds. \qquad (7.38)$$

The Laplacian ∇^2 has the general form

$$\nabla^2 = \sum_{\alpha,\beta} \frac{1}{\sqrt{g}} \frac{\partial}{\partial \omega^\alpha} \sqrt{g}\, g^{\alpha\beta} \frac{\partial}{\partial \omega^\beta} \qquad (7.39)$$

in curvilinear coordinates. The contravariant tensor $g^{\alpha\beta}$ is related to the inverse of $g_{\alpha\beta}$ so that $g^{\alpha\gamma}g_{\gamma\beta} = \delta^\alpha_\beta$. Here the repeated index γ is summed, and the quantity δ^α_β is unity if $\alpha = \beta$ and zero if $\alpha \neq \beta$. This result for the Laplacian can be used to find ∇^2 in the time-dependent (u, s) coordinate space, noting that $g_{11} = 1$, $g_{12} = g_{21} = 0$, $g_{22} = (1 + uK)^2$, and $g^{11} = 1$, $g^{12} = g^{21} = 0$, $g^{22} = (1 + uK)^{-2}$. To obtain the interface dynamics from the field equation (7.2), the mapping $(x, y) \to (u, s)$ must be used only within the boundary layer of the interface: i.e. $-\xi/2 < u < \xi/2$. During the late stages of the evolution $\xi \ll K^{-1}$. Therefore, one needs to evaluate ∇^2 only to first order in uK, so that the determinant $g = (1 + uK)^2 \approx 1$, but $\sqrt{g} = (1 + uK)$ and $g^{\alpha\beta} \approx \delta^{\alpha\beta}$ is a unit tensor. This leads to the result

$$\frac{\partial}{\partial u}\sqrt{g}\frac{\partial}{\partial u} = \sqrt{g}\frac{\partial^2}{\partial u^2} + \left(\frac{\partial\sqrt{g}}{\partial u}\right)\frac{\partial}{\partial u} = \frac{\partial^2}{\partial u^2} + K\frac{\partial}{\partial u}, \qquad (7.40)$$

and

$$\nabla^2 \approx \frac{\partial^2}{\partial u^2} + \frac{\partial^2}{\partial s^2} + K\frac{\partial}{\partial u}, \qquad (7.41)$$

since $uK\, \partial^2/\partial u^2$ is a higher-order term.

In the linearization using $\psi = \psi_o + \delta\psi$, both $\delta\psi$ and $(\partial u/\partial\tau)$ are small. Thus,

$$\frac{\partial\psi}{\partial\tau} = \frac{\partial\psi_o}{\partial\tau} + \frac{\partial\delta\psi}{\partial\tau} = \frac{\partial\psi_o}{\partial u}\frac{\partial u}{\partial\tau} + \frac{\partial\delta\psi}{\partial u}\frac{\partial u}{\partial\tau}$$

$$\approx \frac{\partial\psi_o}{\partial u}\frac{\partial u}{\partial\tau} = -v\frac{\partial\psi_o}{\partial u}. \tag{7.42}$$

The third and final ingredient needed to obtain Eq. (7.17) concerns $\delta f/\delta\psi -$ $\xi^2\nabla^2\psi$. For the equilibrium profile ψ_o, $\frac{\delta f}{\delta\psi_o} = \xi^2\frac{d^2\psi_o}{du^2}$. Using this relation and $\psi = \psi_o + \delta\psi$, to linear order in $\delta\psi$ we have

$$\frac{\delta f}{\delta\psi} = \frac{\delta f}{\delta\psi_o} + \frac{\delta^2 f}{\delta\psi_o^2}\delta\psi, \tag{7.43}$$

and

$$\nabla^2\psi = \left(\frac{\partial^2}{\partial u^2} + \frac{\partial^2}{\partial s^2} + K\frac{\partial}{\partial u}\right)(\psi_o + \delta\psi)$$

$$\approx \frac{d^2\psi_o}{du^2} + K\frac{d\psi_o}{du} + \left(\frac{\partial^2}{\partial u^2} + \frac{\partial^2}{\partial s^2}\right)\delta\psi, \tag{7.44}$$

since ψ_o is independent of s and $K\frac{\partial}{\partial u}\delta\psi$ is of higher order and may be neglected in the linearization procedure. Putting all these results together, we get

$$-\frac{\partial\psi_o}{\partial u}v = \nabla^2\left[\hat{\Omega}_2\delta\psi - \xi^2 K\frac{d\psi_o}{du}\right], \tag{7.45}$$

where

$$\hat{\Omega}_2 = \frac{\delta^2 f\{\psi_o\}}{\delta\psi_o^2} - \xi^2\left(\frac{\partial^2}{\partial u^2} + \frac{\partial^2}{\partial s^2}\right). \tag{7.46}$$

This is Eq. (7.17).

8

Domain growth and structure factor for model B

As a phase-separating mixture evolves towards equilibrium, domains of coexisting phases grow in size and the domain structure coarsens. During this coarsening process the system exhibits self-similarity, which can be exploited to deduce general features of the late-stage domain growth for model B. If the dynamics is self-similar then, on average, the pattern formed by domains at a given time τ can be made to match that at an earlier time τ' by a uniform rescaling of the system size (see Fig. 6.1). This scaling factor is

$$l = \frac{R(\tau)}{R(\tau')},\tag{8.1}$$

where R is some measure of the average domain size. The dynamics also becomes invariant if time is rescaled in accordance with the growth law, $R(\tau) \sim \tau^n$. Thus, on average, the equations describing the system should be invariant to a renormalization in which lengths are scaled by a factor of l and time is scaled by a factor of $l^{1/n}$, with n the unknown growth exponent.

8.1 Domain growth law

The analysis of the model B domain growth law begins with the nonlocal relation between interface curvature and the interface velocity in Eq. (7.27), which we reproduce here:

$$\sigma\xi^2 K(s) = \int du d\mathbf{x}' \, G(\mathbf{x}, \mathbf{x}') \frac{d\psi_o(u')}{du'} \frac{d\psi_o(u)}{du} v(\mathbf{x}').\tag{8.2}$$

Even though this result was deduced in Chapter 7 for two-dimensional systems, it is also true for $d \geq 2$ if s is replaced by a $(d-1)$-dimensional surface vector s. At late times we expect interfaces to be sharp. The interfacial profile $\psi_o(u)$ can then be approximated by a step function with infinitesimal width. Then, if $\Theta(u)$ is

60

the Heaviside function, which is zero for negative arguments and unity for positive arguments, one has $\psi_o(u) = \psi_o(-\infty)(1 - \Theta(u)) + \psi_o(+\infty)\Theta(u) = \Delta\psi\Theta(u) + \psi_o(-\infty)$, where $\Delta\psi = [\psi_o(+\infty) - \psi_o(-\infty)]$. From this relation it follows that $d\psi_o/du = \Delta\psi \, \delta(u)$. Substituting these expressions into Eq. (8.2), and noting that $\mathbf{x}' \equiv (u', \mathbf{s}')$, we get

$$\sigma\xi^2 K(\mathbf{s}) = |\Delta\psi|^2 \int d^{(d-1)}s' \, G(\mathbf{R}(\mathbf{s}), \mathbf{R}(\mathbf{s}'))v(\mathbf{s}'). \tag{8.3}$$

Applying the scaling operation to this equation yields

$$\sigma\xi^2 K^l(\mathbf{s}) = |\Delta\psi|^2 \int d^{(d-1)}s'^l \, G^l(\mathbf{R}(\mathbf{s}^l), \mathbf{R}(\mathbf{s}'^l))v^l(\mathbf{s}'^l), \tag{8.4}$$

where the superscript l denotes a scaled quantity. Here we assumed that $\xi^l \to \xi$: i.e. the interface width is a time-invariant *nonscaling* length. Next we use dimensional analysis to find the dependence of $\mathbf{s}$, K, and v on l. We find: $d^{(d-1)}s'^l \to l^{(d-1)}d^{(d-1)}s'$, $K^l \to l^{-1}K$, and $v^l \to l^{1-\frac{1}{n}}v$. Since the Green function satisfies Eq. (7.20), it must scale with l such that

$$[\nabla^l]^2 G^l = \delta(l\mathbf{x} - l\mathbf{x}'). \tag{8.5}$$

In d dimensions, $[\nabla^l]^2 \to l^{-2}\nabla^2$, and $\delta(l\mathbf{x} - l\mathbf{x}') \to l^{-d}\delta(\mathbf{x} - \mathbf{x}')$, so that $G^l \to l^{2-d}G$. Substituting these scaling results into Eq. (8.4) gives

$$\sigma\xi^2 K(\mathbf{s}) = l^{3-\frac{1}{n}} |\Delta\psi|^2 \int d^{(d-1)}s' \, G(\mathbf{R}(\mathbf{s}), \mathbf{R}(\mathbf{s}'))v(\mathbf{s}'), \tag{8.6}$$

which should be identical to Eq. (8.3) if self-similarity is to hold. Thus, invariance under the scaling operation requires that $n = \frac{1}{3}$. The origin of this growth exponent is related to the nonlinear structure of the model B field theory,

$$\frac{\partial\psi}{\partial\tau} = \nabla^2\mu, \tag{8.7}$$

with $\mu = -\psi + \psi^3 - \nabla^2\psi$. All three terms in the expression for the chemical potential are needed to construct the interfacial profile. The result, as discussed in Chapter 7 (see the analysis between Eqs. (7.2) and (7.27)), is that the order parameter field ψ decomposes into a kink-like solution $\psi_o(u)$ describing the interface and a bulk contribution $\delta\psi(u, \mathbf{s})$. The growth exponent, $n = \frac{1}{3}$, emerges from the time independence of the interfacial width ξ inherent in the kink solution ψ_o and the coupling of the bulk dynamics to the interfacial curvature through the surface tension. These two features are also the key ingredients of the well-known Lifshitz–Slyozov–Wagner theory of coarsening (see Chapter 12).

The emergence of sharp interfaces is a significant aspect of the late stage coarsening in model B. Because the interface is coupled to the bulk through the surface tension, the interfacial width remains an important length scale in the problem, even during the late stages. This is a particularly subtle point, since ξ is an asymptotically irrelevant length scale with respect to the shape of the scaling function discussed later in Chapter 9. By contrast, in model A, where the bulk and the interfacial dynamics are not coupled, ξ can be ignored in the late-stage dynamics. For model B, the correct late-stage theory must be such that the surface tension emerges in a natural way.

8.2 Porod's law and other consequences of sharp interfaces

Sharp interfaces in systems with a scalar order parameter ψ are examples of topological defects. More generally, for a system in which the order parameter is an n-component vector ψ, other types of stable topological defect that depend on the space dimensionality d can also be generated. For $n = 1$ (a scalar ψ), the topological defect is a $(d - 1)$-dimensional domain wall; for $n = d = 2$ it is a vortex; it is a string if $n = 2, d = 3$, and a monopole or a "hedgehog" if $n = d = 3$. In general all the n components of ψ must vanish at the defect core, which defines a surface of dimension $(d - n)$. The existence of such defects therefore requires $d \geq n$. For $n = 1$, a domain wall is the defect core where ψ is identically zero at all points.

For a system undergoing phase ordering or separation, the existence of topological defects has important consequences for the short-distance correlations of the order parameter: i.e. the small x behavior of $G(\mathbf{x}, \tau)$ or the large k behavior of $S(\mathbf{k}, \tau)$. Short distance means $\xi \ll x \ll R(\tau)$ for two points $\mathbf{x_0}$ and $(\mathbf{x_0} + \mathbf{x})$. Consider a scalar order parameter ψ in a system in the late stages of the evolution following a quench to a point within the coexistence region, such that sharp domain walls have been established and $\psi = \pm 1$ within the bulk regions. Then, the product $\psi(\mathbf{x_0})\psi(\mathbf{x_0} + \mathbf{x})$ will be -1 if a wall passes between the two points and will be $+1$ otherwise. Since R is the typical inter-defect distance, for low defect densities $x = |\mathbf{x}| \ll R$ and the probability of finding more than one wall can be neglected. The probability that a randomly placed rod of length x will cut a domain wall is of the order of x/R, so that, for $x \ll R$, we have

$$G(\mathbf{x}) \equiv \langle \psi(\mathbf{x_0})\psi(\mathbf{x_0} + \mathbf{x}) \rangle \approx (-1)\frac{x}{R} + (+1)(1 - \frac{x}{R}),$$

$$= 1 - 2\frac{x}{R} + \cdots,$$

$$= \langle \psi(\mathbf{x_0})\psi(\mathbf{x_0}) \rangle - \frac{1}{2}|\Delta\psi|^2\frac{x}{R} + \cdots, \tag{8.8}$$

where $R = R(\tau)$ and $\Delta\psi$ is the change in ψ across a domain wall. A more accurate calculation for $n = 1$, $d = 2$ leads to

$$G(\mathbf{x}, \tau) = G(\mathbf{x} = 0, \tau) - |\Delta\psi|^2 \frac{L}{\pi A} x + |\Delta\psi|^2 \frac{L}{24\pi A} \langle K^2 \rangle x^3 + \cdots, \quad (8.9)$$

where L is the total interface length in a system of area A (Rogers, 1989).

A number of observations follow from an examination of Eq. (8.9). For $d > 1$, $G(\mathbf{x}, \tau)$ is nonanalytic in $\mathbf{x}$ at $\mathbf{x} = 0$, since it is linear in $x = |\mathbf{x}|$. This presents no difficulty, since the result is valid only for $x > \xi$ and breaks down within the defect core. The coefficient of x^2 is zero in the expansion (8.9). This is due to the smoothness of the interface, as was first pointed out in 1952 by Porod (1952, 1982) and later by Tomita (1984). The fact that the coefficient of x^2 is zero is often referred to as the Tomita sum rule. The coefficient of x^3 contains $\langle K^2 \rangle$, which is an average of the square of the curvature K over the length of the interface. Both the linear and cubic terms are proportional to the interface perimeter density L/A, which is proportional to $1/R(\tau)$.

Generalization of Eq. (8.9) to a d-dimensional system with an n-component vector order parameter has been made by Bray and Puri (1991) and independently by Toyoki (1992), and the leading singular term takes the form

$$G_{sing}(\mathbf{r}) = \begin{cases} -2\frac{r}{R(\tau)} & \text{for } n = 1, \\ \mathcal{B}_{d,n} \, \rho_d \, |\mathbf{x}|^n & \text{for } n \le d, \end{cases} \quad (8.10)$$

where ρ_d is the defect density and $\mathcal{B}_{d,n}$ is a universal constant depending only on d and n.

Equation (8.9) was obtained using only the geometrical properties of the system and the existence of random sharp interfaces. Consequently, the result is valid for both models A and B, and for any other universality class in which defects are created. The nonanalytic form of $G(\mathbf{x}, \tau)$ in Eq. (8.9) implies a power law tail in its Fourier transform $S(\mathbf{k}, \tau)$. Using the fact that the defect density L/A typically scales as $1/R(\tau)$, a simple power-counting argument for large k leads to

$$S(\mathbf{k}, \tau) \equiv \int d^d x \, e^{i\mathbf{k}\cdot\mathbf{r}} G(\mathbf{x}, \tau) \sim \frac{1}{R(\tau)k^{d+1}}, \quad (8.11)$$

for a scalar order parameter. This result is known as Porod's law, since it is implicit in the early work of Porod (see also Debye *et al.*, 1957).

Using a similar power-counting argument on Eq. (8.10), the Fourier transform of $G_{sing}(\mathbf{r})$ leads to

$$S(\mathbf{k}, \tau) \sim \frac{\rho_d}{k^{d+n}} = \frac{1}{[R(\tau)]^n k^{d+n}}. \quad (8.12)$$

The expressions in Eqs. (8.11) and (8.12) are consistent with scaling, since

$$S(\mathbf{k}, \tau) = k^{-d} F(kR(\tau)), \tag{8.13}$$

when the domain size $R(\tau)$ is the dominant length.

8.3 Small-k behavior of $S(\mathbf{k}, \tau)$

For a scalar order parameter, $S(\mathbf{k}, \tau) \equiv \langle |\psi_k(\tau)|^2 \rangle$ has two peaks at late times. One peak is a rather broad Lorentzian $\sim k_B T/(k^2 + \xi^{-2})$ arising from the equilibrium Ornstein–Zernicke structure. This peak is rather difficult to see in experiments (see, however, Nagler *et al.*, 1988). The other Bragg peak arises from growth and ordering processes. This peak evolves in time and takes the form $\psi_o^2 (2\pi)^d \delta(\mathbf{k} - \mathbf{k_0})$ at $\tau = \infty$. For model A the Bragg peak is centered at $k = 0$ and its width is of order $1/R(\tau)$. For a scalar conserved order parameter the Bragg peak is qualitatively different: for small k one has

$$\frac{\partial \psi_k}{\partial \tau} = -k^2 \mu_k(\tau),$$

due to the conservation law. Yeung (1988) and Furukawa (1989) have argued that $S(\mathbf{k}, \tau) \sim k^4$ for small k, implying that $\mu_k(\tau)$ contains a k-independent term. These arguments have been put on a firm basis by Fratzl *et al.* (1991).

Another consequence of sharp interfaces is Nozieres' estimate for the order parameter autocorrelation function $G(\tau) \equiv \langle \psi(\mathbf{x_0}, \tau) \psi(\mathbf{x_0}, \tau) \rangle$. At late times $\psi(\mathbf{x}, \tau) = \pm \psi_o$ except near interfaces, where it is approximately zero. The "volume" in which it is zero is proportional to $R^{d-1} \xi / R^d$, so that

$$G(\tau) = \psi_o^2 \left(1 - \mathcal{C} \frac{\xi}{R(\tau)} + \cdots \right), \tag{8.14}$$

where $\mathcal{C}$ is a constant of order unity. This result is true regardless of the explicit time dependence of $R(\tau)$: e.g. $\tau^{1/2}$ for model A and $\tau^{1/3}$ for model B.

9

Order parameter correlation function

The model A Langevin equation without thermal noise is known as the time-dependent Ginzburg–Landau (TDGL) equation. In dimensionless form this equation is

$$\frac{\partial \psi}{\partial \tau} = \xi^2 \nabla_x^2 \psi - \frac{\delta f}{\delta \psi}. \tag{9.1}$$

The interface dynamics in a system governed by this equation is described most conveniently by transforming to the interface-specific, space-and-time-dependent coordinates (u, s). The coordinate u is zero on the interface. If $\hat{n}$ is a unit vector normal to the interface pointing in the direction of increasing ψ, then u is positive (negative) in the $+ (-)$ phase. The $(d - 1)$-dimensional vector s is tangent to the interface, such that $x = R(s) + u\hat{n}(s)$, with R the position of a point on the interface. In Chapter 7 we used the broken symmetry due to the interface to expand ψ as $\psi = \psi_o(u(x, \tau)) + \delta\psi$, where $\delta\psi = \sum_{n=1}^{\infty} a_n(s)\tilde{\zeta}_n(u)$. The sum starts with $n = 1$ since the $n = 0$ coefficient $a_o = 0$ on account of the orthogonality of the bulk degrees of freedom, $\delta\psi$, to the Goldstone mode $\zeta_o(u) \equiv d\psi_o(u)/du$.

A simple way to determine the velocity of the interface from Eq. (9.1) is as follows: close to the domain wall, $\nabla\psi = \left(\frac{\partial \psi}{\partial u}\right)_\tau \hat{n}$ so that

$$\nabla^2 \psi = \left(\frac{\partial^2 \psi}{\partial u^2}\right)_\tau + \left(\frac{\partial \psi}{\partial u}\right)_\tau \nabla \cdot \hat{n}. \tag{9.2}$$

Also, since $\left(\frac{\partial \psi}{\partial \tau}\right)_u = -\left(\frac{\partial \psi}{\partial u}\right)_\tau \left(\frac{\partial u}{\partial \tau}\right)_\psi$, Eq. (9.1) becomes

$$-\left(\frac{\partial \psi}{\partial u}\right)_\tau \left(\frac{\partial u}{\partial \tau}\right)_\psi = \xi^2 \left(\frac{\partial \psi}{\partial u}\right)_\tau \nabla \cdot \hat{n} + \xi^2 \left(\frac{\partial^2 \psi}{\partial u^2}\right)_\tau - \frac{\delta f}{\delta \psi}. \tag{9.3}$$

Assuming that the profile is given by the equilibrium condition

$$\xi^2 \left(\frac{\partial^2 \psi}{\partial u^2}\right)_\tau \approx \frac{\delta f}{\delta \psi}, \tag{9.4}$$

the last two terms in Eq. (9.3) cancel and one gets the Allen–Cahn result,

$$\left(\frac{\partial u}{\partial \tau}\right)_\psi \equiv v = -\xi^2 \nabla \cdot \hat{\mathbf{n}} \equiv -\xi^2 K, \tag{9.5}$$

where $K \equiv \nabla \cdot \hat{\mathbf{n}}$ is $(d - 1)$ times the mean curvature H. For $2d$ systems, the interface is a one-dimensional curve, and there is only one curvature K. For $3d$ systems, the mean and Gaussian curvatures were introduced in Chapter 6, and are useful in describing surfaces, interfaces, and membranes (Safran, 2003). In model A the bulk and interface motions decouple, and since in dimensionless units $\xi = 1$, one gets $v = -K$. Note that here $K = (d - 1)H$ and should not be confused with the Gaussian curvature.

9.1 Dynamic scaling and Ohta–Jasnow–Kawasaki theory

In approximate theories the physical field $\psi(\mathbf{x}, \tau)$ is often replaced by an auxiliary field $m(\mathbf{x}, \tau)$, which is akin to, but not exactly the same as, the curvilinear coordinate $u(\mathbf{x}, \tau)$. Such theories are exemplified by the work of Ohta, Jasnow, and Kawasaki (1982) (OJK). Since $m(\mathbf{x}, \tau)$ is smooth on all length scales, standard approximation methods that do not work for $\psi(\mathbf{x}, \tau)$ may work for $m(\mathbf{x}, \tau)$. The replacement of the physical field is carried out by introducing the approximate nonlinear relation

$$\psi(m) \approx \tanh(m/\sqrt{2}), \tag{9.6}$$

for a sigmoid-shape profile or

$$\psi(m) \approx \text{sign}(m), \tag{9.7}$$

for a step function shape profile. Since $K \equiv \nabla \cdot \hat{\mathbf{n}}$ and $\hat{\mathbf{n}} \sim \nabla u$ is a unit vector normal to the interface, we have $\hat{\mathbf{n}} = \nabla m/|\nabla m|$, which leads to

$$v = -K = -\nabla \cdot \hat{\mathbf{n}} = -\nabla \cdot \frac{\nabla \cdot m}{|\nabla \cdot m|}$$
$$= \left(-\nabla^2 m + (\hat{\mathbf{n}}.\nabla)(\hat{\mathbf{n}}.\nabla)m\right)/|\nabla \cdot m|. \tag{9.8}$$

In a frame of reference co-moving with the interface,

$$\frac{dm}{d\tau} = 0 = \frac{\partial m}{\partial \tau} + \mathbf{v} \cdot \nabla m. \tag{9.9}$$

Since $\mathbf{v}$ is parallel to ∇m, $\mathbf{v} \cdot \nabla m = v|\nabla m|$, and Eq. (9.9) then gives $v = -\frac{\partial m}{\partial \tau} \frac{1}{|\nabla m|}$. Substituting this result into Eq. (9.8) to eliminate v, we obtain an equation for m:

$$\frac{\partial m}{\partial \tau} = \nabla^2 m - (\hat{\mathbf{n}} \cdot \nabla)(\hat{\mathbf{n}} \cdot \nabla)m. \tag{9.10}$$

This equation was first obtained by Ohta *et al.* (1982). Since $\hat{\mathbf{n}}$ depends on m, this is a nonlinear equation. In the OJK theory $(\hat{\mathbf{n}} \cdot \nabla)(\hat{\mathbf{n}} \cdot \nabla)$ is replaced by its (isotropic) spherical average, $n_\alpha n_\beta \approx \delta_{\alpha\beta}/d$, where $\alpha, \beta = 1, 2, \ldots, d$. This yields $(\hat{\mathbf{n}} \cdot \nabla)(\hat{\mathbf{n}} \cdot \nabla) \approx \frac{1}{d}\nabla^2$ and, using this approximation, Eq. (9.10) reduces to a linear diffusion equation,

$$\frac{\partial m}{\partial \tau} = D\nabla^2 m, \tag{9.11}$$

with $D = \left(1 - \frac{1}{d}\right)$. Provided there are no long-range correlations present in the system at $\tau = 0$, random initial conditions should not play an important role in late stages of the evolution. This justifies the assumption that $m(\mathbf{x}, \tau)$ is a random Gaussian field initially:

$$\langle m(\mathbf{x}, 0)\rangle = 0,$$
$$\langle m(\mathbf{x}, 0)m(\mathbf{x}', 0)\rangle = \Delta\delta(\mathbf{x} - \mathbf{x}'). \tag{9.12}$$

Since Eq. (9.11) is linear, an initial Gaussian field will continue to remain Gaussian at all later times. Thus, Eq. (9.11) may be solved and averaged over initial conditions using Eq. (9.12) (equivalent to a nonequilibrium average) to obtain the equal time correlation function of m. We obtain

$$\langle m(\mathbf{x}_1, \tau)m(\mathbf{x}_2, \tau)\rangle = \frac{\Delta}{(8\pi D\tau)^{d/2}} \exp\left(-\frac{r^2}{8D\tau}\right), \tag{9.13}$$

where $\mathbf{x}_2 \equiv \mathbf{x}_1 + \mathbf{r}$. If we denote $m(\mathbf{x}_i, \tau)$ by $m(i)$, $i = 1, 2$, we also get

$$s_o(1) \equiv \langle m(1)^2\rangle = \frac{\Delta}{(8\pi D\tau)^{d/2}} = s_o(2) \equiv \langle m(2)^2\rangle. \tag{9.14}$$

The normalized correlation function $\gamma(12)$ is

$$\gamma(12) = \frac{\langle m(1)m(2)\rangle}{(s_o(1)s_o(2))^{\frac{1}{2}}} = \exp\left(-\frac{r^2}{8D\tau}\right). \tag{9.15}$$

The joint probability for $m(1)$ and $m(2)$ may be expressed in terms of γ, $s_o(1)$ and $s_o(2)$, since $m(i)$ is a Gaussian field:

$$P\big(m(1), m(2)\big) = N \exp\left(-\frac{1}{2(1-\gamma^2)}\left[\frac{m(1)^2}{s_o(1)} + \frac{m(2)^2}{s_o(2)} - 2\gamma\frac{m(1)m(2)}{\sqrt{s_o(1)s_o(2)}}\right]\right).$$

(9.16)

The normalization constant N is given by

$$N = (2\pi)^{-1}\left[(1-\gamma^2)s_o(1)s_o(2)\right]^{-\frac{1}{2}}.$$

(9.17)

Assuming the step function form for $\psi(m)$ given in Eq. (9.7), the order parameter correlation function takes the form

$$G(r,\tau) = \langle\psi(m(1))\psi(m(2))\rangle = \langle\text{sign}(m(1))\text{sign}(m(2))\rangle$$
$$= \frac{2}{\pi}\ \sin^{-1}\big(\gamma(12)\big),$$

(9.18)

where the average $\langle\ldots\rangle$ is over the distribution $P\big(m(1), m(2)\big)$ in Eq. (9.16). Since $\gamma(12)$ is a function of $r^2/(D\tau)$, G is also a function of this combination of variables. Thus, the domain size $R(\tau)$ can be calculated as

$$R(\tau) = \langle r\rangle(\tau) = \int d^d r\ rG\left(\frac{r^2}{D\tau}\right)\bigg/\int d^d r,$$
$$= (D\tau)^{\frac{1}{2}}\int d^d x\ xG(x^2)\bigg/\int d^d x.$$

(9.19)

From OJK theory one obtains dynamic scaling, the scaled correlation function, and the power law domain growth, $R(\tau) \sim (D\tau)^{\frac{1}{2}}$, with $D = (d-1)/d$. Since an analytic expression for the order parameter correlation function G is obtained in the theory, the proportionality constant in the expression for the domain size is also obtained in this theory, as seen from Eq. (9.19). The scaling function compares well with the results from numerical simulations for $d = 2$ and 3.

9.2 Other theories

The domain growth law $R(\tau) \sim \tau^{\frac{1}{2}}$ for model A is firmly based on results from experiments and computer simulation. It was also among the results obtained by Kawasaki, Yalabik, and Gunton (1978) (KYG) using a weak coupling, long-time approximation. The KYG theory used earlier ideas of Suzuki (1976a, 1976b) to carry out a resummation of an approximate infinite order perturbation theory for the TDGL equation. For the quartic potential $f(\psi) = (1 - \psi^2)^2/4$ this equation takes the form

$$\frac{\partial \psi}{\partial \tau} = \nabla^2 \psi + \psi - g\psi^3,$$ (9.20)

where g is an expansion parameter which is set equal to unity at the end of calculation. The basic idea in this approach is to treat g as a small parameter and extract the leading asymptotic behavior in τ of each term in the series. The resulting series may then be resummed to obtain the result. The KYG solution can be expressed as

$$\psi(m) = \frac{m}{(1 + m^2)^{\frac{1}{2}}},$$ (9.21)

with m satisfying the equation

$$\frac{\partial m}{\partial \tau} = \nabla^2 m + m.$$ (9.22)

This equation contains exponential growth and diffusion terms akin to the "reactive" and "diffusive" elements in Eq. (9.20). These equations may be contrasted with those in the OJK theory by comparing Eq. (9.22) with Eq. (9.11), and Eq. (9.21) with Eq. (9.7). If Eq. (9.7) is used instead of Eq. (9.21), one obtains Eq. (9.18) for the correlation function $G(r, \tau)$, with γ as in Eq. (9.15) but with $D = 1$. If Eq. (9.21) is substituted into Eq. (9.20), one finds that m satisfies

$$\frac{\partial m}{\partial \tau} = \nabla^2 m + m - 3\frac{m(\nabla m)^2}{1 + m^2}.$$ (9.23)

The third nonlinear term cannot be neglected. A weakness of the KYG theory is a lack of self-consistency, which arises from the difference between Eqs. (9.23) and (9.22).

Other similar approaches have been attempted by Mazenko (1989, 1990, 1991) and Bray and Humayun (1993). Dynamic scaling emerges naturally in the Bray and Humayun approach.

9.3 Extension to model B

Thus far we have discussed theories for model A with a scalar order parameter. Using similar ideas and approximations, attempts have been made to obtain the functional form of the structure factor and the correlation function for model B. The motions of interfaces are strongly correlated as a consequence of the existence of the conservation law. This feature introduces major challenges for theoretical descriptions. The Gaussian closure technique, which was quite successful for model A, does not appear to work as well for model B. Yeung *et al.* (1994) compared the Gaussian closure theory with simulation results for the $3d$ model B system of Shinozaki and Oono (1991, 1993). The theory compares very well with the simulation results for the scaled correlation function $\tilde{G}(X)$ up to its second zero, but the agreement becomes progressively worse for larger values of X. The theory gives the incorrect small-k behavior of the structure factor $S(k, \tau)$: $S \sim k^2$ instead of the correct k^4 behavior. With the assumption that $m(\mathbf{x}, \tau)$ is a random Gaussian field, the OJK relation between G and γ (Eq. (9.18)) is independent of the dynamical model, and remains true for model B. Using this relation the spectral density γ was extracted from the simulation data. By definition, $\gamma(k)$ is positive for all k, but for $k < 0.5$ and $1.5 < k < 4$ it was found to be negative. This leads to the inevitable conclusion that the Gaussian approximation cannot be used to recover either the correlation function $\tilde{G}(X)$ or the structure factor $S(k, \tau)$ over the entire X or k ranges.

10

Vector order parameter and topological defects

Domain walls or interfaces are the simplest forms of topological defect, and occur in systems with a scalar order parameter. The link between interfaces and the Goldstone mode was pointed out earlier. A planar interface is topologically stable, since local changes in the order parameter may move the interface but cannot destroy it. For curved interfaces with an intrinsic width ξ, at late times the local radius of curvature $K^{-1} \sim R(\tau)$. Since $R(\tau) \gg \xi$, the interface is locally planar to a good approximation, and curved interfaces are also topologically stable.

Now consider vector fields. The $O(n)$ model is appropriate for systems with an n-component vector field $\psi(\mathbf{x}, \tau)$. The generalization of the free energy functional is

$$\mathcal{F}[\psi(\mathbf{x})] = \int d^d x \left[f(\psi) + \frac{1}{2}(\nabla \psi)^2 \right], \tag{10.1}$$

where

$$(\nabla \psi)^2 = \sum_{i=1}^{d} \sum_{\alpha=1}^{n} \left(\frac{\partial \psi^\alpha}{\partial x_i} \right)^2; \tag{10.2}$$

i.e. a scalar product over both spatial and internal coordinates. Here

$$f(\psi) = (1 - \psi^2)^2, \tag{10.3}$$

is the "mexican hat"-like double well potential shown in Fig. 10.1(a). The free energy functional $\mathcal{F}[\psi]$ is invariant under a global rotation of ψ (a continuous symmetry) for $n > 1$. This should be contrasted with the inversion symmetry $(\psi \rightarrow -\psi)$, which is a discrete symmetry for the scalar $(n = 1)$ case.

The generalized TDGL model A is

$$\frac{\partial \psi}{\partial \tau} = \nabla^2 \psi - \frac{\delta f}{\delta \psi}. \tag{10.4}$$

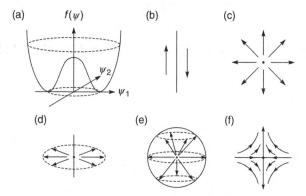

Fig. 10.1. For the $O(n)$ model, (a) the double well free energy with $n = 2$, (b)–(f) different types of topological defect: (b) $n = 1$, domain wall; (c) $n = d = 2$, vortex; (d) $n = 2$, $d = 3$, string; (e) $n = d = 3$, monopole or "hedgehog"; (f) antivortex. After Bray (1994).

For model B (conserved vector fields), the generalization is obtained by acting on the right-hand side with $(-\nabla^2)$. As in the case of a scalar order parameter field, the stationary solutions of Eq. (10.4) with appropriate boundary conditions lead to stable topological defects for vector order parameter fields. In order to accommodate the global rotation symmetry, the order parameter field configurations are radially symmetric with respect to the defect core. In d dimensions all n components of ψ vanish at the defect core. This constraint generates a surface of dimension $(d - n)$: e.g. a planar interface (domain wall) is a surface of dimension $(d - 1)$ (Fig. 10.1(b)).

The existence of such defects requires that $d \geq n$. For $d = n$ the defects are point defects. For $n = 2$ they are vortices if $d = 2$ (Fig. 10.1(c)) and strings or vortex lines if $d = 3$ (Fig. 10.1(d)). For $d = n = 3$ the defects are point monopoles or hedgehogs (Fig. 10.1(e)).

For $d > n$, ψ vanishes only in the n dimensions "orthogonal" to the defect core; it is uniform in the remaining $(d - n)$ dimensions "parallel" to the core. The defects are spatially extended, and curvature plays an important role. Coarsening occurs by the removal of sharp features from the defect surface and the shrinkage and disappearance of small domain bubbles or vortex loops. Such processes reduce the total "area" of the spatially extended defects.

For $d = n$ only point defects exist, and coarsening occurs by the annihilation of defect–antidefect pairs (see Figs. 10.1(c) and (f), which show examples of a vortex defect and an antivortex defect). Vortex and antivortex defects have different topological charges. On going along a loop around the defect, the order parameter ψ rotates by $\pm 2\pi$, respectively, for vortex and antivortex defects. Thus, an antivortex cannot be obtained from a vortex by simply reversing the signs of all the components of the ψ field. On the other hand, an antimonopole is obtained by reversing the arrows of the monopole. In general an antidefect cannot be obtained by simple rotation of a defect.

The defects shown in Fig. 10.1 (c), (d), and (e) are radially symmetric, and one can write $\psi(\mathbf{x}) = \hat{\mathbf{x}}\zeta(r)$, where $\hat{\mathbf{x}}$ is a unit vector in the radial direction and $\zeta(r)$ is the profile function. Putting $\partial \psi / \partial \tau = 0$ and using this symmetric form in Eq. (10.4), one gets

$$\frac{d^2\zeta(r)}{dr^2} - \frac{(n-1)}{r}\frac{d\zeta(r)}{dr} - \frac{(n-1)}{r^2}\zeta(r) - \frac{\delta f(\zeta)}{\delta \zeta} = 0, \tag{10.5}$$

subject to the boundary conditions $\zeta(0) = 0$ and $\zeta(\infty) = 1$. For a scalar order parameter $(n = 1)$ the profile approaches its asymptotic value exponentially. For $n > 1$ a different behavior is obtained. Let $\zeta(r) = 1 - \epsilon(r)$ and assume that $\epsilon(r)$ is small. Substituting this expression for $\zeta(r)$ into Eq. (10.5) and expanding in powers of $\epsilon(r)$, one obtains the first-order result that

$$\epsilon(r) \approx \frac{(n-1)}{f''(\zeta = 1)}\frac{1}{r^2}, \quad \text{as } r \to \infty. \tag{10.6}$$

A useful definition of the core size ξ (analogous to the interface width) can be found using the relation

$$\zeta(r) \to 1 - \frac{\xi^2}{r^2}, \tag{10.7}$$

which gives

$$\xi = \left(\frac{(n-1)}{f''(1)}\right)^{\frac{1}{2}}, \quad \text{for } n > 1. \tag{10.8}$$

In Chapter 7, the fact that interface velocity v at a point is proportional to the local curvature K was deduced and discussed for model A with a scalar order parameter. This result can be generalized to the case of an n-component vector field (Bray and Puri, 1991; Toyoki, 1992), and is an extension of the approximate KYG theory. The argument can be cast in terms of two phenomenological parameters: the surface tension σ and friction constant η. The force due to curvature is σK per unit "area" of defect, and the frictional retarding force is ηv. Equating these two forces, identifying $K \sim 1/R(\tau)$ and $v \sim dR/d\tau$, one gets

$$R(\tau) \approx A_d \left(\frac{\sigma \tau}{\eta}\right)^{\frac{1}{2}}, \tag{10.9}$$

where A_d is a universal constant depending only on d and the definition of $R(\tau)$. For example, one can define $R(\tau)$ from the relation $G(R(\tau), \tau) = 1/2$.

The case of a vector *conserved* order parameter ($O(n)$ model) has also been studied (Puri, 2004). The Gaussian closure technique has been used by Puri *et al.* (1995) and Rojas *et al.* (2001) for the XY and Heisenberg models. Coniglio and Zannetti (1989) showed that the structure factor exhibits multiple-length scaling rather than single-length scaling in the limit $n \to \infty$. Bray and Humayun (1992) showed that this multi-scaling is a singular property of the $n = \infty$ limit. Qian and Mazenko (2004) studied vortex kinetics for $n = d = 2$ using and extending a heuristic scaling treatment for the large speed tails of defect velocity distribution developed by Bray (1997).

11

Liquid crystals

Liquid crystals are ubiquitous. They are in silk, snail slime, and crude oil. They are in mantles of neutron stars, and provide models for cosmic strings. They are in our food (gluten) and drinks (milk). The behavior of hair cells in the inner ear and the function of DNA are affected by them. The insulating coating of the axons of nerve cells is a liquid crystal called myelin. Liquid crystals are very responsive to excitations, which has led to many useful applications, such as liquid crystal displays. A great deal is known and understood about liquid crystalline materials (Chandrasekhar, 1992; de Gennes and Prost, 1993).

Liquid crystalline materials are orientationally ordered soft matter (Palffy-Muhoray, 2007). These materials are composed of large organic molecules, which have a long and rigid core, typically consisting of several linked benzene rings, terminated by a flexible alkyl chain. Such a molecular structure is then often modeled by disk-like or rod-like entities, depending on the cylindrical aspect ratio. Such model molecules have a head–tail symmetry. Thus, at high densities, liquid crystals can naturally create local orientational order. Onsager (1949) showed that hard rods tend to align at volume fractions larger than about four times their breadth-to-length ratio. Many liquid crystal phases can exist, depending on the temperature and solvent concentration. Some of these phases are shown in Fig. 11.1.

An isotropic disordered liquid phase exists at high temperatures. As the temperature is lowered, there is a competition between the positional and orientational entropies: the former favors a random location for a rod and the latter a random orientation. At low temperatures or high concentrations, maximum positional entropy is favored and the rods become nearly parallel, forming the nematically ordered phase. This is the simplest of the ordered phases, in which the molecules acquire a partial orientational order where the long axes of the molecules tend to align along a spontaneously selected direction. Anisotropic molecular polarizability in liquid crystal molecules leads to van der Waals forces, which also contribute

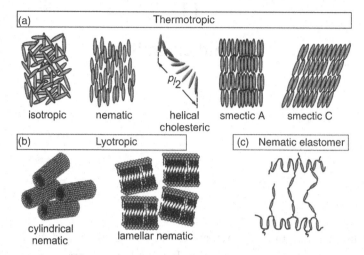

Fig. 11.1. Structures of various liquid crystal phases. For the helical cholesteric phase, half a rotation period is shown. The phase transitions of (a) thermotropic and (b) lyotropic liquid crystals are, respectively, functions of temperature and solvent concentration. (c) Liquid crystal elastomers are rubbers whose constituent molecules are orientationally ordered. Reprinted with permission from Palffy-Muhoray (2007). Copyright 2007, American Institute of Physics.

to the creation of parallel alignment below a critical temperature. In the nematic phase the liquid crystalline material has some orientational order but no translational order.

11.1 Nematic liquid crystals

Nematic liquids are almost always uniaxial. Their orientational order is described by a tensorial order parameter $\mathbf{Q}$. The orientation of a molecule l, with its center of mass located at $\mathbf{x}^{(l)}$, is defined through a unit vector field, the director $\boldsymbol{v}^{(l)}$, which points along its long axis. Since the two ends of the molecule are indistinguishable due to inversion symmetry, the order parameter is a symmetric traceless second-rank tensor which is even in the director field,

$$Q_{ij} = \frac{V}{N} \sum_{l=1}^{N} \left(v_i^{(l)} v_j^{(l)} - \frac{1}{3} (\boldsymbol{v}^{(l)} \cdot \boldsymbol{v}^{(l)}) \delta_{ij} \right) \delta(\mathbf{x} - \mathbf{x}^{(l)}). \qquad (11.1)$$

The volume V contains N molecules, $v_i^{(l)}$ is the ith component of $\boldsymbol{v}^{(l)}$, and Q_{ij}, as defined, is dimensionless. In the high-temperature isotropic phase, the average of the symmetric traceless tensor $\langle \mathbf{Q} \rangle$ over directions is zero. Its average is nonzero in the ordered nematic phase. Assuming that the nematic phase is uniaxial, one can choose the coordinate system with one axis along the direction of molecular alignment to obtain

$$\langle \mathbf{Q} \rangle = S(\hat{\mathbf{n}}\hat{\mathbf{n}} - \frac{1}{3}\mathbf{I}), \tag{11.2}$$

where $\mathbf{I}$ is a unit tensor and the unit vector $\hat{\mathbf{n}}(\mathbf{x})$ specifies the direction of the principal axis of $\langle \mathbf{Q} \rangle$. The unit vector $\hat{\mathbf{n}}$ is called the director. Computing $\langle \mathbf{Q} \rangle : \hat{\mathbf{n}}\hat{\mathbf{n}}$, one gets the scalar order parameter

$$S = \frac{1}{2}\langle 3(\mathbf{v}^{(l)} \cdot \hat{\mathbf{n}})^2 - 1 \rangle. \tag{11.3}$$

The Landau free energy density f for a nematic liquid crystal must be invariant under all rotations. Since $\mathbf{Q}$ transforms like a tensor under the rotation group, f can only be a function of the scalar combinations $\mathrm{Tr}\langle \mathbf{Q} \rangle^p$. The term with $p = 1$ is zero since $\mathrm{Tr}\mathbf{Q}$ vanishes by definition. Then, up to fourth order in $\langle \mathbf{Q} \rangle$ (Chaikin and Lubensky, 1995),

$$f = \frac{1}{2}a_2\left(\frac{3}{2}\mathrm{Tr}\langle \mathbf{Q} \rangle^2\right) - a_3\left(\frac{9}{2}\mathrm{Tr}\langle \mathbf{Q} \rangle^3\right) + a_4\left(\frac{3}{2}\mathrm{Tr}\langle \mathbf{Q} \rangle^2\right)^2$$
$$= \frac{1}{2}a_2 S^2 - a_3 S^3 + a_4 S^4, \tag{11.4}$$

where a_3 and a_4 are temperature-independent and, in analogy to the Landau theory for a scalar order parameter, $a_2 = a_{20}(T - T^*)$, where T^* is the mean field transition temperature. For a 3×3 traceless symmetric tensor, $\mathrm{Tr}\langle \mathbf{Q} \rangle^4 = (1/2)(\mathrm{Tr}\langle \mathbf{Q} \rangle^2)^2$, so that a $\mathrm{Tr}\langle \mathbf{Q} \rangle^4$ term is implicitly included in the last term proportional to the a_4 factor. The third-order term is allowed since the constituent molecules have inversion (head–tail) symmetry. The quadrupolar symmetry of the rod-like molecules leads to the tensorial order parameter, and rotational invariance does not rule out odd terms in the free energy density. In this mean field theory, the presence of a third-order term leads to a first-order transition at a temperature T_c which is greater than T^*. At high temperatures, $f(S) = \frac{1}{2}a_2 S^2 - a_3 S^3 + a_4 S^4$ has a single minimum at $S = 0$. As the temperature is decreased, two additional minima occur at symmetric points $\pm S_o$. At a temperature T_c these minima, which now occur at $\pm S_c$, and the minimum at $S = 0$ have equal values of free energy. Thus, $(\partial f / \partial S)|_{S_C} = 0$ and $f(S_c) - f(0) = 0$. Using these two relations, one can show that T_c is related to the parameters by $a_{2c} = a_{20}(T_c - T^*) = a_3^2/2a_4$ and $S_c = a_3/2a_4$. At temperatures below T_c the minima at nonzero S_o have a lower free energy than at $S = 0$. The phase transition at T_c is thus a first-order transition with a discontinuous jump in S.

A square gradient term can be added to the Landau free energy density to obtain a simple free energy functional (de Gennes and Prost, 1993),

$$\mathcal{F}[\langle \mathbf{Q} \rangle] = \int d^d x \left[f(\langle \mathbf{Q} \rangle) + \frac{\kappa}{2} \text{Tr} |\nabla \langle \mathbf{Q} \rangle|^2 \right]. \tag{11.5}$$

Using Eq. (11.2) and the relation $\nabla (\hat{\mathbf{n}} \cdot \hat{\mathbf{n}}) = 0$, the gradient term can be expressed as being proportional to $\sum_{i,j} (\partial \hat{n}_j / \partial x_i)^2$, where $j = 1, \ldots, n$ and $i = 1, \ldots, d$ for an $O(n)$ model for a system in d spatial dimensions. The above form of the gradient term assumes that the internal and spatial dimensions can be considered as distinct. However, in real nematic liquid crystals these spaces are coupled, since rotations of the spatial coordinates and the order parameter are produced by the same rotation operator. Thus, the two vectors $\hat{\mathbf{n}}$ and ∇ are not independent. The gradient free energy should be invariant under uniform rotations of the entire system, and under the symmetry operations $\hat{\mathbf{n}} \rightarrow -\hat{\mathbf{n}}$ and $\mathbf{x} \rightarrow -\mathbf{x}$. Also, since $\hat{\mathbf{n}}$ is a unit vector, $\hat{n}_i \nabla_j \hat{n}_i = 0$. These requirements result in three possible allowed scalar combinations of $\hat{\mathbf{n}}$ and ∇ in three dimensions: $(\nabla \cdot \hat{\mathbf{n}})^2$, $[\hat{\mathbf{n}} \cdot (\nabla \times \hat{\mathbf{n}})]^2$, and $[\hat{\mathbf{n}} \times (\nabla \times \hat{\mathbf{n}})]^2$, which respectively correspond to splay (nonzero $\nabla \cdot \hat{\mathbf{n}}$), twist (nonzero $\hat{\mathbf{n}} \cdot (\nabla \times \hat{\mathbf{n}})$), and bend (nonzero $\hat{\mathbf{n}} \times (\nabla \times \hat{\mathbf{n}})$) distortions of the director $\hat{\mathbf{n}}$. The resulting gradient free energy density is the Frank energy density (Oseen, 1933; Frank, 1958)

$$\mathcal{F}_n = \frac{1}{2} \int d^d r \left[K_1 (\nabla \cdot \hat{\mathbf{n}})^2 + K_2 [\hat{\mathbf{n}} \cdot (\nabla \times \hat{\mathbf{n}})]^2 + K_3 [\hat{\mathbf{n}} \times (\nabla \times \hat{\mathbf{n}})]^2 \right], \tag{11.6}$$

where the Frank elastic constants K_1, K_2, and K_3 are associated, respectively, with splay, twist, and bend of the director $\hat{\mathbf{n}}$. Figure 11.2 schematically shows the bend, splay, and twist distortions of the director $\hat{\mathbf{n}}$ of a nematic liquid crystal. The constants K_i have dimensions of energy/length, and are of the order of $k_B T_{ni}/a$, where T_{ni} is the nematic–isotropic transition temperature and a is a typical molecular length. A simple approximation where the constants K_i are taken to be equal, $K_1 = K_2 = K_3$, leads to the simpler form of square gradient term in Eq. (11.5). For nematic phases this approximation is often reasonable.

The relaxational dynamics of model A can be explored using the simple free energy functional in Eq. (11.5). The order parameter dynamics is nonconserved and ignores coupling to hydrodynamic variables. The equation of motion for $\langle \mathbf{Q} \rangle$ must include the constraint that it is a traceless tensor at all times. This leads to

$$\frac{\partial \langle \mathbf{Q} \rangle}{\partial t} = -\frac{\delta \mathcal{G}[\langle \mathbf{Q} \rangle]}{\delta \langle \mathbf{Q} \rangle}, \tag{11.7}$$

where

$$\mathcal{G}[\langle \mathbf{Q} \rangle] = \mathcal{F}[\langle \mathbf{Q} \rangle] - \int d^d x \lambda(\mathbf{x}) \langle \mathbf{Q} \rangle(\mathbf{x}), \tag{11.8}$$

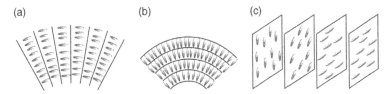

Fig. 11.2. (a) Bend distortion of the director leads to a large deviation of the layer spacing from its preferred value. (b) Splay distortion of director at constant layer spacing. (c) Twist distortion of the director requires a third dimension. After Chaikin and Lubensky (1995). Reprinted with the permission of Cambridge University Press.

and λ is a Lagrange multiplier introduced to maintain the condition $\mathrm{Tr}\langle \mathbf{Q} \rangle = 0$. Substitution of Eq. (11.5) for $\mathcal{F}$ and Eq. (11.4) for f into Eq. (11.7) leads to

$$\frac{\partial \langle \mathbf{Q} \rangle}{\partial t} = \kappa \nabla^2 \langle \mathbf{Q} \rangle - \frac{3}{2} a_2 \langle \mathbf{Q} \rangle + \frac{27}{2} a_3 \langle \mathbf{Q} \rangle^2 - 9 a_4 \langle \mathbf{Q} \rangle \mathrm{Tr} \langle \mathbf{Q} \rangle^2 + \lambda \mathbf{I}. \quad (11.9)$$

Taking the trace of this equation and imposing the constraint that $\mathrm{Tr}\langle \mathbf{Q} \rangle = 0$, one obtains $\lambda = -\frac{9}{2} a_3 \mathrm{Tr} \langle \mathbf{Q} \rangle^2$. Substituting this value of λ into Eq. (11.9), the equation satisfied by $\langle \mathbf{Q} \rangle$ is found to be

$$\frac{\partial \langle \mathbf{Q} \rangle}{\partial t} = \kappa \nabla^2 \langle \mathbf{Q} \rangle - \frac{3}{2} a_2 \langle \mathbf{Q} \rangle + \frac{27}{2} a_3 \left(\langle \mathbf{Q} \rangle^2 - \frac{1}{3} \mathbf{I} \, \mathrm{Tr} \langle \mathbf{Q} \rangle^2 \right) - 9 a_4 \langle \mathbf{Q} \rangle \mathrm{Tr} \langle \mathbf{Q} \rangle^2. \quad (11.10)$$

A nematic liquid crystal is a nearly perfect experimental system with a tensor order parameter. When such systems are rapidly quenched from the disordered phase to the nematic phase, topological defects are generated. It is of interest to extract scaling laws observed during the coarsening defect dynamics. The rate at which the quench takes place determines the initial characteristic length over which the symmetry-breaking direction is correlated. The interfaces between the patches with different symmetry directions cost gradient energy, which the system attempts to minimize during coarsening. However, topological constraints prevent the smooth merging of different symmetry patches and lead to regions with singularities in the gradient energy. These are the topological defects.

Since $\hat{\mathbf{n}}$ has inversion (head–tail) symmetry, $\langle \mathbf{Q} \rangle$ is traceless and the kinetics need not be in the same universality class as that of a vector order parameter system discussed in the previous chapter. The symmetry of the isotropic phase is the full rotation group $O(3)$. In the transition to the nematic phase, the $O(3)$ symmetry

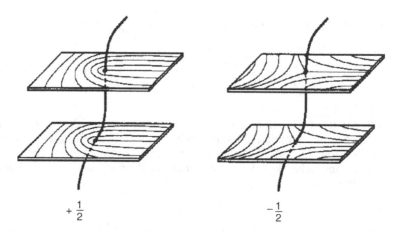

Fig. 11.3. Examples of $+\frac{1}{2}$ and $-\frac{1}{2}$ disclination lines. Reprinted from Yurke *et al.* (1992). Copyright 1992, with permission from Elsevier.

is spontaneously broken to the symmetry of a cylinder, $D_{\infty h}$. Due to the existence of inversion symmetry, it is sufficient to consider the symmetry breaking of the $SO(3)$ subgroup of $O(3)$ to $O(2)$. A number of defect types are supported by nematic liquid crystals (Kléman, 1983). Point monopole defects arise from the global symmetry under the rotation of $\hat{\mathbf{n}}$. The local inversion symmetry under $\hat{\mathbf{n}} \rightarrow -\hat{\mathbf{n}}$ leads to $\pm\frac{1}{2}$ string defects or disclinations. The director $\hat{\mathbf{n}}$ rotates through $\pm\pi$ when a path which encloses a $\pm\frac{1}{2}$ string is traversed. These strings, which are illustrated in Fig. 11.3, are topologically stable. Texture defects are also possible. These defects are nonsingular, except during a brief time when their unwinding occurs. Among the monopoles, strings, and textures, the $\pm\frac{1}{2}$ strings are the most dominant defects.

The approximate KYG theory, discussed in Section 9.2, has been extended to obtain the pair correlation function and the structure factor from Eq. (11.10) (Bray *et al.*, 1993) in order to describe the experimental measurements of Yurke *et al.* (1992). For nematic phases, a k^{-5} Porod tail is obtained for $d = 3$, implying that the behavior is similar to that for $n = 2$ vector order parameter systems. A $t^{1/2}$ time dependence of the characteristic length $R(t)$ is also obtained. Since the dimension of the defect is $d - n$, the number of defects per unit volume ρ behaves as R^{-n} and the total string length, with $n = 2$, decreases as t^{-1}. Experiments show that the density of monopole defects ($n = 3$) decays faster than R^{-3}, and the $\pm\frac{1}{2}$ disclination lines occur with the greatest abundance.

The disclination line tension and the viscous dissipation within the system are weak logarithmic functions of R. Both are proportional to $\ln(R/R_c)$, where R_c is the defect core radius. Consequently, the dynamics of a string can be understood qualitatively in terms of a constant line tension σ and a constant mobility Γ. If r is the local radius of curvature of the string, its equation of motion is

$$\Gamma \frac{dr}{dt} = -\frac{\sigma}{r},$$
(11.11)

which integrates to

$$r^2 = \frac{2\sigma(t_o - t)}{\Gamma},$$
(11.12)

where t_o is the time it takes for the defect to collapse. A similar simple argument for the decay of the string density can also be constructed as follows. If there is only one characteristic length ξ, the string density is given by

$$\rho \propto \xi^{-2}.$$
(11.13)

The characteristic radius of curvature is proportional to ξ and the characteristic line tension force per unit length is $f_\sigma \propto \xi^{-1}$. The frictional force per unit length is $f_\Gamma = -\Gamma v$, where v is the characteristic velocity of the string. Equating the two forces gives $v \propto \xi^{-1}$. The elastic energy per unit volume, E, stored in the disclination lines is

$$E = \frac{\sigma \xi}{\xi^3} \propto \sigma \rho.$$
(11.14)

The energy dissipation rate per unit volume due to viscous friction is

$$W = \frac{f_\Gamma v \xi}{\xi^3} = -\frac{\Gamma v^2}{\xi^2} \propto \Gamma \rho^2.$$
(11.15)

The equation of motion for ρ is obtained by equating W and the rate of change of elastic energy dE/dt:

$$\frac{d\rho}{dt} = -\frac{c\Gamma}{\sigma} \rho^2,$$
(11.16)

where c is a constant of proportionality. The solution of this equation is

$$\rho = \frac{\sigma}{c\Gamma}(t - t_o)^{-1},$$
(11.17)

where t_o is a constant of integration. Thus, at late times, the string density scales as t^{-1} and the characteristic length scales as $t^{1/2}$.

11.2 Smectic liquid crystals

Smectic liquid crystals have molecules in well-defined layers with a spacing that is of the order of the molecular length. In each layer, molecules move as in liquids and have no spatial correlation with molecules in other layers. There is, on average, some positional order such as a periodic density variation in directions normal to the layer, in addition to the orientational order akin to that in nematic liquid crystals. A system of microphase-separated block-copolymer microdomains has the same symmetry as a two-dimensional smectic phase and has been extensively studied (Harrison *et al.*, 2002). A single layer of such microdomains in a cylindrical phase of an asymmetric copolymer can be considered as a model 2*d* smectic phase. Two-dimensional smectic phases have a liquid-like positional order along one axis and a mass density wave along an orthogonal axis.

While the transition from a disordered phase to a nematic phase breaks rotational symmetry, the layering of molecules in a smectic phase breaks rotational and translational symmetries. Disclination defects also occur in the smectic phase, and the development of orientational order dominates the pattern coarsening kinetics. Annealing of disclination defects leads to the growth of orientational order. Figure 11.4 shows a 2*d* section of $\pm\frac{1}{2}$ disclination defects, along with a scanning electron microscopy image of such defects in an annealed sample of a 2*d* asymmetric diblock copolymer film. Such defects arise from the topological constraints present in these systems.

Stripe patterns with disclination defects have a striking similarity to dermatoglyphic prints on our palms and soles (Fig. 11.4). Both are director field patterns and so have similar topological constraints (Penrose, 1965). Dermatoglyphs increase one's ability to grip, act as a stitching to secure the epidermis to the dermis layer, and probably arise due to some process involving a field with a tensorial character, such as strains arising from surface curvature of the palms, fingers, soles, and toes. In contrast to smectic phases, the dermatoglyphic patterns are in a steady state, originate at isolated regions (center of finger tips), and proceed from the distal to proximal regions of a limb. In dermatoglyphic prints, a $+\frac{1}{2}$ disclination is called a loop and a $-\frac{1}{2}$ disclination a triradius. The arch, loop, and whorl describe typical configurations on digits. On each palm there are at least four triradii, which are usually in the positions as shown at a, b, d, and t in Fig. 11.4. For every loop in the palm there is an additional triradius. Thus, over the whole hand, including the fingers, the number of triradii T and the number of loops L are related. Since every digit is equivalent to one loop, the relation is $T + 1 = L + D$, where D is the number of digits, with the wrist excluded. Conditions which do not satisfy this simple rule have been described by Penrose (1965).

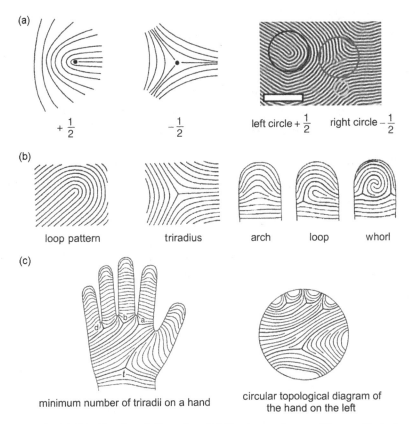

(a)

$+\frac{1}{2}$ $-\frac{1}{2}$ left circle $+\frac{1}{2}$ right circle $-\frac{1}{2}$

(b)

loop pattern   triradius   arch   loop   whorl

(c)

minimum number of triradii on a hand   circular topological diagram of the hand on the left

Fig. 11.4. (a) Disclination defects in a *2d* block copolymer. The small circle on the right contains an edge dislocation. The large circles enclose the disclination defects shown schematically on the left. Reprinted Figure 3b with permission from Harrison *et al.* (2002). Copyright 2002 by the American Physical Society. (b), (c) Disclination defects in dermatoglyphic patterns. Reprinted by permission from Penrose (1965). Copyright 1965 by Macmillan Publishers Ltd.

Returning to smectic phases, the molecular layering can be defined through a phase field u of the density wave. The molecular density can be written as

$$\rho(\mathbf{x}) = \rho_o + \sum_m Re[\langle\psi_m\rangle e^{im\mathbf{q}_o\cdot\mathbf{x}}], \qquad (11.18)$$

where the complex Fourier coefficient is related to the phase $u(\mathbf{x})$ by

$$\langle\psi_m\rangle = |\langle\psi_m\rangle|e^{-imq_o u}. \qquad (11.19)$$

Here $\mathbf{q}_o = (2\pi/d)\mathbf{n_o}$, where d is the smectic layer spacing, $\mathbf{n_o}$ points in the direction normal to the layers (chosen as z axis), and u is the translational elastic variable.

The free energy density of smectic phases depends on the spatial gradient of u in addition to that of the director $\hat{\mathbf{n}}$. The constraint that $\hat{\mathbf{n}}$ is normal to the layers makes the twist and bend distortions of $\hat{\mathbf{n}}$ energetically much more costly than splay distortions, since the latter can be produced at constant layer spacing (see Fig. 11.2).

Two-dimensional systems do not have twist ($K_2 = 0$), and the highly costly bend distortion implies $K_3 \gg K_1$. The anisotropy in elastic constants $\epsilon = (K_3 - K_1)/(K_3 + K_1)$ is therefore nearly unity for smectic phases, while it is nearly zero for nematic phases.

If the director fluctuations are written as $\hat{\mathbf{n}} = \mathbf{n_o} + \delta\hat{\mathbf{n}}$, then the splay distortions are described by $\delta\hat{\mathbf{n}} = -\nabla_\perp u$, the negative gradient of u in the direction perpendicular to the layer normal. Thus, the low-energy rotational distortions $\delta\hat{\mathbf{n}}$ are completely determined by $\nabla_\perp u$ and the long-wavelength elastic distortions in smectic phases are fully described by u. The gradient energy density associated with splay distortions then reduces to $K_1(\nabla \cdot \hat{\mathbf{n}})^2 = K_1(\nabla_\perp^2 u)^2$, since $\nabla_\parallel \cdot \hat{\mathbf{n}} = -\nabla_\parallel \cdot \nabla_\perp u = 0$. The free energy density expression which contains these low-energy excitations is given by (Chaikin and Lubensky, 1995)

$$\mathcal{F}_{el}[u] = \frac{1}{2} \int d^d x [B(\nabla_\parallel u)^2 + K_1(\nabla_\perp^2 u)^2], \qquad (11.20)$$

where B is related to the energy cost of compressing or stretching the layers. Such a term is specific to smectic phases. For a $2d$ system of block copolymer cylindrical microdomains, at low excitation energies (long wavelengths), the distortion field $u(x, z)$ has the form $u_o \sin(\mathbf{q} \cdot \mathbf{x})$ for a splay distortion and $u_o \sin(\mathbf{q} \cdot \mathbf{z})$ for the distortions due to compression and dilation, where the smectic layer normal is in the z direction. This leads to a splay energy density of $K_1 u_o^2 q^4$ and to a compression/expansion energy density of $B u_o^2 q^2$. The ratio of these two energies is $(K_1/B)q^2$. For $2\pi/q$ comparable to the orientational correlation length ξ_θ, this ratio is about 10^{-3} for copolymer systems, which implies that the splay distortion is three orders of magnitude lower than a compressibility distortion. This large ratio results in molecular splay patterns with very small distortions in the microdomain spacing.

An orientation field $\theta(\mathbf{r})$ at a point $\mathbf{r}$ in a stripe pattern can be constructed, taking account of the twofold degeneracy of the cylindrical microdomains. The associated order parameter field $\psi_2(\mathbf{r}) = \exp[2i\theta(\mathbf{r})]$, and the correlation function $g_2(\mathbf{r}) = \langle \psi(0)\psi(\mathbf{r}) \rangle$ can then be computed. The orientational correlation length ξ_θ is obtained by fitting $g_2(\mathbf{r})$ to the exponential form $g_2(\mathbf{r}) = \exp(-r/\xi_\theta)$. A quantitative characterization of the degree of microdomain order can be obtained through the measurement of ξ_θ as a function of the annealing time. Harrison *et al.* (2002) found that $\xi_\theta(t) \sim t^{0.25 \pm 0.02}$ for a $2d$ system of block copolymer cylindrical microdomains. In this smectic system, $\pm\frac{1}{2}$ disclinations are among the relevant topological defects associated with orientational disorder: as these defects anneal, the orientational order increases. The increase in the inter-disclination spacing with time is also consistent with the exponent of $\frac{1}{4}$ for $\xi_\theta(t)$.

The interaction and annihilation of defects play a key role in understanding this coarsening process. As a result of broken rotational symmetry, disclination defects occur in both nematic and smectic phases. In smectic phases, broken translational symmetry leads to the possibility of edge dislocations. Some dislocations are trapped in the strain field of each disclination but most are relatively free. In smectic phases, the dislocation density is typically higher than the disclination density. The strength of a dislocation in a periodic (crystal) or a quasi-periodic (smectic liquid crystal) structure is quantified through its Burgers vector (Chaikin and Lubensky, 1995). The Burgers vector is an extra vector displacement needed to close a loop which encloses a dislocation line. For edge dislocations encountered in smectic phases, the Burgers vector is perpendicular to the dislocation line. A disclination pair has an associated Burgers vector with magnitude equal to twice the separation distance.

The evolution of stripe patterns in smectic phases has an essential topological constraint, which requires production or absorption of dislocations during the motion of disclinations. Since the Burgers vector is conserved, the annihilation of a disclination pair (disclination dipole) produces a number of dislocations equal to the magnitude of the original Burgers vector divided by the layer spacing d. In contrast, the annihilation of a dislocation pair, or a disclination quadrupole, leads to a real decrease of defect density since the net Burgers vector for such defects is zero. Since creation of a dislocation costs energy, the process of annihilation of a disclination pair alone is not favored during coarsening in comparison with multi-disclination annihilation. In the latter case third and other disclinations can act as a source or sink of dislocations, relieving the topological constraint. Figure 11.5 shows a sequence of atomic force microscopy images during the coarsening of a striped pattern of a single layer of cylindrical block copolymer microdomains in a thin film, and illustrates the annihilation of a disclination quadrupole.

The additional conservation constraint on the Burgers vector is responsible for the lower kinetic exponent of $\frac{1}{4}$ in smectic phases, compared with $\frac{1}{2}$ in nematic phases. The $\xi_\theta(t) \sim t^{1/4}$ behavior can be qualitatively understood as follows. In a multi-disclination annihilation event, consider a pair of disclinations which will annihilate. Oppositely charged disclinations which are a distance r apart attract each other with a potential varying as $\ln(r)$. This is equivalent to a force varying as r^{-1}. The topological constraint prevents disclinations from moving in response to this force. A coherent move by a $\pm\frac{1}{2}$ disclination pair by one layer requires an associated motion of a dislocation from this disclination pair to another disclination. Suppose a dislocation must move a distance r in order for the disclinations in the pair to move one unit distance. The free energy of the disclination strain field is then decreased by

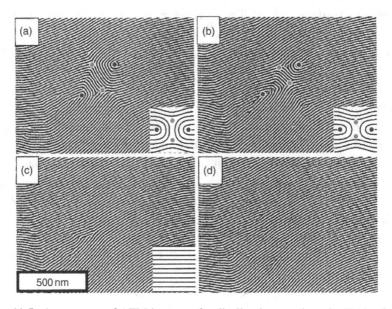

Fig. 11.5. A sequence of AFM images of a disclination quadrupole. Each of the three schematic insets shows the idealized quadrupole configuration at that time. From Harrison *et al.* (2000). Reprinted with permission from AAAS.

an amount $\Delta E \sim r^{-1}$. The force f which drives the movement of the dislocation is $\Delta E/r$ so that $f \sim r^{-2}$. If the motion of the dislocation is overdamped, its speed v is proportional to this force, and then $v \sim r^{-2}$. The dislocation travels a distance r in time dt so that $dt \sim r/v \sim r^3$. During this time the disclinations in the pair move towards one another by a distance $-dr \sim 1$ and therefore $dr/dt \sim -1/r^3$. This integrates to $r \sim (t_f - t)^{1/4}$, where, at time t_f, the disclination pair annihilates. This implies that a typical spacing between the remaining disclinations at time t grows as $\xi(t) \sim t^{1/4}$, leading to the kinetic exponent of $\frac{1}{4}$ in smectic phases. These results are consistent with a renormalization group analysis by Bray (see Fig. 24 in Bray, 1994).

12

Lifshitz–Slyozov–Wagner theory

The late stages of model B dynamics for asymmetric quenches, where the initial condition places the post-quench system just inside and quite near to the coexistence curve, exhibit characteristic features. For an asymmetric system, ψ_o is a measure of the extent of off-criticality of the system. For $\psi_o > 0$ the majority phase equilibrates at $\psi_+ = +1$ and the minority phase at $\psi_- = -1$. At late times, the minority-phase clusters have a characteristic radius $R(\tau)$ which is much larger than the interface width ξ. An important coupling exists between the interface and the majority phase through the surface tension σ. The conservation law dictates that the minority phase will occupy a much smaller "volume" fraction than the majority phase in the final equilibrium state. The dynamics is governed by interactions between the different domains of the minority phase. At late times, these domains have spherical and circular shapes for three- and two-dimensional systems, respectively. Late-stage coarsening is referred to as Ostwald ripening.

The late-stage dynamics may be mapped onto a diffusion equation with sources and sinks (domains) whose boundaries are time dependent. The classic papers by Lifshitz and Slyozov (1961) and Wagner (1961) form the theoretical cornerstone for the description of domain coarsening dynamics for model B. The Lifshitz–Slyozov–Wagner (LSW) theory of coarsening is based on the assumption that each interface between a minority phase domain and the majority phase background is infinitely sharp. It describes the diffusive interactions between the domains through a mean-field treatment with precisely defined boundary conditions at each of the interfaces. The LSW analysis is based on the premise that the clusters of the minority phase compete for growth through an evaporation–condensation mechanism, whereby larger clusters grow at the expense of smaller ones. Material of the minority phase evaporates from a small cluster (source), diffuses through the majority phase background, and condenses on a large cluster (sink). The dominant growth mechanism is the transport of the material from regions of high curvature to regions of low curvature by diffusion through the intervening bulk phase. Throughout this chapter we use dimensionless quantities. These are defined in Chapter 6.

The basic Cahn–Hilliard equation (7.2) can be linearized by using $\psi = 1 + \delta\psi$ and keeping terms up to first order in $\delta\psi$. This linearization is performed around the majority phase equilibrium $\psi_+ = +1$. The result is

$$\frac{\partial}{\partial\tau}\delta\psi = -\xi^2\nabla^4\delta\psi + \left(\frac{\delta^2 f}{\delta\psi^2}\right)_{\psi=1}\nabla^2\delta\psi. \tag{12.1}$$

Since the characteristic length scales are large compared with ξ at late times, the ∇^4 term is negligible and $\delta\psi$ satisfies a diffusion equation,

$$\frac{\partial}{\partial\tau}\delta\psi = f''(1)\nabla^2\delta\psi. \tag{12.2}$$

Due to the existence of a conservation law, the diffusion field $\delta\psi$ relaxes on a time scale that is shorter than the time it takes for significant interface motion. If the domain size is $R(\tau)$, the diffusion field relaxes on a time scale $\tau_D \sim R^2$. However, as we shall see, the typical interface velocity scales as $\sim R^{-2}$. Thus, in a time τ_D, interfaces move a distance of ~ 1, which is much smaller than R. This implies that the diffusion field $\delta\psi$ is always in approximate equilibrium with the interfaces and, thus, obeys Laplace's equation,

$$\nabla^2\delta\psi = 0, \tag{12.3}$$

in the bulk.

12.1 Gibbs–Thomson boundary condition

The coupling between the interface and the majority phase enters through the Gibbs–Thomson boundary condition. To derive the boundary condition, it is convenient to work with the chemical potential instead of the diffusion field. We have

$$\frac{\partial\psi}{\partial\tau} = -\nabla\cdot\mathbf{j}, \tag{12.4}$$

where

$$\mathbf{j} = -\nabla\mu, \quad \mu = f'(\psi) - \xi^2\nabla^2\psi. \tag{12.5}$$

In the bulk, linearization of μ leads to $\mu = f''(\psi_+)\delta\psi - \xi^2\nabla^2\delta\psi$. The term containing ∇^2 is again negligible, so that μ is proportional to $\delta\psi$. Thus, μ also obeys Laplace's equation,

$$\nabla^2\mu = 0, \tag{12.6}$$

in the bulk. Consider the chemical potential μ near an interface. As in Eqs. (7.18) and (7.19), the Laplacian in the curvilinear coordinates $(u, \mathbf{s})$ can be written such that μ in Eq. (12.5) becomes (near the interface)

$$\mu = f'(\psi) - \xi^2 \left(\frac{\partial \psi}{\partial u} \right)_\tau K - \xi^2 \left(\frac{\partial^2 \psi}{\partial u^2} \right)_\tau, \tag{12.7}$$

where $K = \nabla \cdot \hat{\mathbf{n}}$ is the total curvature and ψ is assumed to be independent of s near the interface. Following the development in Chapter 7, the value of μ at the interface can be obtained by multiplying Eq. (12.7) with $(\partial \psi / \partial u)_\tau$ (which is sharply peaked at the interface) and integrating over u across the interface. Since μ and K vary smoothly through the interface, one obtains

$$\mu \Delta \psi = \Delta f - \xi^2 \sigma K, \tag{12.8}$$

where $\Delta \psi$ is the change in ψ across the interface, Δf is the difference in the minima of the free energy f for the two bulk phases, and the expression for σ is given in Eq. (7.26). Equation (12.8) is referred to as the Gibbs–Thomson boundary condition. For a symmetric double-well free energy, $\Delta f = 0$ and $\Delta \psi = 2$. Thus,

$$\mu = -\frac{1}{2} \xi^2 \sigma K, \tag{12.9}$$

at the interface. This calculation can also be carried out when the free energy minima have unequal depths (Bray, 1994). The supersaturation $\epsilon \equiv \delta \psi(\infty)$ is defined as the mean value of $\delta \psi$, and reflects the presence of other subcritical clusters in the system. Far from the interface, $\mu = f''(\psi_+) \delta \psi$ and takes the value $2 \delta \psi$ for $f(\psi) = \psi^4 / 4 - \psi^2 / 2$. Then one has

$$\delta \psi(\infty) = - \lim_{u \to \infty} \delta \psi(u) = -\frac{\mu}{2} = +\frac{\xi^2 \sigma K}{4} \tag{12.10}$$

for the supersaturation.

Equation (12.9) determines μ on the interfaces in terms of the curvature. In the bulk of the majority phase, μ satisfies Laplace's equation (12.6). Since $\mathbf{j} = -\nabla \mu$, an interface will move due to the imbalance between the currents flowing into and out of it. Therefore, the interface velocity is given by

$$j_{out} - j_{in} = v \Delta \psi. \tag{12.11}$$

From Eq. (12.5),

$$j_{out} - j_{in} = -\left[\frac{\partial \mu}{\partial u} \right] = -[\hat{\mathbf{n}} \cdot \nabla \mu], \tag{12.12}$$

where $[\hat{\mathbf{n}} \cdot \nabla \mu]$ denotes the discontinuity in $\hat{\mathbf{n}} \cdot \nabla \mu$ across the interface. Therefore, the interface velocity is

$$v = -[\hat{\mathbf{n}} \cdot \nabla \mu]/\Delta \psi. \tag{12.13}$$

Equations (12.6), (12.9), and (12.13) together determine the interface motion.

12.2 LSW analysis for evaporating and growing droplets

Consider a single spherical domain of minority phase, $\psi_- = -1$, in an infinite sea of majority phase, $\psi_+ = +1$. From the definition of μ in Eq. (12.5), $\mu = 0$ at infinity. Let $R(\tau)$ be the domain radius. The solution of Laplace's equation (12.6) for $d > 2$, with boundary condition $\mu(\infty) = 0$ and Eq. (12.9) at $r = R$ with $K = (d - 1)/R$, is

$$\mu = \begin{cases} -\frac{(d-1)\sigma \xi^2}{2r}, & \text{for } r \geq R, \\ -\frac{(d-1)\sigma \xi^2}{2R}, & \text{for } r \leq R. \end{cases} \tag{12.14}$$

Then using Eq. (12.13) with $\Delta \psi = 2$, we get

$$\frac{dR}{d\tau} = v = -\frac{1}{2} \left[\frac{\partial \mu}{\partial r} \right]_{R-\epsilon}^{R+\epsilon} = -\frac{(d-1)\xi^2 \sigma}{4R^2}. \tag{12.15}$$

Integrating Eq. (12.15) and setting $\xi = 1$, we obtain

$$R^3(\tau) = R^3(0) - \frac{3}{4}(d-1)\sigma \tau, \tag{12.16}$$

which leads to $\tau \propto R^3$ for the time dependence of the evaporating domain.

Again, consider a single spherical droplet of minority phase, $\psi_- = -1$, of radius R immersed in a sea of majority phase, but now suppose that the order parameter of the majority phase at infinity is slightly smaller than $+1$: i.e. $\psi(\infty) \equiv \psi_o < 1$. The majority phase is now "supersaturated" with the dissolved minority species. The supersaturation is $\epsilon = (1 - \psi_o)$. If the minority droplet is large enough, it will grow by absorbing material from the majority phase. Otherwise it will evaporate as above. The two regimes are separated by a critical radius R_c.

Let $f(\pm 1) = 0$ by convention. Then, the Gibbs–Thomson boundary condition, Eq. (12.8), becomes

$$(1 + \psi_o)\mu = f(\psi_o) - \frac{(d-1)\sigma}{R} \tag{12.17}$$

at $r = R$. From the last equality in Eq. (12.5),

$$\mu = f'(\psi_o) \tag{12.18}$$

at $r = \infty$. For $d = 3$ the solution of Laplace's equation (12.6) with these boundary conditions is

$$\mu = \begin{cases} f'(\psi_o) + \left(\frac{f(\psi_o)}{1+\psi_o} - f'(\psi_o)\right)\frac{R}{r} - \frac{2\sigma}{(1+\psi_o)}\frac{1}{r}, & r \geq R, \\ \frac{f(\psi_o)}{1+\psi_o} - \frac{2\sigma}{(1+\psi_o)}\frac{1}{R}, & r \leq R. \end{cases} \tag{12.19}$$

Using Eqs. (12.13) and (12.19), one finds the interface velocity $v \equiv dR/d\tau$ to be

$$\frac{dR}{d\tau} = \left(\frac{f(\psi_o)}{(1+\psi_o)^2} - \frac{f'(\psi_o)}{(1+\psi_o)}\right)\frac{1}{R} - \frac{2\sigma}{(1+\psi_o)^2}\frac{1}{R^2}. \tag{12.20}$$

For small supersaturation, $\psi_o = 1 - \epsilon$ with $\epsilon \ll 1$. To leading (nontrivial) order in ϵ, Eq. (12.20) reduces to

$$v(R) \equiv \frac{dR}{d\tau} = \frac{\sigma}{2R}\left(\frac{1}{R_c} - \frac{1}{R}\right) \tag{12.21}$$

with $R_c = \sigma/(f''(1)\epsilon)$ as the critical radius.

The form of $v(R)$ in Eq. (12.21) is valid only for $d = 3$. If we write it as

$$\frac{dR}{d\tau} = \frac{\alpha_d}{R}\left(\frac{1}{R_c} - \frac{1}{R}\right), \tag{12.22}$$

then the general expression for α_d is $\alpha_d = (d-1)(d-2)\sigma/4$ (Yao et al., 1993, 1994). For $d = 2$, α_d vanishes due to the singular nature of the Laplacian in two-dimensional systems. In this case, in the limit of small (zero) volume fraction of the minority phase, we have (Rogers and Desai, 1989)

$$\frac{dR}{d\tau} = \frac{\sigma}{4R\ln(4\tau)}\left(\frac{1}{R_c} - \frac{1}{R}\right), \tag{12.23}$$

with $R_c = \sigma/(2f''(1)\epsilon)$. A change of variable $\tau^* = \tau/\ln(4\tau)$ converts Eq. (12.23) into the same form as Eq. (12.21), but now the time-like variable has a logarithmic form.

An assembly of drops is considered in the LSW analysis. Growth proceeds by evaporation from drops with $R < R_c$, followed by diffusion of material within the majority phase, culminating with condensation onto drops with $R > R_c$. The supersaturation ϵ changes in time, so that $\epsilon(\tau)$ plays the role of a mean field due to all other droplets. This also implies that the critical radius, $R_c(\tau) = \sigma/(f''(1)\epsilon(\tau))$, is time dependent.

For any dimension d, the cluster size distribution function $n(R, \tau)$ is defined such that $n(R, \tau)dR$ is the number of clusters per unit "volume" with radius between R and $R + dR$. Assuming no nucleation of new clusters and no cluster coalescence occurs, $n(R, \tau)$ satisfies the continuity equation

$$\frac{\partial n}{\partial \tau} + \frac{\partial}{\partial R}(vn) = 0, \tag{12.24}$$

where $v \equiv dR/d\tau$ is given by Eq. (12.21). Finally, the conservation law is imposed on the entire system as follows. Let the spatial average of the conserved order parameter be $(1 - \epsilon_o)$. At late times the supersaturation $\epsilon(\tau)$ tends to zero, giving the constraint

$$\epsilon_o = \epsilon(\tau) + V_d \int_0^\infty dR R^d n(R, \tau) \sim V_d \int_0^\infty dR R^d n(R, \tau), \tag{12.25}$$

where V_d is the volume of the d-dimensional unit sphere. Equations (12.22) with $R_c(\tau)$, (12.24), and (12.25) constitute the LSW problem for the cluster size distribution function $n(R, \tau)$. The LSW analysis of these equations starts by introducing a scaled distribution of droplet sizes:

$$n(R, \tau) = R_c^{-(d+1)} \mathcal{N}\left(\frac{R}{R_c}\right). \tag{12.26}$$

Using this relation and denoting the scaled droplet size by $x = R/R_c$, Eq. (12.25) becomes

$$\epsilon_o = V_d \int_0^\infty dx \, x^d \, \mathcal{N}(x), \tag{12.27}$$

which fixes the normalization of $\mathcal{N}(x)$. If Eq. (12.26) is substituted into Eq. (12.24), and the velocity equation (12.22) is used, we obtain

$$\frac{\dot{R}_c}{R_c^{d+2}}\left((d+1)\mathcal{N}(x) + x\frac{d\mathcal{N}}{dx}\right) = \frac{\alpha_d}{R_c^{d+4}}\left(\left(\frac{2}{x^3} - \frac{1}{x^2}\right)\mathcal{N}(x) + \left(\frac{1}{x} - \frac{1}{x^2}\right)\frac{d\mathcal{N}}{dx}\right). \tag{12.28}$$

To be consistent with the scaled form of the droplet size distribution in Eq. (12.26), the R_c dependence should drop out of Eq. (12.28). This implies that

$$R_c^2 \dot{R}_c = \alpha_d \gamma, \tag{12.29}$$

which integrates to

$$R_c(\tau) = (3\alpha_d\gamma\tau)^{\frac{1}{3}}. \tag{12.30}$$

Equation (12.28) can then be simplified to give

$$\left[\frac{2}{x^3} - \frac{1}{x^2} - \gamma(d+1)\right]\mathcal{N}(x) = \left[\gamma x - \frac{1}{x} + \frac{1}{x^2}\right]\frac{d\mathcal{N}}{dx}. \tag{12.31}$$

Integration of this equation yields

$$\ln\mathcal{N}(x) = \int^x \frac{dy}{y} \frac{(2 - y - \gamma(d+1)y^3)}{(\gamma y^3 - y + 1)}. \tag{12.32}$$

For large y the integrand on the right-hand side of this equation behaves as $-(d-1)/y$, which implies that $\mathcal{N}(x) \rightarrow x^{-(d+1)}$ for large x. This would create a diverging integral in the normalization constraint on $\mathcal{N}(x)$ (see Eq. (12.27)), unless it vanishes for x greater than some cutoff value x_o. Lifshitz and Slyozov show that the only stable solution for $\mathcal{N}(x, \gamma)$ is such that it and all its derivatives vanish at the cutoff value x_o, whereby $\mathcal{N}(x)$ vanishes for $x \geq x_o$. Equations (12.22) and (12.30) together yield an equation for $x = R/R_c$:

$$\frac{dx}{d\tau} = \frac{1}{3\gamma\tau}\left(\frac{1}{x} - \frac{1}{x^2} - \gamma x\right) \equiv \frac{1}{3\gamma\tau}h(x). \tag{12.33}$$

The quantity γ (introduced in Eq. (12.29)) may in general be dependent on τ, even though it is assumed to be a constant above. There are three possibilities for the asymptotic behavior of $\gamma(\tau)$ as $\tau \rightarrow \infty$: $\gamma(\tau) \rightarrow \infty$, 0, or a nonzero constant. Consider the form of $h(x)$ which is shown in Fig. 12.1. For $\gamma > \gamma_o$, $h(x)$ is negative for all x, all drops evolve to $x = 0$, and the conservation condition cannot be satisfied since the total amount of material goes to zero. For $\gamma < \gamma_o$, $h(x)$ is positive for values of x between its two positive real roots at x_1 and x_2. For this case, drops with $x < x_1$ will evolve to $x = 0$ and drops with $x > x_1$ will asymptotically approach $x = x_2$. In both cases the conservation condition cannot

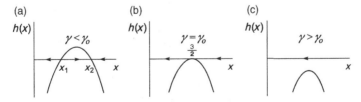

Fig. 12.1. $h(x)$ as a function of x for three possible classes of γ. After Bray (1994).

be satisfied: the total amount of material goes to zero in the former case and to infinity in the latter case. If $\gamma = \gamma_o = \frac{4}{27}$, then at $x = \frac{3}{2}$, $h(x) = 0$ but is negative for other values of x. In this case, all points with $x < \frac{3}{2}$ evolve to $x = 0$ and again the conservation condition cannot be satisfied. Points with $x > \frac{3}{2}$ move to reach $x = \frac{3}{2}$. These observations lead to the result that the only allowed possibility which is consistent with the conservation condition, Eq. (12.27), is that $\gamma(\tau)$ asymptotically approaches γ_o from above, and this approach should take an infinite time. (If γ reaches γ_o in a finite time, all drops with $x > \frac{3}{2}$ would eventually arrive at $x = \frac{3}{2}$ and cease to evolve. One would have a repeat of the $\gamma < \gamma_o$ case.) Lifshitz and Slyozov (1961) find

$$\gamma_o = \gamma(\tau)\left(1 - \frac{3}{4\tau^2}\left(1 + \frac{1}{(\ln \tau)^2}(1 + \cdots)\right)\right). \tag{12.34}$$

The asymptotic form of the scaled distribution $\mathcal{N}(x)$ consistent with the conservation condition can be then obtained by choosing $\gamma = \gamma_o$, and evaluating the integral in Eq. (12.32). For $\gamma = \gamma_o = \frac{4}{27}$, the integrand in Eq. (12.32) has a double pole at $x = \frac{3}{2}$ and a simple pole at $x = -3$. One can use the calculus of residues to evaluate the integral to find that the cutoff value x_o is the double pole of the integrand in Eq. (12.32): i.e. the cutoff occurs at $x_o = \frac{3}{2}$. The evaluation of the integral in Eq. (12.32) leads to the asymptotic form of the scaled distribution for $d = 3$:

$$\mathcal{N}(x) = \begin{cases} Cx^2(3+x)^{-(1+\frac{4d}{9})}(\frac{3}{2}-x)^{-(2+\frac{5d}{9})}e^{(-\frac{d}{3-2x})}, & \text{for } x < \frac{3}{2}, \\ 0, & \text{for } x \geq \frac{3}{2}. \end{cases} \tag{12.35}$$

The normalization condition in Eq. (12.27) can be used to determine $\mathcal{C}$. For $d = 3$, $\mathcal{C} = 3^4 e / 2^{\frac{5}{3}}$. The distribution $\mathcal{N}(x)$ in Eq. (12.35) is shown in Fig. 12.2. The asymptotic scaled distribution function $\mathcal{N}(x)$ can be now employed to obtain $n(R, \tau)$ using Eq. (12.26), where $R_c(\tau)$ is given by Eq. (12.30). Together these three quantities constitute the asymptotic solution for the LSW problem defined through Eqs. (12.22), (12.24), and (12.25). One can find a family of self-similar distribution functions, consistent with the scaling form (Eq. (12.26)) and the τ-dependence of $R_c(\tau)$ in Eq. (12.30), where each function is localized on a finite interval $[0, x_m]$, can be parameterized by σ, and may depend on the initial conditions (Giron *et al.*, 1998). However, all such solutions are unstable with respect to addition of an infinite tail. Only the LSW solution given above is globally stable.

The LSW theory is a mean field theory. The reason why and the manner in which a system undergoing Ostwald ripening selects the globally stable solution from the family of allowed solutions have been investigated by considering the role of fluctuations (Meerson, 1999; Meerson *et al.*, 2005). A natural extension of the LSW

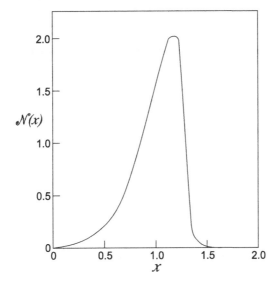

Fig. 12.2. The asymptotic cluster size distribution $\mathcal{N}(x)$ from the LS analysis, for $d = 3$.

mean-field continuum model is found in the mean-field rate equations of cluster models for the cluster size distribution function $N_s(\tau)$, where s is the number of atoms in a cluster (Gunton *et al.*, 1983; Binder, 1987). Treating the number of monomers N_1 separately, regarding s as a continuous variable, and using a second-order Taylor expansion in $1/s$, one can obtain a Fokker–Planck equation for $N_s(\tau)$. Terms up to first order in $1/s$ lead to the mean field equation (12.26), and the second-order contribution adds a diffusion term in which the diffusion coefficient in s space is determined in terms of attachment and detachment rates of single atoms to a cluster of size s. This analysis shows that even when the initial state of the system is localized on a finite interval, fluctuations produce an infinite tail in the self-similar distribution function and drive it towards the globally stable LSW solution.

The LSW theory is designed for Ostwald ripening in dense mixtures where the molecular mean free path is very small compared with some minimum cluster radius. This is not the case for liquid–vapor or solid–vapor systems which arise during the gas phase production of powders made up of nano-size particles. Ostwald ripening in such a rarefied system can lead to qualitatively different cluster size distribution functions (Burlakov, 2006).

13

Systems with long-range repulsive interactions

New features appear in the kinetics of phase ordering and phase separation in systems where long-range repulsive interactions (LRRI) compete with the short-range attractive interactions considered earlier. Competing interactions can lead to the emergence of modulated phases, where a particular symmetry, wavelength, and amplitude are selected (Seul and Andelman, 1995). Both in equilibrium and nonequilibrium systems such modulated phases have domain structures with various shapes, patterns, and morphologies. Figure 13.1 shows some domain structures seen in systems displaying modulated phases. Modulated phases in materials are important in technological applications (Park *et al.*, 1997; Black *et al.*, 2000). An understanding of such phases is crucial in order to be able to design materials with specific properties and control their morphology.

Many systems in nature can be modeled through the inclusion of long-range interactions. Examples of such systems are uniaxial ferromagnetic films, ferromagnetic surface layers, ferrofluid films, ferroelectrics, Langmuir (lipid) monolayers, block copolymers, and cholesteric liquid crystals. A uniaxial ferromagnetic film in the presence of an external magnetic field can be modeled by augmenting the standard scalar order parameter model A with an additional long-range interaction arising from the parallel orientation of magnetic dipoles (Roland and Desai, 1990). This repulsive interaction competes with the attractive domain wall energy. An external magnetic field makes the film's magnetization a nonconserved quantity so that a description based on model A is appropriate. Block copolymers and Langmuir monolayers are examples of conserved order parameter (model B) systems where the connectivity between the covalently bonded blocks of the polymer chains results in an effective LRRI (Sagui and Desai, 1994).

At a coarse-grain level the contribution to the free energy arising from long-range repulsive interactions can be expressed as

$$\mathcal{F}_{LR}[\phi(\mathbf{r})] = \frac{\alpha}{2} \int d^d r d^d r' \, \phi(\mathbf{r}) G(|(\mathbf{r} - \mathbf{r}')|)\phi(\mathbf{r}'). \qquad (13.1)$$

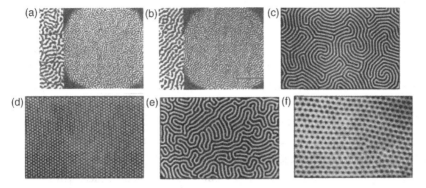

Fig. 13.1. Partially aligned (a) and globally aligned (b) stripes in Langmuir films of a DMPC–cholesterol mixture at 10 °C with 27% cholesterol (the bar marks 100 μm). Reprinted Figure 3 with permission from Seul and Chen (1993). Copyright 1993 by the American Physical Society. Stripe (c) and bubble (d) phases in ferromagnetic garnet film with characteristic length ∼ 10 μm; ferrofluid confined between two glass plates in a magnetic field normal to the fluid layer displaying (e) labyrinthine (period ∼ 1 cm) and (f) bubble (period ∼ 4 μm) states. From Seul and Andelman (1995). Reprinted with permission from AAAS.

Here α is the strength of the long-range (nonlocal) interaction whose spatial dependence is given through the *dimensionless* kernel G. In general, G may depend on the order parameter ϕ: for instance, when elastic forces are present. If G is independent of ϕ, then the interaction term is harmonic. This is the case for Langmuir monolayers (Andelman *et al.*, 1987) and uniaxial ferromagnetic films (Garel and Doniach, 1982), where the interaction arises from dipolar forces. We now consider Langmuir monolayers and block copolymers in more detail to illustrate how the long-range interactions arise in these models.

13.1 Langmuir monolayers

Amphiphilic organic molecules such as surfactants, fatty acids, or lipids can form insoluble Langmuir monolayers at the water/air interface, provided their concentration is not so high that they form micelles. The molecular head of the amphiphile is invariably positioned slightly under the water surface due to strong bonding forces. The tail part is typically a nonpolar hydrocarbon, which experiences a weaker hydrophobic interaction and, due to entropic forces, is excluded from water. The molecule is localized at the interface and is oriented with its tail pointing away from water. In order to understand various equilibrium and kinetic properties of monolayers and other more complex structures formed by such amphiphiles (see Figs. 1.1 and 1.2), electrostatic interactions need to be considered. These interactions have at least two possible origins. Most neutral surfactant molecules carry a permanent electric dipole moment, which has a preferential orientation perpendicular to the interface. In a mesoscopic description, thermal averaging of the molecular dipole moment vectors causes the component parallel to the water/air interface to vanish, and only the component normal to the interface remains nonzero. Charged monolayers create an electrical double layer, which sometimes can be viewed as a layer of permanent dipoles. For both of these cases Andelman *et al.* (1987) have shown that the free energy can be expressed in the form given above for $\mathcal{F}_{LR}$.

Consider a neutral monolayer composed of two partially incompatible amphiphiles, say A and B. The amphiphiles can diffuse laterally and form an incompressible film that fully covers a $2d$ domain. The relative concentration of A, $\phi(\mathbf{r})$, is the order parameter. The free energy of such a system has three terms: two of these, the local free energy and the inhomogeneity term related to the line tension of the interface between A-rich and B-rich coexisting regions, have the forms given in Eq. (3.2). The line tension expresses the energy cost incurred in forming the linear interface. It favors the minimization of the total interface length, which results in the growth of the A-rich and B-rich regions. However, the growth is tempered if the two species carry dipole moments. For simplicity, let only species A have an electric dipole moment μ. If all molecular dipoles point in the direction normal to the water/air interface, they interact through pairwise repulsive dipole–dipole electrostatic interactions. A physical dipole can be viewed as a pair of equal and opposite charges of magnitude Q separated by a distance l. A mathematical dipole results from the limit where the dipole moment $\mu = Ql$ is held fixed while $l \to 0$ and $Q \to \infty$. In the continuum limit, a physical dipolar monolayer can be viewed as a pair of parallel sheets with opposite charge separated by a distance l. The magnitude of charge density in each of the sheets is $Q\phi(\mathbf{r})$. The electrostatic energy of such a charge distribution, which describes the overall electrically neutral monolayer, may be written as

$$\mathcal{F}_{LR}[\phi(\mathbf{r})] = \frac{\mu^2}{4\pi\epsilon_o l^3} \int d^2r d^2r' \, \phi(\mathbf{r})G(|\mathbf{r} - \mathbf{r}'|)\phi(\mathbf{r}'), \qquad (13.2)$$

where

$$G(|\mathbf{r} - \mathbf{r}'|) = \frac{l}{|\mathbf{r} - \mathbf{r}'|} - \frac{l}{\left((\mathbf{r} - \mathbf{r}')^2 + l^2\right)^{\frac{1}{2}}},$$

$$= \int d^2(lk) \, \frac{(1 - e^{-lk})}{lk} e^{i\mathbf{k}.(\mathbf{r}-\mathbf{r}')}. \qquad (13.3)$$

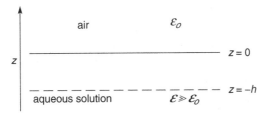

Fig. 13.2. Schematic structure of a Langmuir film. Molecular dipoles are located in the plane $z = -h$ and are fully immersed in the aqueous solution.

In the limit of small l, $G(|\mathbf{r} - \mathbf{r}'|)$ approaches $l^3/(2|\mathbf{r} - \mathbf{r}'|^3)$, which leads to a repulsive dipolar kernel

$$\mathcal{F}_{LR}[\phi(\mathbf{r})] = \frac{\mu^2}{8\pi\epsilon_o} \int d^2r d^2r' \, \phi(\mathbf{r})|\mathbf{r} - \mathbf{r}'|^{-3}\phi(\mathbf{r}'). \tag{13.4}$$

This result does not take into account the dielectric properties of water and air. Let ϵ be the dielectric constant of water and ϵ_o that of air. The dielectric nature of the medium may be accounted for as follows. Consider the Langmuir monolayer sketched in Fig. 13.2, where a flat interface at $z = 0$ separates air (dielectric constant ϵ_o) from an aqueous solution (dielectric constant $\epsilon \gg \epsilon_o$). The molecular dipoles with dipole moment μ are confined to the plane $z = -h$ and are fully immersed in the solution. Let $\phi(\mathbf{r})$ be the in-plane monolayer concentration. The local electric polarization points in the z direction and has a magnitude $\mu\phi(\mathbf{r})$. The electrostatic potential V satisfies Laplace's equation $\nabla^2 V = 0$ and has a discontinuity at $z = -h$: $\Delta V = \mu\phi/\epsilon$. Also, V vanishes at $z = +\infty$ and $z = -\infty$. Furthermore, $\epsilon(z)\partial V/\partial z$ is continuous at $z = 0$. If one assumes that the surface concentration oscillates with wavevector $\mathbf{k}$, $\phi(\mathbf{r}) = \phi_o + \phi_k e^{i\mathbf{k}\cdot\mathbf{r}}$, Laplace's equation can be solved in the limit of vanishing h. The solution is

$$V = \begin{cases} \frac{\mu\,\phi_k}{(\epsilon_o+\epsilon)}e^{-|k||z|+i\mathbf{k}\cdot\mathbf{r}}, & z > 0, \\ \frac{-\epsilon_o\mu\phi_k}{\epsilon(\epsilon_o+\epsilon)}e^{|k||z|+i\mathbf{k}\cdot\mathbf{r}}, & z < 0. \end{cases} \tag{13.5}$$

The electrostatic free energy of the dipoles per unit area, subject to this potential V, may be written in terms of the electric field $\mathbf{E}$ and the electric polarization $\mathbf{P}$, and is given by

$$\frac{F_{el}}{A} = -\frac{1}{2A}\int_A d^2r \, \mathbf{P}\cdot\mathbf{E} = \frac{1}{2A}\int_A d^2r \, \mu\phi(\mathbf{r})\left(\frac{\partial V}{\partial z}\right)_{z=0^-}$$

$$= -\frac{1}{2}|k||\mu|^2\phi_k^2\frac{\epsilon_o}{\epsilon(\epsilon_o + \epsilon)}. \tag{13.6}$$

This result can be understood by noting that a given dipole μ below the surface at $z = -h$ has an image dipole, $\mu' = \mu(\epsilon_o - \epsilon)/(\epsilon_o + \epsilon)$, located at $z = h$. Another real dipole located at a distance r from the first one sees the field created by both dipoles, $(\mu + \mu') = 2\epsilon_o\mu/(\epsilon_o + \epsilon)$. For $h \ll r$, the interaction energy between the two real dipoles is then given by

$$G(r) = \frac{\mu^2\epsilon_o}{2\pi\epsilon(\epsilon_o + \epsilon)}\frac{1}{r^3}. \tag{13.7}$$

For a layer of dipoles at $z = 0^-$ with concentration $\phi(\mathbf{r})$, one can add the energy contributions from each of the dipole pairs to obtain

$$F_{el} = \frac{\mu^2\epsilon_o}{4\pi\epsilon(\epsilon_o + \epsilon)} \int d^2r d^2r' \, \phi(\mathbf{r}) \, |\mathbf{r} - \mathbf{r}'|^{-3}\phi(\mathbf{r}'). \tag{13.8}$$

In Fourier space this result is the same as that obtained in Eq. (13.6). It also has the same form as Eq. (13.4). Thus, for Langmuir monolayers, one can identify the strength of the long-range repulsive interaction in Eq. (13.1) as

$$\alpha = \frac{\mu^2\epsilon_o}{\pi l^3\epsilon(\epsilon_o + \epsilon)}, \tag{13.9}$$

and the dimensionless kernel $G(r)$ by Eq. (13.3). The analysis given above also applies to magnetic garnet films, where the role of electric dipoles is replaced by magnetic dipoles (Garel and Doniach, 1982; Roland and Desai, 1990).

13.2 Block copolymers

A block copolymer is a linear polymer chain consisting of at least two subchains of incompatible polymers joined by covalent bonds. Consider a diblock system of X and Y subchains. If the repulsion between X and Y subchains is sufficiently strong, segregation can occur; however, due to covalent bonding, only microphase separation is possible. Below the microphase separation transition (MST) temperature the system self-organizes into an inhomogeneous periodic structure of alternating X and Y domains (modulated phase).

The block copolymer free energy functional (for a d-dimensional system) can be written in dimensionless form as (Leibler, 1980; Ohta and Kawasaki, 1986)

$$\mathcal{F}[\psi] = \int d^d x \, [-\frac{1}{2}\psi^2 + \frac{1}{4}\psi^4 + \frac{1}{2}(\nabla\psi)^2]$$
$$- \frac{B}{2} \int d^d x d^d x' \psi(\mathbf{x}')G(\mathbf{x} - \mathbf{x}') \, \psi(\mathbf{x}), \tag{13.10}$$

where B is a parameter proportional to N^{-2}, N being the degree of polymerization (total number of monomeric units of the X and Y subchains), and ψ is the difference in the local volume fractions of the X and Y monomers. The kernel G is the diffusion Green function, which satisfies the equation $\nabla^2 G(\mathbf{x}, \mathbf{x}') = \delta(\mathbf{x} - \mathbf{x}')$. In Fourier space, one has $\hat{G}(\mathbf{k}) = -1/k^2$. Assuming a relaxational dynamics for the microphase separation of diblock copolymers, one has

$$\frac{1}{M} \frac{\partial \psi}{\partial \tau} = \nabla^2 \frac{\delta \mathcal{F}[\psi]}{\delta \psi} \tag{13.11}$$

$$= \nabla^2 \left[(\psi^3 - \psi - \nabla^2 \psi) - B \int d^d x' G(\mathbf{x} - \mathbf{x}')[\psi(\mathbf{x}') - \psi_o] \right],$$

or

$$\frac{1}{M} \frac{\partial \psi}{\partial \tau} = \nabla^2 (\psi^3 - \psi - \nabla^2 \psi) - B[\psi(\mathbf{r}) - \psi_o], \tag{13.12}$$

where ψ_o is the average value of ψ and M is the kinetic mobility coefficient.

Equation (13.11) can be compared with Eq. (6.4). It is then clear that microphase separation in block copolymers is an example of kinetics leading to modulated phases which can be described by model B, where the free energy functional contains a long-range repulsive interaction. If a system consists of an incompressible blend of X and Y polymer chains (polymer melt) instead of X–Y block copolymers, then model B equations *without* the LRRI term would apply. For such a polymer melt the form of the free energy functional that is appropriate is the Flory–Huggins–de Gennes functional (Glotzer, 1995).

The static solution of Eq. (13.12) obeys the equation

$$0 = \nabla^2 \left(\psi^3 - \psi - \nabla^2 \psi \right) - B(\psi(\mathbf{x}) - \psi_o). \tag{13.13}$$

The critical point for the system is at $\psi_o = 0$, $B = 1/4$. The static solution corresponds to a homogeneous, disordered phase for $B > 1/4$. Microphase separation occurs for $B < 1/4$. There are two interesting regimes, depending on the value of B: for values of B very close to the onset of microphase separation, the size of the patterns grows as $B^{-1/4}$ (Leibler, 1980; Ohta and Kawasaki, 1986). This regime is called the weak segregation regime. Far from onset, one has a strong segregation regime where the patterns grow as $B^{-1/3}$ (Hashimoto *et al.*, 1980; Ohta and Kawasaki, 1986; Liu and Goldenfeld, 1989). In this strong segregation regime, one typically observes a disordered lamellar structure unless the overall direction of the lamellae is fixed by either a flow or a boundary condition, or if the system is very slowly annealed (Liu and Goldenfeld, 1989; Bahiana and Oono, 1990; Chakrabarti and Gunton, 1993).

13.3 Langevin models A and B including LRRI

The Langevin equations for models A and B can be constructed by adding the long-range repulsive interaction to the GLW free energy functional so that $\mathcal{F} = \mathcal{F}_{GLW} + \mathcal{F}_{LR}$. The long-range repulsive interaction term $\mathcal{F}_{LR}$ is given by Eq. (13.1). Recall from Chapter 3 that the GLW free energy functional $\mathcal{F}_{GLW}$ consists of the Landau free energy density term $f(\phi^*)$ and the Ginzburg square gradient term. From Eqs. (3.2) and (3.14), choosing $f_c = 0$ as the zero level of energy,

$$\mathcal{F}_{GLW}[\phi(\mathbf{r},t)] = \int d^d r \left[(\frac{1}{2}a_2\phi^{*2} + \frac{1}{4}a_4\phi^{*4} - \mathcal{H}\phi^*) + \frac{\kappa}{2}(\nabla\phi)^2 \right]. \quad (13.14)$$

Here $\phi^* = (\delta\phi + \phi_o)$, where the constant ϕ_o is the deviation of the average order parameter $\bar{\phi}$ from the critical order parameter ϕ_c. The linear term $(-\mathcal{H}\phi^*)$ couples ϕ^* to an external field $\mathcal{H}$. Since $\phi = \bar{\phi} + \delta\phi$, at any given time $(1/V)\int d^d r \phi(\mathbf{r},t) = \bar{\phi}(t)$ and $\int d^d r \delta\phi(\mathbf{r},t) = 0$. Since the order parameter is conserved in model B, $\bar{\phi}$ is a time-independent constant. This is not the case in model A, and $\bar{\phi}(t)$ has its own equation of motion in which $\mathcal{H}$ plays a role.

The field equations for models A and B with LRRI are obtained by using the total free energy functional $\mathcal{F}$ in the model A (Eq. (5.1)) and model B (Eq. (6.4)) Langevin equations. The Langevin equations for models A and B are often expressed as a single equation,

$$\frac{\partial \delta\phi(\mathbf{r},t)}{\partial t} = -M(-\nabla^2)^n \frac{\delta\mathcal{F}}{\delta\phi} + \zeta(\mathbf{r},t), \quad (13.15)$$

with

$$\langle \zeta(\mathbf{r},t)\zeta(\mathbf{r}',t') \rangle = 2kTM(-\nabla^2)^n \delta(\mathbf{r}-\mathbf{r}')\delta(t-t'), \quad (13.16)$$

where $n = 0$ for model A and $n = 1$ for model B. For critical quenches, $\delta\phi(\mathbf{r},t)$ is ϕ^* and for off-critical quenches it is $(\phi^* - \phi_o)$.

The Langevin equations now contain an additional term arising from $\delta\mathcal{F}_{LR}/\delta\phi$, which introduces the physics of the competing long-range interactions in the field theory description of models A and B. Using Eq. (13.1) for $\mathcal{F}_{LR}$, one obtains

$$\frac{\delta\mathcal{F}_{LR}}{\delta\phi} = \alpha \int d^d r' \, G(|\mathbf{r}-\mathbf{r}'|)\phi(\mathbf{r}'). \quad (13.17)$$

It is useful to introduce a dimensionless quantity

$$\beta = \alpha \frac{\xi^d}{a_o|T_c - T|} = \alpha \frac{\kappa^{(d/2)}}{(a_o|T_c - T|)^{(d+2)/2}}, \tag{13.18}$$

which is the strength of the long-range repulsive interaction relative to that of the short-range square gradient interaction. Then, in terms of scaled variables, the additional term in the model A Langevin equation due to LRRI reduces to $\beta \int d^d x' \, G(|\mathbf{x} - \mathbf{x}'|) \psi(\mathbf{x}')$. Since this is a convolution, its spatial Fourier transform is $\beta \hat{G}(\mathbf{k}) \hat{\psi}(\mathbf{k})$.

For Langmuir and ferromagnetic films, the kernel in Eq. (13.3) can be rewritten in scaled variables as

$$G(|\mathbf{x} - \mathbf{x}'|) = \frac{L}{|\mathbf{x} - \mathbf{x}'|} - \frac{L}{((\mathbf{x} - \mathbf{x}')^2 + L^2)^{\frac{1}{2}}}$$

$$= \int d\mathbf{k} \frac{(1 - e^{-Lk})}{k} e^{i\mathbf{k}\cdot(\mathbf{x}-\mathbf{x}')}, \tag{13.19}$$

where the dimensionless thickness of the film l/ξ is denoted by L. Notice that for $|\mathbf{x}-\mathbf{x}'|$ large compared with L, $G(|\mathbf{x}-\mathbf{x}'|)$ approaches $L^3/(2|\mathbf{x}-\mathbf{x}'|^3)$, which is like a repulsive dipolar kernel. The spatial Fourier transform of the LRRI kernel is $\hat{G}(k) = (1 - e^{-Lk})/k$; its limit for $k = 0$ is the dimensionless film thickness L, which enters in the equation for $\overline{\psi}(\tau)$ (Eq. (13.24) below). An important dimensionless number, the Bond number $\mathcal{B}$, characterizes many systems, such as ferromagnetic and Langmuir films, where electric or magnetic dipoles play a role at a macroscopic level. For quasi-two-dimensional systems such as Langmuir films, $\mathcal{B} = (\Delta p)^2/\sigma$, where Δp is the dipolar density and σ is the line tension. In our notation, $\mathcal{B} = (\beta L^2)/2$.

For an off-critical quench in a general system with long-range interactions, the Langevin equations in dimensionless variables take the form

$$\frac{\partial \psi(\mathbf{x}, \tau)}{\partial \tau} = \frac{(-\nabla^2)^n}{2} \left[(\nabla^2 + q_c^2(\tau))\psi(\mathbf{x}, \tau) - 3\overline{\psi}(\tau)\psi^2(\mathbf{x}, \tau) - \psi^3(\mathbf{x}, \tau) \right.$$

$$\left. - \beta \int d^d x' G(|\mathbf{x} - \mathbf{x}'|)\psi(\mathbf{x}', \tau) \right] + \sqrt{\epsilon}\eta(\mathbf{x}, \tau), \tag{13.20}$$

where

$$\langle \eta(\mathbf{x}, \tau)\eta(\mathbf{x}', \tau')\rangle = (-\nabla^2)^n \delta(\mathbf{x} - \mathbf{x}')\delta(\tau - \tau') \tag{13.21}$$

and

$$q_c^2(\tau) = 1 - 3\overline{\psi(\tau)}^2. \tag{13.22}$$

The scaled variables are defined as (see Chapters 5 and 6): $\psi(\mathbf{x}, \tau) = \delta\phi/\phi^*_{min}$, $\mathbf{x} = \mathbf{r}/\xi$, where $\xi = \left(\kappa/(a_o|T_c - T|)\right)^{1/2}$, $\tau = [(2Ma_o|T_c - T|)(a_o|T_c - T|/\kappa)^n\, t]$, and ϵ is defined in Eq. (5.6). The average order parameter is

$$\overline{\psi(\tau)} = \overline{\phi^*(t)}/\phi^*_{min}. \tag{13.23}$$

For model B, $\overline{\psi(\tau)} = \psi_o = \phi_o/\phi^*_{min} = $ constant, so $q_c^2 = (1 - 3\psi_o^2)$ and is independent of time.

Equation (13.20) describes the time evolution of the order parameter fluctuations. This is the only relevant equation for model B. For model A, Eq. (13.20) must be augmented by an equation describing the time variation of the average order parameter $\overline{\psi(\tau)}$:

$$\frac{\partial\overline{\psi(\tau)}}{\partial\tau} = \frac{1}{2}\left[(1 - \beta\hat{G}(0))\overline{\psi(\tau)} - \overline{\psi(\tau)}^3 + h\right], \tag{13.24}$$

where

$$h = \left(\frac{a_4}{a_o|T_c - T|}\right)^{\frac{1}{2}}\frac{\mathcal{H}}{a_o|T_c - T|}. \tag{13.25}$$

The role of the field is to produce and maintain a net overall magnetization $\overline{\psi(\tau)}$, which rapidly reaches equilibrium and remains constant so that the subsequent dynamics is governed solely by Eq. (13.20). For $h \neq 0$ this constant, $\overline{\psi(\tau = \infty)}$, is the real positive root of the cubic equation

$$\left[\overline{\psi}^3 - (1 - \beta\hat{G}(0))\overline{\psi} - h\right] = 0. \tag{13.26}$$

Solving the cubic for its roots, one gets

$$\overline{\psi(\infty)} = \begin{cases} 2[(1 - \beta\hat{G}(0))/3]^{(1/2)}\cos(\phi/3), & \text{if } h^2/4 \leq [(1 - \beta\hat{G}(0))^3/27], \\ (A_+ + A_-), & \text{otherwise,} \end{cases} \tag{13.27}$$

where

$$\cos(\phi) = [(h/2)/[(1 - \beta\hat{G}(0))^3/27]^{(1/2)}], \tag{13.28}$$

and

$$A_\pm = [(h/2) \pm [(h^2/4) - (1 - \beta\hat{G}(0))^3/27]^{(1/2)}]^{(1/3)}. \tag{13.29}$$

For $h = 0$, the three roots of the cubic are $0, \pm(1 - \beta\hat{G}(0))^{(1/2)}$. The first root is unstable, and the equilibrium system consists of a random distribution of "up" and "down" domains corresponding to the two nonzero roots, leading to an overall average $\overline{\psi} = 0$. Also, for $h = 0$, the order parameter is conserved and its average remains zero for all times (model B).

The stochastic integro-differential equations described by Eqs. (13.20)–(13.25) are characterized by five dimensionless parameters: the off-criticality ψ_o, which defines the asymmetry between the two phases; the relative strength of the LRRI, β; the dimensionless film thickness, $L = l/\xi$; the strength of the thermal noise, ϵ; and the external field, h. In the absence of long-range interactions ($\beta = 0$), these equations reduce to the following equations for models A and B (compare with Eq. (6.11)):

$$\frac{\partial\psi(\mathbf{x}, \tau)}{\partial\tau} = \frac{(-\nabla^2)^n}{2}\left[(\nabla^2 + q_c^2)\psi(\mathbf{x}, \tau) - 3\overline{\psi}\,\psi^2(\mathbf{x}, \tau) - \psi^3(\mathbf{x}, \tau)\right]$$
$$+ \sqrt{\epsilon}\eta(\mathbf{x}, \tau). \tag{13.30}$$

The nonlinear terms in Eqs. (13.20) and (13.30) are the same; however, the linear terms are different. In both equations the outer Laplacian $(-\nabla^2)^n$ represents the presence (model B, $n = 1$) or absence (model A, $n = 0$) of a conservation law. The quadratic nonlinearity is present for asymmetric off-critical quenches: when it is dominant, one obtains nucleation-like droplet growth. The cubic nonlinearity, which is present for all quenches, including critical quenches, treats both phases in a symmetric manner, leading to spinodal-like labyrinthine patterns. The cubic term also leads to a saturation of $\psi(\mathbf{x}, \tau)$ that curbs the exponential growth arising from the linear terms.

The complexity of the nonlinear terms and the relative simplicity of the linear terms are revealed when Eq. (13.20) is transformed to Fourier space using

$$\psi(\mathbf{x}, \tau) = \sum_{\mathbf{k}} e^{-i\mathbf{k}\cdot\mathbf{x}}\psi_{\mathbf{k}}(\tau) = \psi_0(\tau) + \sum_{\mathbf{k}\neq 0} e^{-i\mathbf{k}\cdot\mathbf{x}}\psi_{\mathbf{k}}(\tau). \tag{13.31}$$

Note $\psi_0(\tau) = \overline{\psi}(\tau)$. One obtains

$$\frac{\partial\psi_{\mathbf{k}}(\tau)}{\partial\tau} = \gamma_k\psi_{\mathbf{k}}(\tau) - \frac{(k^2)^n}{2}\left[3\psi_0(\tau)\sum_{\mathbf{k}'}\psi_{\mathbf{k}'}(\tau)\psi_{\mathbf{k}-\mathbf{k}'}(\tau)\right.$$
$$\left. + \sum_{\mathbf{k}'}\sum_{\mathbf{k}''}\psi_{\mathbf{k}'}(\tau)\psi_{\mathbf{k}''}(\tau)\psi_{\mathbf{k}-\mathbf{k}'-\mathbf{k}''}(\tau)\right] + \sqrt{\epsilon}\eta_{\mathbf{k}}. \tag{13.32}$$

The linear dispersion relation is

$$\gamma_{\mathbf{k}} = \frac{(k^2)^n}{2}[q_c^2 - k^2 - \beta\hat{G}(\mathbf{k})], \tag{13.33}$$

where the dependence of q_c^2 on time (see Eq. (13.22)) can be neglected. The time dependence of $\overline{\psi}$ in q_c^2 arises from the external field, and rapidly reaches the constant equilibrium value ψ_o. Then $q_c^2 = (1 - 3\psi_o^2)$. The values of k for which $\gamma_{\mathbf{k}}$ is positive correspond to unstable modes. The origins of various terms in Eq. (13.33) are of interest: $k^{2n}/2$ is a consequence of the conservation law, q_c^2 arises from the linearized local free energy, $-k^2$ from the attractive square gradient interaction, and $-\beta\hat{G}(\mathbf{k})$ from the nonlocal long-range interactions.

Experimental studies of quenches where the temperature is varied can be mapped onto this model. Equation (13.18) gives the temperature dependence of β. For most experimental systems α and κ, which are the strengths of the long-range repulsive and short-range attractive forces, respectively, are fixed. The long-range repulsive forces suppress the fluctuations near T_c, so that the onset of modulated patterns occurs for temperatures slightly lower than T_c. In a single-mode approximation of wavenumber k_{SM}, the new ordering temperature T_o is given by $T_o = T_c[1 - k_{SM}^2 - \beta\hat{G}(k_{SM})]$. Near this ordering temperature, $|a_2| = a_o(T_c - T)$ reaches its minimum value a_{2c} and β its maximum β_c. As the temperature is lowered, $|a_2|$ increases and β decreases. Quenches to high temperatures (shallow quenches) are mimicked by quenches with high β while quenches to low temperatures (deep quenches) are mimicked by quenches with low β. In this way β controls the depth of the quench while the noise term accounts for the thermal fluctuations. These considerations are applicable quite generally for any $\hat{G}(\mathbf{k})$.

14

Kinetics of systems with competing interactions

In the presence of competing interactions, phase-ordering kinetics and phase separation processes exhibit features that are distinct from those in systems with short-range attractive interactions. The equilibrium ground states of such systems are also qualitatively different. Depending on the values of $\psi_0(\tau = \infty)$ and β, the equilibrium ground state for a $2d$ film of thickness L can be a homogeneous disordered phase (D), a modulated stripe or lamellar phase (S), or a modulated hexagonal or bubble phase (H). In the hexagonal phase, circular-shaped minority domains (bubbles) are arranged in a hexagonal lattice. In the two ordered phases, S and H, the ground state is infinitely degenerate, since either the normal vector in the stripe phase or one of the three symmetry axes in the hexagonal phase can be oriented along any direction within the plane of the film. This is a consequence of the isotropy of the $2d$ system. Figure 14.1 shows a stripe pattern which is globally isotropic but has a well-defined local orientation. The competition between the square gradient term and the long-range repulsive interaction in the free energy results in a nonzero optimal wavevector whose direction is arbitrary in the plane of the film.

14.1 Equilibrium phase diagram

Figure 4.1 gives a pictorial summary of how the equilibrium phase diagram can be obtained from a model free energy. In order to obtain the ground states and the equilibrium phase diagram for a system with competing interactions, we must consider the solutions of the equation $\delta\mathcal{F}/\delta\psi = 0$. The equilibrium states within a single phase minimize the free energy $\mathcal{F}$ and, in the region of a pair of coexisting phases, e.g. $i = D, H$, the equilibrium states correspond to the absolute minima of $\mathcal{F}_i - \mu\psi_o$. In the Ginzburg–Landau approach the free energy functional is that for an effective $2d$ system. The square gradient term is the bulk $3d$ term averaged

Fig. 14.1. Magnetic stripe domains in a ferromagnetic garnet film. Up and down domains form a labyrinth of dark and light stripes. Each cluster is coded according to orientation of its line segments. From Seul *et al.* (1991) (cover image, 13 December 1991). Reprinted with permission from AAAS.

over the film thickness. The long-range interaction term arises from surface interactions: it is a surface demagnetizing term for ferromagnetic films and a surface polarization term for Langmuir films. The order parameter for a ferromagnetic film is the component of magnetization along the film-normal, averaged over the film thickness. It is the amphiphilic concentration for a Langmuir film.

In the stripe and hexagonal phases, the equilibrium system is inhomogeneous and the order parameter varies in space in a periodic manner. The equilibrium analog of Eq. (13.31) is

$$\psi(\mathbf{x}) = \sum_{\mathbf{k}} e^{-i\mathbf{k}\cdot\mathbf{x}} \psi_{\mathbf{k}} = \psi_0(\tau = \infty) + \sum_{\mathbf{k}\neq 0} e^{-i\mathbf{k}\cdot\mathbf{x}} \psi_{\mathbf{k}}. \tag{14.1}$$

For model B, the space-averaged order parameter $\psi_0(\tau = \infty) = \psi_o$. For model A, it depends on h. For uniaxial ferromagnetic films, it is the average magnetization produced by the field h and is given by Eqs. (13.27)–(13.29). In the homogeneous disordered phase, $\psi(\mathbf{x}) = \psi_0(\tau = \infty)$.

Since $\psi(\mathbf{x})$ is real and the S and H lattices have inversion symmetry, it follows that $\psi_{-\mathbf{k}} = \psi_{\mathbf{k}}$. Also, since all of the neighbors at an equal distance from a given stripe or bubble are indistinguishable, $\psi_{\mathbf{k}}$ depends only on the magnitude of $\mathbf{k}$, and we may write $\psi_{\mathbf{k}} = \psi_k$. Thus, one has

$$\sum_{\mathbf{k}\neq 0} e^{-i\mathbf{k}\cdot\mathbf{x}} \psi_{\mathbf{k}} = \sum_{|\mathbf{k}|\neq 0} \psi_k \sideset{}{'}\sum_{\hat{\mathbf{k}}} e^{-i\mathbf{k}\cdot\mathbf{x}}, \tag{14.2}$$

where $\sum'_{\mathbf{k}}$ denotes the sum over all vectors $\mathbf{k}$ such that $|\mathbf{k}| = k$ and $\sum'_{\hat{\mathbf{k}}} = n_k$ is the number of such vectors.

For a smectic phase, which is a stripe phase that is periodic in one dimension, $n_k = 2$. Let the normal to the film be in the z direction. Suppose the stripes are parallel to y so that the normal to the stripes in the plane of the film is in the x direction. Then

$$\sum_{\mathbf{k} \neq 0} e^{-i\mathbf{k} \cdot \mathbf{x}} \psi_{\mathbf{k}} = \sum_{k} 2\psi_k \cos(kx). \tag{14.3}$$

Here the allowed discrete values of k are $k_n = nk_e$, where $k_e = \pi/a$ is the equilibrium wavenumber of the $1d$ stripe lattice with a periodicity of $2a$, and n takes positive integer values. The sum over k can also be written as a sum over n.

For the hexagonal phase, the system has three symmetry directions, and the number of vectors n_k with the same magnitude $|\mathbf{k}| = k$ is not invariant but varies with k. This should be contrasted with the stripe phase, where $n_k = 2$. In Fourier space the allowed discrete wavevectors can be enumerated through $\mathbf{k_n} = k_e(n_1\hat{\mathbf{i}} + n_2\hat{\mathbf{j}})$, where the equilibrium wavenumber is $k_e = 4\pi/(\sqrt{3}a)$, with a the lattice constant of the hexagonal lattice. The direction of $\mathbf{k_n}$ is determined from the coordinate pair (n_1, n_2), and $\hat{\mathbf{i}}$ and $\hat{\mathbf{j}}$ are unit vectors along two of the three symmetry directions with $\cos(\hat{\mathbf{i}} \cdot \hat{\mathbf{j}}) = \frac{1}{2}$. The pairs $(0, 1)$, $(1, 0)$, $(1, -1)$ and their symmetrical counterparts give the sixfold Bragg peak. The magnitude of the wavevector is $|\mathbf{k_n}| = k_e\sqrt{n_1^2 + n_1 n_2 + n_2^2}$, and the sum over k can be equivalently written as a double sum over n_1 and n_2, each of which takes integer values.

In Fourier space, the free energy functional $\mathcal{F}$ can be written in dimensionless form as

$$\frac{\mathcal{F}}{A} = -\frac{1}{2} \sum_{\mathbf{k}} [q_c^2 - k^2 - \beta G(k)] \psi_{-\mathbf{k}} \psi_{\mathbf{k}} + \psi_o \sum_{\mathbf{k}} \sum_{\mathbf{k}'} \psi_{-\mathbf{k}} \psi_{\mathbf{k}'} \psi_{\mathbf{k}-\mathbf{k}'}$$

$$+ \frac{1}{4} \sum_{\mathbf{k}} \sum_{\mathbf{k}'} \sum_{\mathbf{k}''} \psi_{-\mathbf{k}} \psi_{\mathbf{k}'} \psi_{\mathbf{k}''} \psi_{\mathbf{k}-\mathbf{k}'-\mathbf{k}''}, \tag{14.4}$$

where A is the area of the film. Equation (14.1) implies that $\delta\psi(\mathbf{x})/\delta\psi_k = \sum'_{\hat{\mathbf{k}}} e^{-i\mathbf{k} \cdot \mathbf{x}}$, and the extremization condition for $\mathcal{F}$ becomes

$$\frac{1}{A} \frac{\delta \mathcal{F}_i}{\delta \psi_k} = -[q_c^2 - k^2 - \beta G(k)] n_k \psi_k + 3\psi_o \sum_{k'} \sum_{\hat{\mathbf{k}}}{}' \psi_{k'} \psi_{|\mathbf{k}-\mathbf{k}'|}$$

$$+ \sum_{k'} \sum_{k''} \sum_{\hat{\mathbf{k}}}{}' \psi_{k'} \psi_{k''} \psi_{|\mathbf{k}-\mathbf{k}'-\mathbf{k}''|} = 0, \tag{14.5}$$

where the subscript i again denotes the S or H phases. Each of the sums should be interpreted as in Eq. (14.2).

The construction of the phase diagram involves three steps:

(1) One must solve the set of coupled equations $\partial \mathcal{F}_i / \partial k_e = 0$ and $\delta \mathcal{F}_i / \delta \psi_k = 0$. The second of these two equations, given by Eq. (14.5), consists of an infinite number of coupled equations for the Fourier components of the order parameter ψ_k. The first equation corresponds to the optimization of k_e and reduces to

$$\sum_{\mathbf{n}} \left(\frac{\partial}{\partial k_e} [q_c^2 - (f(\mathbf{n})k_e)^2 - \beta G(f(\mathbf{n})k_e)] n_k \psi_k^2 \right) = 0, \qquad (14.6)$$

where $k = f(\mathbf{n})k_e$, and $f(\mathbf{n}) = \sqrt{n_1^2 + n_1 n_2 + n_2^2}$ for the hexagonal phase and is $f(\mathbf{n}) = n$ for the stripe phase. Equation (14.6) is coupled to Eq. (14.5) through ψ_k. This coupled set of equations must be solved numerically to obtain the Fourier coefficients ψ_k and the equilibrium wavenumber k_e for specified values of ψ_o, β, and L. The infinite set of equations is truncated in practice to a finite number of modes. At order s, one has $(s + 1)$ equations in the $(s + 1)$ unknowns: k_e and the s amplitudes ψ_k of the lowest s modes. The number of modes needed to accurately describe the system depends on β. Since the limit $\beta \to 0$ corresponds to a square-well profile, a large number of modes is required for small β. On the other hand, only a few modes, or even a single mode, suffice if β is close to the critical value β_c.

(2) The resulting ψ_k are used in Eqs. (14.1) and (14.2) to obtain an approximation to the order parameter $\psi(\mathbf{x})$, which is then used to determine the free energy functional.

(3) The double-tangent construction (Fig. 4.1) is used to obtain the different regions of the phase diagram. One seeks the absolute minimum of $\mathcal{F}_i - \mu \psi_o$, where $\mathcal{F}_i = \mathcal{F}_S$, $\mathcal{F}_H$, or $\mathcal{F}_D$, and μ is the chemical potential or the magnetic field magnitude, which is coupled to the average order parameter ψ_o.

Using this procedure, the phase diagram has been constructed analytically in the one-mode approximation (Garel and Doniach, 1982; Andelman et al., 1987) and numerically for an arbitrary number of modes (Sagui and Desai, 1994). In the (β, ψ_o) plane of the phase diagram, the stripe and hexagonal phases are separated by a region of stripe-hexagonal coexistence (see Fig. 14.2(a)). Similarly, the hexagonal and disordered phases are separated by a region of hexagonal-disorder coexistence.

Two types of fluctuation could give rise to melting of the bubble and stripe lattices: directional fluctuations of the Brazovskii type (Brazovskii, 1975), which lead to first-order transitions, and dislocations leading to second-order Kosterlitz–Thouless melting (Kosterlitz and Thouless, 1973, 1978). The stripe–bubble transition is expected to be first order.

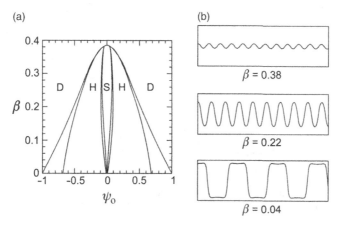

Fig. 14.2. (a) The phase diagram in the (β, ψ_o) plane for $L = 10$. The stripe phase meets the disordered phase at the point $(\beta_c = 0.385, \psi_o = 0)$. (b) Order parameter profiles for $\psi_o = 0, L = 10$ and three different values of β. Reprinted Figure 1 with permission from Sagui and Desai (1994). Copyright 1994 by the American Physical Society.

The profiles of the order parameter $\psi(\mathbf{x})$ for the symmetric stripe phase in Fig. 14.2(b) show that the profile approaches a square-well form as $\beta \to 0$, indicating the existence of sharp interfaces. The amplitude of the profile increases towards unity and its characteristic wavelength becomes progressively longer. At $\beta = 0$ the amplitude is precisely unity and its wavelength is infinite. The stripe region reduces to a point $\psi_o = 0$ at $\beta = 0$. As $\beta \to \beta_c$, the profile adopts a more sinusoidal form, while both its amplitude and its characteristic wavelength continuously decrease. The amplitude is the saturation value of the order parameter ψ_{sat}, and in Fig. 14.3(a) we observe that it decreases monotonically from 1 to 0 as β increases from 0 to β_c.

The equilibrium wavenumber k_e, which is inversely proportional to the characteristic wavelength, increases monotonically with β (Fig. 14.3(b)). The competition between the attractive square gradient interaction and the long-range repulsive interaction plays an important role in the determination of k_e. This competition is clearly seen in the single-mode approximation, as seen from Eq. (14.6). The spatial modulation of the order parameter is given by $\psi_{SM} \sim \cos(k_{SM}x)$. The single-mode approximation k_{SM} to the equilibrium wavevector k_e is the solution of

$$\frac{d}{dk}\left(k^2 + \beta G(k)\right) = \frac{d}{dk}\left(k^2 + \beta \frac{(1 - e^{-kL})}{k}\right) = 0. \tag{14.7}$$

For thick films, $kL \gg 1$, $k_{SM} = (\beta/2)^{1/3}$, while for thin films $kL \ll 1$ and $k_{SM} = \beta L^2/4$. For films with arbitrary thickness, k_{SM} is the same as the maximally unstable mode k_m of the model A linear dispersion relation γ_k in Eq. (13.33).

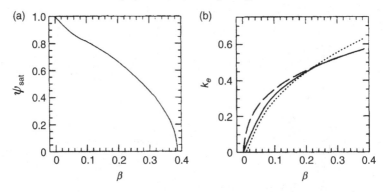

Fig. 14.3. (a) Variation of ψ_{sat} with β and (b) variation of k_e (solid line), k_{SM} (long dashed line) and k_{HW} (dotted line) with β for $L = 10$ and $\psi_o = 0.2$.

The single mode approximation overestimates the "wall" energy, and k_{SM} is an upper bound to the equilibrium wavenumber k_e. Once k_{SM} is obtained, the amplitude ψ_{SM} may be obtained from the absolute minimum of $\mathcal{F}_i - \mu\psi_o$ (Garel and Doniach, 1982; Andelman *et al.*, 1987).

In the hard-wall approximation, the order parameter profile is approximated by zero-width sharp interfaces. In this approximation, which is valid for small β, the equilibrium wavenumber k_{HW} is given by the solution of (Kooy and Enz, 1960)

$$\frac{d}{dk}\left[k + \beta \sum_{n=0}^{\infty} \frac{1}{(2n+1)^3} \frac{(1 - e^{-(2n+1)kL})}{k} \right]_{k=k_{HW}} = 0. \qquad (14.8)$$

The equilibrium wavenumber k_{HW} provides a lower bound to k_e. The variations of k_e, k_{SM} and k_{HW} with β are compared in Fig. 14.3(b) (see Sagui and Desai, 1994).

14.2 Linear stability analysis

Phase separation and phase ordering are nonlinear processes because of the double-well structure of the local free energy functional. A naive linearization of the problem replaces the double well by an inverted harmonic function which has no equilibrium state. Nevertheless, a linear stability analysis of the Langevin equation (13.32) helps to identify the initially unstable modes of the system. The linearized form of Eq. (13.32),

$$\frac{\partial \psi_\mathbf{k}(\tau)}{\partial \tau} = \gamma_k \psi_\mathbf{k}(\tau) + \sqrt{\epsilon}\eta_\mathbf{k}, \qquad (14.9)$$

has the solution

$$\psi_\mathbf{k}(\tau) = e^{\gamma_k \tau}\,\psi_\mathbf{k}(0) + \sqrt{\epsilon}\int_0^\tau d\tau^* \eta_\mathbf{k}(\tau^*)e^{\gamma_k(\tau-\tau^*)}. \qquad (14.10)$$

This solution can be used to obtain the linear structure factor, which is valid for very early times. Recall that the pre-quench system described by $\psi_{\mathbf{k}}(0)$ is in the single-phase region (see also Chapter 6). The corresponding stationary solution for the structure factor is the initial ($\tau = 0^+$) post-quench structure factor, which is given by

$$\langle |\psi_{\mathbf{k}}(0)|^2 \rangle = \frac{\epsilon_o}{q_c^2 + k^2 + \beta \hat{G}(k)}. \tag{14.11}$$

The thermal noise strength $\epsilon_o = \epsilon T_o (a_2(T))^{2-\frac{d}{2}} / [(2\pi)^d T (a_2(T_o))^{2-\frac{d}{2}}]$, where T_o and T are pre- and post-quench temperatures. (See Eq. (5.6) and discussion in Chapter 6.) The dimensionless analog of the time-dependent structure factor introduced in Chapter 4 is $\langle |\psi_{\mathbf{k}}(\tau)|^2 \rangle$. It is related to the two-point correlation function by

$$\langle \psi_{\mathbf{k}}(\tau) \psi_{\mathbf{k}'}(\tau) \rangle = \delta(\mathbf{k} + \mathbf{k}') \langle |\psi_{\mathbf{k}}(\tau)|^2 \rangle. \tag{14.12}$$

Using Eq. (14.10), the result for the linear structure factor is

$$\langle |\psi_{\mathbf{k}}(\tau)|^2 \rangle = \langle |\psi_{\mathbf{k}}(0)|^2 \rangle \, e^{2\gamma_k \tau} + \frac{(k^2)^n}{(2\pi)^d} \left[\frac{e^{2\gamma_k \tau} - 1}{2\gamma_k} \right] \epsilon. \tag{14.13}$$

In the linear regime, the dispersion function γ_k determines the exponential growth of $\psi_{\mathbf{k}}$ and the structure factor. If γ_k is positive, the corresponding Fourier mode fluctuations grow in time: i.e. these are the linearly unstable modes of the system. From Eqs. (13.33) and (13.19), one has

$$\gamma_k = \frac{(k^2)^n}{2} \left(q_c^2 - k^2 - \frac{\beta(1 - e^{-Lk})}{k} \right), \tag{14.14}$$

for either a Langmuir monolayer or uniaxial ferromagnetic film. As usual, $n = 0$ for model A and $n = 1$ for model B. The dispersion function γ_k is shown in Fig. 14.4 for models A (ferromagnetic film) and B (Langmuir monolayer) (see Sagui and Desai, 1994).

14.2.1 Analysis in the absence of LRRI

For $\beta = 0$ there are no long-range interactions, and γ_k takes the form $\gamma_k = k^{2n}(q_c^2 - k^2)/2$. The band of unstable wavenumbers lies in the range $0 < k < q_c$ for both models A and B. Since the range of the band extends to zero, domains will grow to macroscopic size. The power-law form of domain growth is an important aspect of the late stages of phase separation kinetics in the absence of long-range interactions

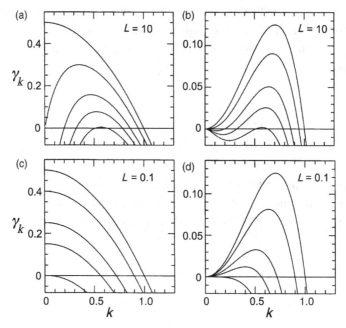

Fig. 14.4. Dependence of the growth exponent γ_k on the wavenumber k for critical quenches ($\psi_o = 0$, $q_c^2 = (1 - 3\psi_o^2) = 1$). (a) and (c) correspond to model A, and (b) and (d) correspond to model B. From top to bottom, (a) and (b) show the curves for a thick film ($L = 10$) for $\beta = 0, 0.1, 0.22, 0.30$, and 0.38; (c) and (d) show the curves for a thin film ($L = 0.1$) for $\beta = 0, 2, 5, 7$, and 10.

(see Chapters 7–12). The maximally unstable mode occurs at $k = 0$ for model A and at $k = k_m \equiv q_c/\sqrt{2}$ for model B. In the linear approximation, the exponential growth of this mode dominates that of all other modes, and the structure factor has a time-invariant maximum at k_m. The existence of a maximally unstable mode at $k = 0$ for model A makes it possible to construct successful approximate nonlinear theories. The presence of a new characteristic wavevector k_m in model B (and in model A when $\beta \neq 0$) makes the construction of nonlinear theories a more challenging task (see Chapter 9).

The early-time experimental observations on binary polymer mixtures (polymer melts) have been fitted to a structure factor form obtained from a linear theory. Model B can be used to describe polymer mixtures. The limitations and range of validity of the linear theory for model B have been discussed by Binder (1984), Elder *et al.* (1988), Glotzer (1995) and Sander and Wanner (1999). As $\psi(\mathbf{k}, \tau)$ increases with time, the linear approximation loses its validity at some crossover time t_{cr}. This occurs roughly at a time when $\langle \psi^2 \rangle$ reaches a value comparable to $(\psi_{sp} - \psi_o)^2 \approx (\psi_{sp}^2 - \psi_o^2) = \psi_{sp}^2 q_c^2$. One can obtain t_{cr} from Eq. (14.10) by replacing k by k_m, since the maximally unstable mode grows exponentially faster than other modes. The dimensionless crossover time, $\tau_{cr} = t_{cr}/t_o \equiv t_{cr}(2M\kappa/\xi_o^4)$, is obtained from

$$(\psi_{sp} - \psi_o)^2 = \langle |\psi(k_m, \tau_{cr})|^2 \rangle = e^{2\gamma_{k_m}\tau_{cr}}\langle |\psi(k_m, 0)|^2 \rangle, \tag{14.15}$$

where the initial fluctuation spectrum is determined from the Ornstein–Zernicke theory at the pre-quench temperature T_o,

$$\langle |\psi(k_m, 0)|^2 \rangle = \frac{\epsilon_o}{(k_m^2 + q_c^2)}. \tag{14.16}$$

Here ϵ_o is given by Eq. (5.6) evaluated at T_o, and can be written as $\epsilon_o = kT_o a_4 \kappa^{-2} \xi_o^{4-\frac{d}{2}}$. Using the values $k_m^2 = q_c^2/2$, $\psi_{sp}^2 = \frac{1}{3}$, and $\xi_o^2 = \kappa/[a_o(T_o - T_c)]$, one obtains

$$2\gamma_{k_m}\tau_{cr} = \frac{d}{2} \ln \kappa + \ln q_c^4 + \ln \left(\frac{[a_o(T_o - T_c)]^{(2-\frac{d}{2})}}{2kT_o a_4} \right). \tag{14.17}$$

Note also that $t_o = (\xi_o^4/2M\kappa) = \kappa/(2Ma_o^2(T_o - T_c)^2)$. As is evident from the form of the square gradient term in the free energy functional in Eq. (3.2), κ plays the role of the square of the effective range of the interaction. Thus, the dimensionless crossover time depends only weakly ($\sim \ln \kappa$) on the range of the interaction. For polymer chains of length N, $\kappa \sim N$. The dimensionless crossover time τ_{cr} for polymeric systems is not very different from that for systems of small molecules. The experimentally accessible crossover time t_{cr} may be scaled to a dimensionless form τ_{cr} using a characteristic time t_o, which itself increases linearly with κ. One has

$$t_{cr} = t_o \tau_{cr} = \frac{2\kappa \left[\frac{d}{2} \ln(\kappa) + \ln(q_c^4) + \ln \left(\frac{[a_o(T_o - T_c)]^{(2-\frac{d}{2})}}{2kT_o a_4} \right) \right]}{q_c^4 Ma_o^2(T_o - T_c)^2}. \tag{14.18}$$

Thus, t_{cr} behaves like $\kappa \ln \kappa \sim N \ln N$ for polymeric systems. The fact that the linear theory is valid for longer times for polymeric systems is due to the factor κ in t_o, which gives rise to the existence of long characteristic times for these systems.

For initial post-quench states in the metastable region that lies between the classical spinodal and coexistence curves, q_c^2 and γ_k are negative for all values of k. A linear stability analysis is not adequate for the metastable region since it predicts that all modes are stable. Nonlinear terms are important and cannot be ignored in kinetics leading to either nucleation or spinodal decomposition. The transition from spinodal decomposition to nucleation is also not well defined because nonlinear instabilities play an increasingly more important role as the classical spinodal is approached from within.

14.2.2 Analysis in the presence of LRRI

A nonzero value of β signals the presence of long-range interactions. If β is sufficiently large, there is a lower-k cutoff on the initially unstable modes. As a result,
after an initial growth phase, the domain size saturates to a finite, time-independent
value. For small k the dispersion relation given in Eq. (14.14) reduces to

$$\gamma_k = \frac{1}{2}(k^2)^n \left[(q_c^2 - \beta L) + \frac{1}{2}\beta L^2 k - (1 + \frac{\beta L^3}{6})k^2 + O(k^3) \right]. \quad (14.19)$$

Consider the three terms in square brackets. The presence of long-range interactions alters the coefficient of $(-k^2)$ from 1 to $(1 + \beta L^3/6)$, which corresponds
to a renormalization of the surface tension. Similarly, the constant term is changed
from q_c^2 to $(q_c^2 - \beta L)$, renormalizing the harmonic term in the bulk free energy. For
model A, if $(q_c^2 - \beta L)$ is positive (negative), the $k = 0$ mode is unstable (stable);
$\beta = q_c^2/L$ is a special value due to the cancellation of the k-independent term
($\beta = 0.1$ curve in Fig. 14.4(a)). For $\beta > q_c^2/L$, the $k = 0$ mode is stable and the
domain size saturates in time to a finite value according to the linear stability analysis. The second term, $\frac{1}{2}\beta L^2 k$, is a direct consequence of long-range interactions.
Note that the two-dimensional Fourier transform of the dipolar interaction kernel
$G(x) \sim 1/x^3$ is $G(\mathbf{k}) \sim -k$. It competes with the third term, which comes from
the renormalized square-gradient attraction and leads to modulated structures as
the system evolves. For $\beta = 0$ the system is characterized by growing domains
which coarsen with time as $\tau^{\frac{1}{3}}$ for model B and $\tau^{\frac{1}{2}}$ for model A. In both cases
the morphology is self-similar at late times. This is qualitatively different from the
modulated structures that one finds for $\beta > q_c^2/L$. For $q_c^2/L > \beta > 0$, domains
coarsen; however, the associated growth exponent gradually decreases from $\frac{1}{3}$ to
zero as β increases from zero to q_c^2/L.

From Eqs. (14.13) and (14.14) it is clear that the initial growth of the pattern
is determined by the linear dispersion γ_k. The system is unstable with respect to
fluctuations of modes with wavenumber k such that $\gamma_k > 0$, and stable with respect
to fluctuations of modes with wavenumber k such that $\gamma_k < 0$. Using Eq. (14.14), the
maximally unstable mode k_m is easily found. Let $\gamma_m = \gamma(k_m)$. As β increases from
zero, γ_m decreases monotonically for models A and B. The maximally unstable
mode k_m increases monotonically from $k = 0$ to k_e for model A and decreases
monotonically from $k = 1/\sqrt{2}$ to k_e for model B.

14.3 Evolution of $\psi(\mathbf{x}, \tau)$ during phase separation

In the presence of long-range repulsive interactions, phase separation involves the
simultaneous segregation into two phases and the creation of a supercrystal ordering. These processes are characterized by very different time scales. In the early and

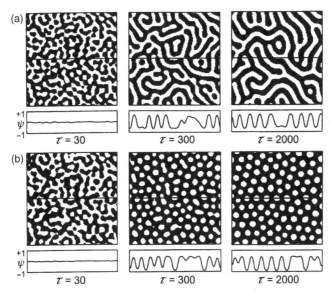

Fig. 14.5. Domain configurations for model B with LRRI for $L = 10$ and $\beta = 0.22$ obtained by integrating Eq. (13.20) with the kernel in Eq. (13.19) and $\epsilon = 0$. (a) Quench into a stripe phase with $\psi_o = 0$. (b) Quench into a hexagonal phase with $\psi_o = 0.2$. Order parameter profile along a horizontal cut in the middle of the configuration is shown below each pattern. Early time stage, $\tau \sim 10\text{--}100$; intermediate time stage, $\tau \sim 100\text{--}1000$.

intermediate stages, processes related to phase segregation and domain formation are predominant. At much later times the domains are essentially monodisperse, and supercrystal ordering dominates. Figure 14.5 shows the patterns that arise from this competition for model B (see Sagui and Desai, 1994). Immediately after the quench, the system acquires a very complicated morphology composed of irregular interpenetrating domains that percolate through the system. The early-time regime corresponds to the amplification of initial fluctuations, the saturation of the order parameter, and the formation of sharp interfaces. It is dominated by modes with wavenumber k_m. During the intermediate-time regime, the domains become progressively more regular and distinct. The variation of the scalar order parameter $\psi(\mathbf{x}, \tau)$ in space is an indicator of the instantaneous domain shapes and the interfaces that separate the domains. Its space and time evolution is associated with how these geometrical degrees of freedom (individual domain shapes and interfaces) of the system evolve in time. The intermediate-time regime is also governed by processes associated with the geometrical degrees of freedom of the system. It is dominated by modes with wavenumber k_e, which determine the equilibrium-modulated structure. The time that the domains require to reach their equilibrium wavelength $2\pi/k_e$ is the characteristic time for this regime. Since the characteristic wavenumber of the system changes from k_m to k_e, the dynamics of this stage of the evolution involves crossover phenomena. The crossover from k_m to k_e is achieved by varying the number of domains. Eventually, the number of domains reaches an essentially constant value, and the distribution function of the stripe width or the bubble radius becomes highly peaked around the equilibrium value. By this time, the shape transition processes are essentially complete, and the system morphology is that of a monodisperse disordered liquid of either stripes or bubbles.

This point marks the onset of the late-time regime. The late stage corresponds to processes associated with the topological degrees of freedom which require two complex order parameters (see Chapter 15). Topological degrees of freedom are associated with the global arrangement of the domains, and are connected to the orientational and translational order of the system. Complex order parameters are useful in describing the evolution of topological order and defects in the system. For model A, the geometrical and topological degrees of freedom are completely decoupled. For model B, this distinction is only an approximation since these degrees of freedom never decouple due to the existence of a conservation law. The kinetics of defect annealing plays an important role during the late stage where supercrystal formation occurs.

The probability that the order parameter has a value between ψ and $\psi + d\psi$ is denoted by $\rho(\psi)d\psi$, and $\rho(\psi)$ is called the one-point distribution function of ψ. It can be extracted from the simulation data at each time instant. For model A, initially $\rho(\psi)$ has a single peak centered at $\psi = 0$ (Fig. 14.6). As time evolves,

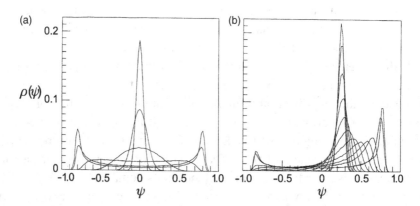

Fig. 14.6. Time evolution of the one-point distribution function $\rho(\psi)$ for model A with $\beta = 0.1$ and $L = 50$. (a) Quench into a stripe phase ($h = 0$): in order of the decreasing peak heights about $\psi = 0$, the curves are at times $\tau = 1, 3, 5, 10, 20$, and 30. (b) Quench into a bubble phase ($h = 1$): in order of the decreasing peak heights about $\psi \approx 0.25$, the curves are at times $\tau = 1, 2, \ldots, 10, 15$, and 30. Reprinted Figure 7 with permission from Roland and Desai (1990). Copyright 1990 by the American Physical Society.

this peak collapses and broadens, and two new peaks grow around the saturation values of the order parameter. The two side peaks are symmetric for the quench into the stripe phase ($h = 0$), but are asymmetric for the quench into the bubble phase ($h = 1$).

The competition between attractive and repulsive interactions not only determines the stability of the shapes of individual domains, it also determines the dynamics of the domain wall fluctuations and shape transitions. For many physical systems the domains exhibit a variety of shapes, which are not always those for an equilibrium system. The response to harmonic distortions for uniaxial ferromagnetic films has been studied for bubble and stripe phases (Garel and Doniach, 1982). Circular shapes have been shown to undergo an elliptical instability. The "stripe–hexagonal" transition in these systems is hindered by a surface energy barrier. This is the origin of topological hysteresis, which is related to nucleation. Langmuir films show instabilities not only to elliptic shapes but also to higher harmonic shapes (McConnell, 1990; Seul *et al.*, 1991). In amphiphilic monolayers, Seul *et al.* (1990, 1991, 1994) have identified the presence of an elliptic instability and a branching instability leading to a melted stripe phase near the consolute point.

15

Competing interactions and defect dynamics

Phase separation in systems with competing interactions involves two dynamic phenomena: segregation into two phases, and the creation of supercrystal (modulated phase) ordering. These two processes occur on very different time scales. The early and intermediate-time regimes were discussed in Chapter 14. In these regimes, all important information about the system may be obtained from the scalar order parameter ψ. During the intermediate-time regime, the domain size reaches its saturation value and the time evolution is ultimately governed by this time independent length scale. Systems with a scalar order parameter form domains of the ordered phase separated by domain walls, the relevant topological defect, and evolve so as to decrease the domain-wall energy.

In the presence of long-range repulsive interactions, the late stage of phase ordering involves the evolution from a disordered liquid of minority phase droplets towards the crystalline (hexagonal) ground state through the gain of orientational and positional order. As discussed in Chapter 9, systems with continuous order parameters have point, line, and other more complex defect structures. The late stages of the phase separation processes are dominated by the motion of these defects and, as time evolves, both their density and energy decrease. This is in contrast to model B in the absence of long-range repulsive interactions, where the late-stage kinetics is curvature driven and the conservation law plays an important role. As long-range repulsive interactions become important, qualitatively different late-stage effects emerge, since dipolar forces compete with forces arising from line tension. These systems present the interesting feature that for the early stages of the evolution domain walls are the relevant topological defects, while in the late stages the defects that exist in two-dimensional solids, dislocations, and disclinations, are the relevant topological defects.

Figure 15.1 shows the late-stage time evolution of droplet morphologies obtained from the simulation of the Langevin equation (13.20) using the kernel given in Eq. (13.19) with different values of β. The Bond number $\mathcal{B} = (\beta \, L^2)/2$ is a useful

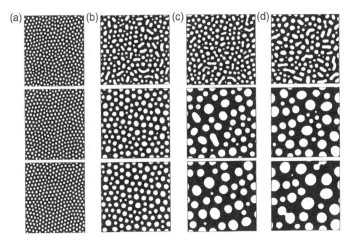

Fig. 15.1. Configurations showing the time evolution for four different sets of parameters: (a) $\beta = 0.22, L = 10$, Bond number $\mathcal{B} = 11$; (b) $\beta = 0.05, L = 10$, $\mathcal{B} = 2.5$; (c) $\beta = 5, L = 0.1, \mathcal{B} = 0.025$; (d) model B (no LRRI, $\beta = 0$). In all cases $\psi_0 = 0.2, \epsilon = 0$. Times, from top to bottom, are $\tau = 1000, 10\,000$, and $20\,000$. Reprinted Figure 12 with permission from Sagui and Desai (1995). Copyright 1995 by the American Physical Society.

correlator for the morphology when long-range repulsive interactions are present. The late-stage morphology is polydisperse in (b), (c), and (d). For large $\mathcal{B}$ in (a), the pattern evolves to a monodisperse-hexagonal crystalline ground state as a result of the domain size saturation expected from the competition between dipolar and line tension forces. The polydisperse morphology which occurs for small $\mathcal{B}$ sometimes becomes frozen or trapped in a metastable state. Different types of morphology can be classified as (a) monodisperse hexagonal (MH), (b) polydisperse frozen (PF), and (c) and (d) polydisperse coarsening (PC).

15.1 Polydisperse systems

The model B ($\beta = 0$) pattern in Fig. 15.1(d) has a Lifshitz–Slyozov–Wagner-type morphology with nonzero volume fraction (see Chapter 12). As β increases, the growth exponent n for the mean droplet size R_m systematically decreases from $\frac{1}{3}$ for model B to zero for the MH morphology. Using the method of matched asymptotic expansions starting from the Langevin description, the LSW analysis has been generalized to describe the PC and PF morphologies in order to understand the Ostwald ripening process in the presence of competing interactions (Sagui and Desai, 1995b). The analysis results in a multidroplet diffusion equation with a Gibbs–Thomson boundary condition that explicitly includes the contribution of the dipolar LRRI to the chemical potential. The normal velocity of the moving interface is proportional to the discontinuity in the chemical potential across the interface, as in the LSW limit. The LRRI effects conveniently break up into intra-droplet and inter-droplet effects, and it is the interplay of the LSW, intra-droplet, and inter-droplet mechanisms that determines the final late-time morphology (PC, PF, or MH). Through a minimization of the energy per unit area, the droplet equilibrium radius R_{eq} can be computed (McConnell, 1989). The effect of the inter-droplet interactions is simply to renormalize the surface tension, which becomes dependent on the volume fraction of the droplet phase and, therefore, on ψ_o. As a result, in certain regions of the β–L plane, only a coarsening solution exists (PC morphology with $R_{eq} = \infty$). The boundary between coarsening (polydisperse coarsening) and noncoarsening (PF morphology) regions occurs when βL^2 is approximately equal to the renormalized surface tension.

15.2 Monodisperse coarsening and defect dynamics

For large values of the Bond number the long-range repulsive interaction is strong and, as discussed above, coarsening leads to a monodisperse hexagonal phase. The formation of a perfect crystalline order in the late stages of the time evolution depends on the depth of the quench (or equivalently, the strength of the dipolar interactions β) and the presence of noise. For this late stage of the coarsening process, where crystalline ordering processes dominate the evolution, it is convenient to characterize the dynamics in terms of the decay of correlations in the system and the development of twofold (for the stripe phase) or sixfold (for the hexagonal phase) symmetry in the structure factor. The new characteristic length scales that appear in the late-stage evolution are related to the sizes of the ordered regions. Since the free energy of the system is degenerate with respect to the direction of the equilibrium wavevector $\mathbf{k}_e$, a quench from the disordered to the ordered phase results in the formation of a modulated phase with defects (disclinations and dislocations). The new length scales grow at a characteristic rate, as different broken symmetry configurations compete to select the ground state.

At the beginning of the late-stage evolution (see the top configuration in Fig. 15.1(a)), the minority phase droplet domains have already attained a circular shape and the scalar order parameter has reached its saturation value. However, the droplets do not have any discernible long-range order in space, and form a liquid-like state. The subsequent time evolution of MH systems is analogous to $2d$ freezing: starting from the liquid state, both the orientational and positional order evolve to that of a perfect crystalline state of circular droplets. Consequently, in the MH regime, it is very useful to describe the ordering process in terms of orientational and translational order parameters, each of which is complex.

The crystal translational order parameter is the local Fourier component of the density, $\rho_K(\mathbf{x}) = e^{i\mathbf{K}\cdot\mathbf{x}}$, where $\mathbf{K}$ is a reciprocal lattice vector corresponding to the first Bragg peak in the structure factor of the crystal. This order parameter is complex, continuous, and Abelian (Nelson, 1983). In $2d$ solids there is no long-range order; only quasi-long-range translational order can exist. The mean translational order parameter of a $2d$ infinite crystal is zero, but its correlation function decays algebraically in space with a temperature-dependent decay exponent. In order to monitor the time evolution of translational order in simulations, it is convenient to define a quantity f_T,

$$f_T = \left|\left\langle \frac{1}{N_T} \sum_{i=1}^{N_T} e^{i\mathbf{K}\cdot\mathbf{X}_i(\tau)} \right\rangle\right|, \tag{15.1}$$

where $\mathbf{X}_i(\tau)$ is the center-of-mass coordinate of droplet i at time τ and the sum is over the centers of the N_T droplets. The angular brackets denote an average over the six symmetry directions of $\mathbf{K}$, and the notation $|\mathcal{R}|$ refers to the modulus of the complex quantity $\mathcal{R}$.

Orientational order in $2d$ is measured by a complex, position-dependent orientational order parameter defined as (Frenkel and McTague, 1979)

$$\psi_6(\mathbf{X}_i(\tau)) = \frac{1}{N_i} \sum_{j=1}^{N_i} e^{i6\theta_j(\mathbf{X}_i)}, \tag{15.2}$$

where N_i is the number of nearest neighbors of droplet i and the sum is over all the nearest-neighbor bonds. The angle made by the bond joining the center of droplet i to that of droplet j, relative to a fixed reference axis, is θ_j. A quantitative measure of orientational order is provided by the correlation function,

$$g_6(x) = \langle \psi_6^*(\mathbf{x})\psi_6(\mathbf{0}) \rangle, \tag{15.3}$$

where the angular brackets represent an average over all pairs of droplets separated by a distance x and an angular average over $\pi/3$ radian segments. The time evolution of the orientational order can be monitored by the quantity f_6, defined as

$$f_6 = \left|\left\langle \frac{1}{N_i} \sum_{j=1}^{N_i} e^{i6\theta_j(\mathbf{X}_i)} \right\rangle\right|, \tag{15.4}$$

where $\langle\ldots\rangle$ and $|\ldots|$ have the same meaning as in Eq. (15.1). Another definition of the orientational order parameter, $f_6' = (g_6(0))^{1/2}$, is often used. It is a measure of local order in the sense that it contains no information about correlations between bonds separated by large distances. Both definitions give $f_6' = f_6 = 1$ in a perfect crystalline solid.

Voronoi constructions can be used to isolate and study the topological defects in the system. The Voronoi cells are hexagons for the MH ground state, and each droplet has exactly six nearest neighbors. For morphologies which are still evolving in time, the coordination number n (number of nearest neighbors) may not be 6. The topological charge of such a droplet is defined to be $q = n - 6$. The graph of droplet centers, which is the *triangulation* construct, is the *dual* structure to the Voronoi partition of space. This implies a one-to-one correspondence between the elements of the Voronoi construct and its dual graph. The Voronoi construction identifies the defects as disclinations or sites with coordination number different from 6. This construction also yields quantities that are useful for the characterization of the orientational order: specifically, bond centers, which are the midpoints of the lines connecting adjacent droplets; and bond angles, which are the angles these lines make with respect to the reference axis. The disclination of the charge q is characterized by a mismatch of $q\pi/3$ in the orientation angle obtained after going around a lattice circuit surrounding the disclination defect. Alternatively, a disclination can be considered to be a site with an atypical number of nearest neighbors determined from the Voronoi polygon construction. In the late stages of the evolution, a majority of the droplets have coordination number $n = 6$. A smaller number of droplets have $n = 5$ (topological charge $q = -1$) and an equal number have $n = 7$ ($q = +1$). In the early stages of the evolution there is a small number of disclinations with higher charge. Isolated disclinations with opposite topological charge attract each other strongly and form more complex topological defects. Fivefold and sevenfold disclinations often occur in local pairs (nearest neighbors separated by a lattice constant) forming edge dislocations. Thus, in a bound pair of dislocations, disclinations are bound in a group of four. An edge dislocation corresponds to two additional half-rows of droplets.

From simulations of the multidroplet diffusion equation discussed in the previous section, one can determine and classify the processes that anneal the topological defects in the MH regime. The experimentally observed defect collision mechanisms correspond to either a T_1 or a T_2 process or combinations of these elementary processes. The T_1 process (see Fig. 15.2(a)) is a neighbor-switching process, which is most easily seen in the Voronoi representation. It can be pictured as taking place when an edge shrinks to zero and is replaced by another edge where the connections to vertices are rearranged (neighbor-switched). In the T_2 process the face of a $2d$ cell vanishes (see the Voronoi representation in Fig. 15.2(b)) and results in the coalescence of two droplets.

The defect analysis is aided by introducing the Burgers vector (BV), which is defined as the amount by which a path around the dislocation core fails to close.

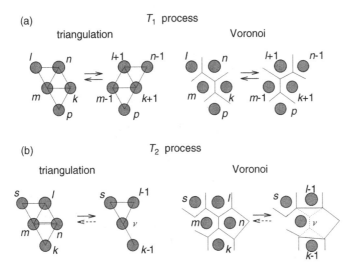

Fig. 15.2. The T_1 and T_2 processes shown in the triangulation and Voronoi representations. For each droplet, its coordination number is shown as l, m, n, etc. For the T_2 case, the reverse process (mitosis) rarely occurs in simulations. Figure 2 with permission from Sagui and Desai (1995). Copyright 1995 by the American Physical Society.

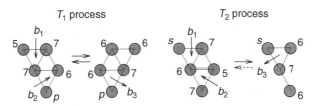

Fig. 15.3. Examples of the T_1 and T_2 processes in the triangular representation. For each droplet, its coordination number is shown. Both processes eliminate one 5-7 pair and correspond to collisions among disclinations, except when $p = 5$ and $s = 5$, in which case they involve collisions between dislocations. Only for the latter case are the corresponding Burgers vectors b_1, b_2, b_3, as shown in the diagram, relevant. Figure 3 with permission from Sagui and Desai (1995). Copyright 1995 by the American Physical Society.

The direction perpendicular to the line joining the bound disclination pair is the glide direction, and the Burgers vector is along this direction. Dislocations move relatively easily in the glide direction, but less easily in the climb direction perpendicular to the Burgers vector. Figure 15.3 illustrates the construction of the Burgers vector. This figure also describes the T_1 and T_2 processes and shows how a 5–7 defect pair is annealed.

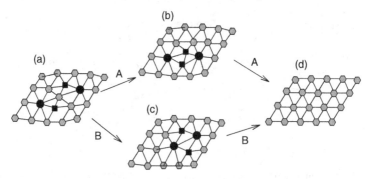

Fig. 15.4. Figure (a), or equivalently figure (b), shows an interstitial defect where there is an extra site in an otherwise perfect lattice. The configuration in (a) can decay through two equivalent processes. In trajectory A, the top dislocation has glided to give (b). A subsequent T_2 process involving the coalescence of the two fivefold droplets (or the elimination of one of the droplets) leads to the perfect lattice in (d). In trajectory B, a T_2 process involving the coalescence of a fivefold droplet and a sixfold droplet (or the elimination of the fivefold droplet) gives configuration (c), with a virtual pair. The virtual pair decays through a T_1 process to configuration (d). (Hexagon $n = 6$; square $n = 5$; circle $n = 7$.) Figure 4 with permission from Sagui and Desai (1995). Copyright 1995 by the American Physical Society.

Figure 15.4 shows how a combination of the T_1 and T_2 processes can lead to the annealing (removal) of an interstitial. The morphologies obtained in the simulation of Langmuir films can also be described in terms of cellular patterns or froths, and the T_1 and T_2 processes are important for the description of the dynamics of these systems. However, there are some important differences. One of the most important differences between froth kinetics and ordering kinetics in Langmuir films is the following: the T_2 processes are dominant for the late stages of froth kinetics, and the T_1 processes are very rare. However, for Langmuir films, in the MH regime, the T_1 processes are by far the dominant mechanism, and defect collisions between two dislocations are prominent. In the MH regime, the area of a "cell" remains fixed, while at smaller values of Bond number in the coarsening regime the area satisfies von Neumann's law, which states that the area A of a froth cell increases with time if $n - 6 > 0$ and decreases if $n - 6 < 0$ so that $\dot{A} = k(n - 6)$, with k a constant.

For the MH regime, the direct integration of the Langevin and multidroplet diffusion equations yields similar results for the time evolution of the orientational and translational order parameters, as shown in Fig. 15.5. Simulations show that in systems with grain boundaries both orientational and translational order are disrupted; in systems without grain boundaries, the onset of quasi-long-range orientational order occurs sooner than the onset of quasi-long-range positional order. In such cases, the system evolves in time through intermediate stages of hexatic

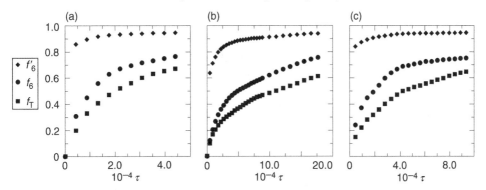

Fig. 15.5. Orientational order parameters f_6, f_6', and translational order parameter f_T for three simulations (a) MH-A, (b) MH-B, and (c) MH-C of the multidroplet diffusion equation. The parameters are: $L = 10$, $\beta = 0.36$ for all three, the area fraction covered by the droplets is 0.04 for MH-A and MH-B, and 0.40 for MH-C. To facilitate visualization the data points for MH-A have been multiplied by 0.6.

order before reaching the crystalline state. The average values of the orientational order parameter f_6 and the translational order parameter f_T shown in this figure indicate that at late times f_6 is higher than f_T. As the orientational order increases, the number of defects (such as 5–7 disclination pairs) decreases. The defect concentration $\rho(\tau)$ can be defined as the ratio of the number of droplets with $z \neq 6$ to the total number of droplets N_T: $\rho(\tau) = 1 - C_6$, where $C_6 = N_6/N_T$ is the concentration of sixfold-coordinated droplets. For each of the three cases in Fig. 15.5, $\rho(\tau)$ decays like a power law with an exponent close to 0.5, which is the value expected when the dislocations are distributed not uniformly in regions of linear dimension L but rather on the surface or the contour of this region.

16

Diffusively rough interfaces

The existence and dynamics of interfaces played a central role in the description of the domain-coarsening phenomena considered in the previous chapters. In the late stages of domain growth the random forces in the order parameter kinetic equations were suppressed and the interface dynamics was treated deterministically. In this chapter we provide a more detailed treatment of the effects of noise and diffusion on the structure of the interface. One may capture the essential physics of diffusively rough interfaces in a general model often called the Kardar–Parisi–Zhang (KPZ) equation (Kardar *et al.*, 1986).

16.1 KPZ equation

Consider a front in a $(d + 1)$-dimensional system extended along $\mathbf{x}$ and moving, on average, in the x_1 direction (see Fig. 16.1). The system is assumed to be infinitely extended along x_1 and has linear dimension L along $\mathbf{x}$. In contrast to the description in Chapter 7, we neglect the intrinsic thickness of the interface and investigate the effects of diffusion and noise on the dynamics of the interfacial profile. Referring to Fig. 16.1, let $h(\mathbf{x}, t)$ be the position of the interface as a function of $\mathbf{x}$ at time t, relative to an arbitrarily selected origin. Its mean position at time t is $\overline{h}(t) = (1/L^d) \int d^d x \; h(\mathbf{x}, t)$. We assume that the front propagates with velocity v in a direction normal to its interface; noise provides a destabilizing influence on the front while diffusion tends to remove any surface roughness.

A sketch of a portion of the interface profile $h(\mathbf{x}, t)$ as a function of $\mathbf{x}$ at time t is shown in Fig. 16.2. The KPZ equation, which describes the dynamics of such fronts, may be motivated using the following physical arguments: assuming that propagation occurs normal to the interface, from the figure we see that the change

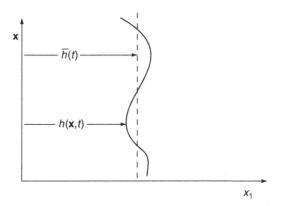

Fig. 16.1. Interfacial profile $h(\mathbf{x}, t)$ as a function of $\mathbf{x}$ at a given time instant. The average front position is $\bar{h}(t)$.

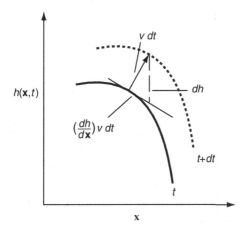

Fig. 16.2. Schematic representation of a curved interface showing the components of the velocity.

dh in the time interval dt is

$$dh = \left[(vdt)^2 + \left(\left(\frac{\partial h}{\partial \mathbf{x}} \right) vdt \right)^2 \right]^{1/2}$$

$$\approx vdt \left(1 + \frac{1}{2} \left(\frac{\partial h}{\partial \mathbf{x}} \right)^2 + \ldots \right). \tag{16.1}$$

Thus, the change in h due to propagation normal to the interface is

$$\frac{\partial h}{\partial t} \approx v + \frac{v}{2} \left(\frac{\partial h}{\partial \mathbf{x}} \right)^2. \tag{16.2}$$

If we then add the changes in h arising from diffusion, $D\partial^2 h/\partial \mathbf{x}^2$, and a Gaussian white noise contribution η with statistical properties

$$\langle \eta(\mathbf{x}, t) \rangle = 0, \tag{16.3}$$

$$\langle \eta(\mathbf{x}, t)\eta(\mathbf{x}', t') \rangle = 2\Gamma\delta(\mathbf{x} - \mathbf{x}')\delta(t - t'), \tag{16.4}$$

we obtain the Kardar–Parisi–Zhang equation,

$$\frac{\partial h}{\partial t} = D\frac{\partial^2 h}{\partial \mathbf{x}^2} + \frac{v}{2}\left(\frac{\partial h}{\partial \mathbf{x}}\right)^2 + \eta. \tag{16.5}$$

In writing this equation we have made the change of variables $h \to h + vt$.

The equation of motion for the average position $\overline{h}(t)$ of the interface at time t can be found by integrating Eq. (16.5) over $\mathbf{x}$ and averaging over realizations of the noise process to obtain

$$\frac{d\overline{h}(t)}{dt} = \tilde{v} = \frac{v}{2L^d} \int d^d x \left\langle \left(\frac{\partial h}{\partial \mathbf{x}}\right)^2 \right\rangle. \tag{16.6}$$

Consequently, even if the planar interface is stationary, an interface following KPZ dynamics will move.

This equation has been used to describe diffusively rough interfaces that arise in a variety of contexts, such as diffusion-limited aggregation and chemical front propagation. The equation is invariant under the transformation $h \to h + \text{constant}$ and under tilting of the interface by an infinitesimal amount, $h \to h + \epsilon \cdot \mathbf{x}$, $\mathbf{x} \to \mathbf{x} + v\epsilon t$. Because of the presence of the $(\partial h/\partial \mathbf{x})^2$ term, the interface does not possess $h \to -h$ symmetry: there is a preferred propagation direction, which typically arises in physical problems from a bias or external force that drives the interface to propagate in a particular direction.

The KPZ equation is related to other well-known evolution equations. If we differentiate Eq. (16.5) with respect to $\mathbf{x}$ and make the change of variables $\mathbf{g} = -\partial h/\partial \mathbf{x}$, we obtain

$$\frac{\partial \mathbf{g}}{\partial t} + v\mathbf{g}\left(\frac{\partial \mathbf{g}}{\partial \mathbf{x}}\right) = D\frac{\partial^2 \mathbf{g}}{\partial \mathbf{x}^2} + \zeta, \tag{16.7}$$

where $\zeta = -\partial \eta/\partial \mathbf{x}$. This is the noisy Burgers' equation.

If we make the transformation $W = e^{vh/2D}$, from Eq. (16.5) we obtain

$$\frac{\partial W}{\partial t} = D\frac{\partial^2 W}{\partial \mathbf{x}^2} + \frac{v}{2D}\eta W, \tag{16.8}$$

which is a diffusion equation subject to parametric noise.

If the nonlinear term in Eq. (16.5) is dropped, we obtain the Edwards–Wilkinson (EW) equation (Edwards and Wilkinson, 1982),

$$\frac{\partial h(\mathbf{x},t)}{\partial t} = D\frac{\partial^2 h(\mathbf{x},t)}{\partial x^2} + \eta(\mathbf{x},t). \qquad (16.9)$$

In the absence of the nonlinear term the interface dynamics possesses an additional $h \to -h$ symmetry characteristic of equilibrium growth processes. We shall consider the consequences of this feature below.

16.2 Interface width scaling

To characterize the interfacial properties, it is useful to consider the width of the interface defined as

$$w(L,t) = \left(\frac{1}{L^d}\int d\mathbf{x}\langle (h(\mathbf{x},t) - \overline{h}(t))^2\rangle\right)^{1/2}, \qquad (16.10)$$

where the angular brackets again signify an average over realizations of the noise process. For a fixed value of L we expect the time variation of the width to take the form shown in Fig. 16.3. Starting from an initially planar interface, after an initial growth phase, the width saturates for times greater than the crossover time t_c to its asymptotic value $w(L) = \lim_{t\to\infty} w(L,t)$.

The sketch in Fig. 16.3 suggests power law behavior for $w(L,t)$ in various limiting regimes. For short times, $t \ll t_c$, and fixed L we define a scaling exponent β such that

$$w(L,t) \sim t^\beta \quad (t \ll t_c, \ L \text{ fixed}).$$

It is found that the width is a function of the system size L, and this system-size dependence also exhibits power-law behavior. The saturated width is observed to

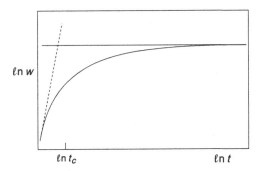

Fig. 16.3. Interfacial width versus time on a logarithmic scale. The crossover time t_c is also indicated in the figure.

scale as

$$w(L) \sim L^\alpha \quad (t \gg t_c),$$

while the crossover time t_c scales as

$$t_c \sim L^z.$$

Using these power laws, a scaling relation among the α, β, and z exponents can be deduced by considering approaches to t_c from above and below. From Fig. 16.3 we have $w(t_c) \sim t_c^\beta = L^{z\beta} = L^\alpha$, which implies $\alpha = z\beta$. Simulations and experiments on a variety of systems expected to possess diffusively rough interfaces indicate that $\beta \approx \frac{1}{3}, \alpha \approx \frac{1}{2}$ and $z \approx \frac{3}{2}$ for systems with $1d$ interfaces. One may now attempt to understand these scaling results on the basis of the interface equations discussed above.

Before considering the KPZ equation we first analyze the EW equation, since it can be solved analytically. We may also obtain the exponent results by a scaling hypothesis. Let ℓ be some scaling length such that $\mathbf{x} \to \mathbf{x}' = \ell\mathbf{x}, h \to h' = \ell^\alpha h$ and $t \to t' = \ell^z t$. We demand that the EW equation be invariant under this scaling,

$$\frac{\partial h'(\mathbf{x}', t')}{\partial t'} = D\frac{\partial^2 h'(\mathbf{x}', t')}{\partial \mathbf{x}'^2} + \eta'(\mathbf{x}', t'). \tag{16.11}$$

First consider the scaled form of the noise. We have

$$\langle \eta'(\mathbf{x}_1', t_1')\eta'(\mathbf{x}_2', t_2')\rangle = 2\Gamma\delta(\mathbf{x}_1' - \mathbf{x}_2')\delta(t_1' - t_2') \tag{16.12}$$
$$= \ell^{-(d+z)}2\Gamma\delta(\mathbf{x}_1 - \mathbf{x}_2)\delta(t_1 - t_2).$$

Thus, comparing with Eq. (16.4), we have $\eta' = \ell^{-(d+z)/2}\eta$. The scaled EW equation then takes the form

$$\frac{\partial h(\mathbf{x}, t)}{\partial t} = +\ell^{z-2}D\frac{\partial^2 h(\mathbf{x}, t)}{\partial \mathbf{x}^2} + \ell^{(-2\alpha-d+z)/2}\eta(\mathbf{x}, t). \tag{16.13}$$

For this equation to be invariant under the transformation, we must have $z = 2$, $\alpha = -d/2 + z/2 = -d/2 + 1$, and $\beta = \alpha/z = -d/4 + \frac{1}{2}$. A direct calculation of $w(L, t)$ for the EW model yields

$$w(L, t)^2 = \frac{\Gamma}{2L}\sum_{n=1}^{\infty}\frac{(1 - e^{-2Dk^2 t})}{2Dk^2}, \tag{16.14}$$

where $k = 2\pi n/L$. The exponents determined from the asymptotic forms of this expression agree with those from the scaling analysis.

If a similar scaling hypothesis is made on the nonlinear KPZ equation, we obtain

$$\frac{\partial h(\mathbf{x}, t)}{\partial t} = \ell^{z-2} \mathcal{D} \frac{\partial^2 h(\mathbf{x}, t)}{\partial \mathbf{x}^2} + \ell^{\alpha-2+z} \frac{v}{2} \left(\frac{\partial h}{\partial \mathbf{x}} \right)^2$$
$$+ \ell^{(-2\alpha-d+z)/2} \eta(\mathbf{x}, t). \tag{16.15}$$

Invariance under this scaling leads to the conditions $z = 2$, $\alpha + z = 2$, and $\alpha = -d/2 + z/2$. This is an overdetermined set of conditions on the exponents. The problem arises from the fact that possible scaling of D and v has been neglected. A full discussion of the correct scaling requires other techniques such as renormalization group methods (Barabási and Stanley, 1995). For $d = 1$ interface dynamics the KPZ scaling exponents are found to be $\alpha = \frac{1}{2}$, $\beta = \frac{1}{3}$, and $z = \frac{3}{2}$. For $d > 2$ interface dynamics, a phase transition exists between two regimes separated by a critical value of the interface velocity $v = v_c$. In the regime with $v < v_c$, termed the weak-coupling regime, the solution flows to the $v = 0$ fixed point corresponding to the linear EW equation. In the strong-coupling regime of the renormalization group equations where $v > v_c$, the nonlinear term becomes relevant. Currently there is no universal agreement on the strong-coupling fixed point and the properties of the KPZ equation in the strong-coupling regime.

16.3 Connection between Langevin and KPZ equations

The KPZ equation was constructed to model a growing interface driven by an external flux of particles. Such a driven nonequilibrium surface can have a roughening transition, which plays an important role in crystal growth. At a fixed temperature, rough or smooth interfaces can be grown by adjusting the external driving force. If an interface grows at a constant rate, it is important to inquire how a constant driving force affects the roughening transition: e.g. the interface roughness above the transition temperature T_R. The KPZ equation is also related to self-organized criticality (SOC) (Bak *et al.*, 1987), which involves states with power-law correlations that are reached without tuning the parameters that control the behavior of the system. A self-organized critical state is characterized by the absence of length and time scales. SOC behavior is different from that near a second-order phase transition (introduced in Chapter 2). The critical point can be reached by tuning parameters such as pressure and temperature in the phase diagram. In contrast, interface equations, such as the KPZ equation, describe a system's behavior after tuning onto a line of first-order phase transitions, such as a solid–liquid coexistence line in the phase diagram of a one-component system. The relationship between SOC and driven growth has been studied extensively (Grossmann *et al.*, 1991). A derivation of the KPZ equation from the model A Langevin equation in the presence of an external driving force is an essential part of this relationship.

To establish the relation between these equations, we begin with the model A Langevin equation (5.1) and the free energy functional $\mathcal{F}_{GLW}$ in Eqs. (3.2) and (3.14),

$$\frac{\partial \delta \phi}{\partial t} = -M \left[-\mathcal{H} - \kappa \nabla^2 \phi^* + \frac{\delta f(\phi^*)}{\delta \phi^*} \right] + \eta^*(\mathbf{r}, t). \qquad (16.16)$$

Here $\delta \phi = (\phi - \overline{\phi}) = (\phi^* - \phi_o)$, with $\phi_o = (\overline{\phi} - \phi_c)$ as defined in Chapter 5; the Landau free energy density $f(\phi^*)$ is given in Eq. (3.15). The thermal noise η^* is a Gaussian random function which obeys the fluctuation–dissipation theorem given in Eq. (5.2). The external field $\mathcal{H}$ plays a central role in the connection between the Langevin and KPZ equations. One can also obtain this Langevin equation from a modification of model C, where a nonconserved order parameter $\phi(\mathbf{r}, t, \mathcal{H})$ is coupled to a conserved field $\mathcal{E}$ in an asymmetric manner, and where the length scale for the diffusion of $\mathcal{E}$ is much larger than all other lengths in the system. With an asymmetric coupling, one has a model for the liquid–solid interface, which is typically used to study the Mullins–Sekerka instability and dendritic growth.

In equilibrium, for a system made up of two coexisting phases separated by a planar diffuse interface located at $x_1 = 0$, ϕ will be time independent and inhomogeneous; we denote it by the function $\phi_o(x_1)$, whose average over x_1 is the average order parameter $\overline{\phi}$. For such a system, equilibrium requires $\mathcal{H} = 0$, and $\phi_o(x_1)$ satisfies the analog of Eq. (7.3) in dimensional form,

$$\kappa \frac{d^2 \phi_o(x_1)}{dx_1^2} = \frac{\delta f[\phi_o(x_1)]}{\delta \phi_o(x_1)}. \qquad (16.17)$$

This result can also be obtained from the average of Eq. (16.16) over the noise. The mean field surface tension is (see Chapter 3)

$$\tilde{\sigma} = \kappa \int_{-\infty}^{\infty} dx_1 \left(\frac{d\phi_o(x_1)}{dx_1} \right)^2. \qquad (16.18)$$

In general, in a far-from-equilibrium driven system, the interface is curved and time dependent. Instead of being located at $x_1 = 0$, the interface is defined by the locus of the zeros of an auxiliary function (see Figs. 7.2 and 16.1),

$$u(\mathbf{r}, t, \mathcal{H}) = 0. \qquad (16.19)$$

The equation of motion for u can be extracted by assuming

$$\phi(\mathbf{r}, t, \mathcal{H}) \simeq \phi_o[u(\mathbf{r}, t, \mathcal{H})]. \qquad (16.20)$$

Implicit in this condition is the assumption that the interface is gently curved and far from other interfaces in the system. We now perform an analysis similar to that in Chapter 7. Using the assumption in Eq. (16.20), the Langevin equation takes the form

$$\frac{\partial u}{\partial t}\frac{\partial \phi_o(u)}{\partial u} = -\dot{M}\left(-\mathcal{H} - \kappa \nabla^2 \phi_o(u) + \frac{\delta f}{\delta \phi_o}\right) + \eta^*. \tag{16.21}$$

If we let $\hat{\mathbf{n}} = \nabla u/|\nabla u| \equiv \ell \nabla u$ be the unit vector normal to the interface, the variation of $\phi_0(u)$ in this normal direction is $\hat{\mathbf{n}} \cdot \nabla \phi_0(u) = \ell^{-1}\partial\phi_0(u)/\partial u$. Equation (16.17) then takes the form

$$\kappa \frac{d^2 \phi_o(u)}{(\ell du)^2} = \frac{\delta f[\phi_o(u)]}{\delta \phi_o(u)}. \tag{16.22}$$

Making use of the fact that the Laplacian of $\phi_0(u)$ may be written as

$$\nabla^2 \phi_o(u) = (\nabla \cdot \hat{\mathbf{n}})\frac{d\phi_0(u)}{d(\ell u)} + \frac{d^2 \phi_0(u)}{d(\ell u)^2}, \tag{16.23}$$

using Eq. (16.17), Eq. (16.21) becomes

$$\frac{\partial u}{\partial t}\frac{\partial \phi_o(u)}{\partial u} = -M\left(-\mathcal{H} - \kappa K \frac{d\phi_o(u)}{(\ell du)}\right) + \eta^*, \tag{16.24}$$

where the curvature is $K = \nabla \cdot \hat{\mathbf{n}}$.

For interfaces with small widths, such as those envisaged in Fig. 16.1, it is appropriate to integrate Eq. (16.24) over the diffuse interface using a projection operator

$$\mathcal{P}(\cdots) = \frac{1}{\Delta\phi_o}\int (\ell du)\left(\frac{d\phi_o}{(\ell du)}\right)(\cdots), \tag{16.25}$$

where $\Delta\phi_o = \phi_o(+\infty) - \phi_o(-\infty)$ is the miscibility gap. The effect of the projection operator on a quantity is to replace it by its average over the diffuse interface (see Fig. 7.1). The analysis follows that in Chapter 7. The result, which uses Eq. (16.18), is

$$\ell\frac{\partial u}{\partial t} = \frac{M\kappa\mathcal{H}\Delta\phi_o}{\tilde{\sigma}} + M\kappa K + \tilde{\eta}, \tag{16.26}$$

which is similar to Eq. (7.30). Here $\tilde{\eta}$ is related to η^* by

$$\tilde{\eta} = (\kappa/\tilde{\sigma})\int (\ell du)\left(\frac{d\phi_o}{(\ell du)}\right)\eta^*. \tag{16.27}$$

Due to the continuity of flux at the interface at $u = 0$, the left-hand side of Eq. (16.26) is the velocity normal to the interface at $u = 0$. The last two terms on the right-hand side correspond to the two terms on the right-hand side of Eq. (7.30). The first term on the right-hand side is new and is proportional to the driving field $\mathcal{H}$. The product $M\kappa$, previously discussed in Chapter 7, must be positive if the motion of the interface is such that it will reduce curvature.

In order to obtain the KPZ equation, we now assume that the interface is almost flat without overhangs: i.e. $u(\mathbf{r}, t) = x_1 - h(\mathbf{x}, t)$. Substituting this expression into Eq. (16.26) one gets

$$\frac{\partial h}{\partial t} = \frac{1}{\ell}\left(v + D\ell^3 \frac{\partial^2 h}{\partial \mathbf{x}^2} + \eta\right),\tag{16.28}$$

where $1/\ell(h) = |\nabla u| = [1 + (\partial h/\partial \mathbf{x})^2]^{1/2}$, $v = -M\kappa\mathcal{H}\Delta\phi_o/\tilde{\sigma}$, $D = M\kappa$, and $\eta = -\tilde{\eta}$. Then expanding to leading order in $(\partial h/\partial \mathbf{x})^2$ and setting $h \to (h - vt)$, we get the KPZ equation

$$\frac{\partial h}{\partial t} = D\frac{\partial^2 h}{\partial \mathbf{x}^2} + \frac{v}{2}\left(\frac{\partial h}{\partial \mathbf{x}}\right)^2 + \eta.\tag{16.29}$$

The above derivation shows that the KPZ equation can be used to describe a nearly flat growing interface in nonconserved model A systems in the presence of a field, or model C systems, such as a solid growing into a supercooled liquid. The self-organized critical states for such systems, which are described by the KPZ equation, are nonequilibrium states within the coexistence curve in an equilibrium phase diagram where a liquid is supercooled or a solid is superheated. Such critical states are reached without tuning. This conclusion imposes strong bounds on the applicable nonequilibrium states: e.g. metastable liquid states. A metastable state eventually decays by droplet nucleation, where bubbles and interacting surfaces become important. The decay rate τ^{-1} of such states, referred to as the nucleation rate, has been calculated (Becker and Doring, 1935; Gunton and Droz, 1983), and is given by

$$\tau \propto \exp\left(\frac{\tilde{\sigma}}{\mathcal{H}^d} + \ln L^{d+1}\right),\tag{16.30}$$

where L^{d+1} is the volume accessible for a large fluctuation needed to nucleate a critical size droplet in a $(d + 1)$-dimensional system. The KPZ equation is useful on time scales $t \ll \tau$.

16.4 Dynamics of curved interfaces

A more general phenomenological description of diffusively rough interfaces that avoids the assumption of small curvature used in the derivation of the KPZ equation can be given as follows (Maritan *et al.*, 1992). Consider an interface in a two-dimensional system. Suppose the interface $\mathbf{R}(s,t) \equiv (x(s,t), y(s,t))$ is parameterized by the continuous variable s. The normal to the interface is $\hat{\mathbf{n}}$. As earlier, we assume that the interface moves in the direction of its normal with speed v. The interface is subjected to noise, and variations along the interface are smoothed by diffusion, characterized by a diffusion coefficient D. Accounting for these contributions, the velocity of the interface is given by

$$\frac{\partial \mathbf{R}(s)}{\partial t} = Dg^{-1/2}\frac{\partial}{\partial s}g^{-1/2}\frac{\partial}{\partial s}\mathbf{R}(s) + v\hat{\mathbf{n}} + \eta, \tag{16.31}$$

where $g = |d\mathbf{R}/ds|^2$. The Laplacian governing diffusion along s is written in curvilinear coordinates (see Chapter 7). The correlation function of the random force η is

$$\langle \eta_\alpha(s,t)\eta_\beta(s',t')\rangle = 2D\delta_{\alpha\beta}g^{-1/2}\delta(s-s')\delta(t-t'). \tag{16.32}$$

This choice of the correlation guarantees that the dynamics is independent of the parametrization adopted. If there are no overhangs, i.e. the interface is a single-valued function of s, this equation reduces to the KPZ equation. To see this, let $\mathbf{R}(s,t) \equiv (h(s,t), y(s,t)) = (h(y,t), y)$, where $h(y,t)$ is a single-valued function of y. Substituting this expression for $\mathbf{R}(s,t)$ into Eq. (16.31) and taking the scalar product of the resulting expression with $\hat{\mathbf{n}} = g^{-1/2}(1, -\partial h/\partial y)$, we obtain

$$\frac{\partial h}{\partial t} = Dg^{-1}\frac{\partial^2 h}{\partial y^2} + vg^{1/2} + g^{1/4}\xi(y,t), \tag{16.33}$$

where

$$\langle \xi(y,t)\xi(y',t')\rangle = 2Dg^{-1/2}\delta(y-y')\delta(t-t'), \tag{16.34}$$

and $g = 1 + (\partial h/\partial y)^2$. In the limit where $(\partial h/\partial y)^2$ is small, so that the g factors can be expanded in this quantity and only lowest-order terms in Eq. (16.33) are retained, we obtain the KPZ equation (16.5) after the transformation $h \to h + vt$.

Equation (16.31) may be used to study the dynamics of disk-shaped nuclei (Kapral *et al.*, 1994). For this purpose it is useful to express the dynamics in a polar coordinate system $x(s,t) = R(s,t)\cos\theta(s,t)$ and $y(s,t) = R(s,t)\sin\theta(s,t)$ (see Fig. 16.4). In this coordinate system, $g = R_s^2 + R^2\theta_s^2$, where the subscript s

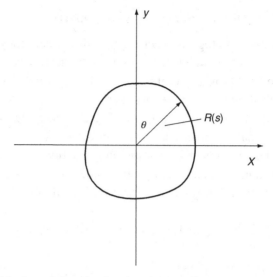

Fig. 16.4. Disk-shaped domain showing the coordinate system used in the text.

indicates differentiation with respect to the arc length. Assuming $R(s,t) = R(\theta,t)$ (no overhangs), we may show that

$$\frac{\partial R(\theta,t)}{\partial t} = D\frac{1}{\sqrt{R_\theta^2 + R^2}}\frac{\partial}{\partial\theta}\frac{1}{\sqrt{R_\theta^2 + R^2}}\frac{\partial}{\partial\theta}R(\theta,t)$$

$$-\frac{D}{R}\left(1 + \frac{R_\theta^2(R_\theta^2 - RR_{\theta\theta})}{(R_\theta^2 + R^2)^2}\right)$$

$$+\frac{v}{R}\sqrt{R_\theta^2 + R^2} + \frac{(R_\theta^2 + R^2)^{1/4}}{R}\xi(\theta,t), \qquad (16.35)$$

where the random force correlation function is

$$\langle \xi(\theta,t)\xi(\theta',t')\rangle = 2D\delta(\theta - \theta')\delta(t - t'). \qquad (16.36)$$

If we assume that $R_\theta/R \ll 1$, we obtain the equation for disk growth:

$$\frac{\partial R}{\partial t} = \frac{D}{R^2}\frac{\partial^2 R}{\partial\theta^2} - \frac{D}{R} + v\left(1 + \frac{R_\theta^2}{2R^2}\right) + \frac{1}{\sqrt{R}}\xi(\theta,t). \qquad (16.37)$$

Even neglecting the KPZ-like term, Eq. (16.37) is nonlinear and the exact solution cannot be found by Fourier transformation.

A perturbative analysis of the angular fluctuations can be carried out as follows. Substituting

$$R(\theta,t) = \sum_{n=-\infty}^{+\infty} R_n(t)e^{in\theta}, \quad \xi(\theta,t) = \sum_{n=-\infty}^{+\infty} \xi_n(t)e^{in\theta} \quad (16.38)$$

in Eq. (16.37) with $\langle \xi_n(t)\xi_n(t') \rangle = \frac{D}{\pi}\delta(t-t')$, and multiplying both sides by R^2, a hierarchy of stochastic differential equations is obtained. This set of equations refers to a fixed reference frame. For nucleation processes it is necessary to determine the radial variable with respect to the actual position of the center of mass. The choice of such a reference frame amounts to setting $R_1(t) = \text{constant} = 0$, and thus the leading corrections to $R_0(t)$ originate from $R_{\pm2}(t)$. We can then truncate the hierarchy of equations at $n = \pm2$. Keeping only the leading terms in the equations for $R_0(t)$, $R_{\pm2}(t)$, we obtain

$$R_0^2\dot{R}_0 + 2R_2R_{-2}\dot{R}_0 + 2R_2R_0\dot{R}_{-2} + 2R_{-2}R_0\dot{R}_2$$
$$= -DR_0 + v(R_0^2 + 6R_2R_{-2})$$
$$+ R_0^{3/2}\left(\xi_0 + \frac{3}{2}\frac{R_2}{R_0}\xi_{-2} - 2\frac{3}{2}\frac{R_{-2}}{R_0}\xi_2 - \frac{3}{8}\frac{R_2R_{-2}}{R_0^2}\xi_0\right),$$

and

$$R_0^2\dot{R}_{\pm2} + 2R_{\pm2}R_0\dot{R}_0 = -5DR_{\pm2} + 2vR_0R_{\pm2} + R_0^{3/2}\xi_{\pm2}. \quad (16.39)$$

Neglecting the angular dependence, i.e. the dependence on $\xi_{\pm2}$ and $R_{\pm2}$, Eq. (16.39) reduces to

$$\dot{R}_0 = -\frac{D}{R_0} + v + \frac{\xi_0}{\sqrt{R_0}}, \quad (16.40)$$

which gives the zeroth order estimate for the critical radius, $R_c = D/v$.

If the noise term is neglected we recover the Allen–Cahn equation for the evolution of a disk-shaped nucleus in model A (see Eq. (7.33) and the discussion that follows this equation). Equation (16.40) can be used to study the effects of fluctuations on the growth of a disk-shaped nucleus, while the more general equation (16.35) provides a KPZ analog for such nucleus growth processes.

17

Morphological instability in solid films

The growth of thin solid semiconductor films is at the heart of the development of modern electronic and optical devices. A key element in strategies for nanoscale fabrication is the exploitation of growth and kinetic instabilities to form surface nanostructures and patterns with desirable functionality.

Epitaxy is a term that is commonly used for the growth of a thin solid layer on top of a substrate. Homoepitaxy denotes the growth of crystals of a material on a crystal face of the same material, while the term heteroepitaxy is used if the materials of the substrate and the growing film are different. Molecular beam epitaxy (MBE) is a common experimental technique that is used to grow such solid films. A film that grows without defects is called a coherently grown film. In such a film the constituent atoms arrange themselves on top of the substrate as its natural extension. The film has the same crystal structure as the substrate.

In the epitaxial growth of a crystal film on another crystal, elasticity plays a dominant role and leads to long-range effective interactions between the adatoms on the surface. These interactions are repulsive and compete with the stronger short-range chemical interactions. The repulsive nature of the long-range interactions can be qualitatively understood as follows. Consider a planar solid surface of a semi-infinite crystal. When an adatom is placed on this surface, its interaction with the atoms in the top layer creates a stress which changes the distance between its nearest neighbors in the top atomic layer of the surface. If the change is an increase in this distance, a compression of the interatomic distances between more distant atomic pairs will occur in the immediate neighborhood of the adatom. This discourages other adatoms from being located close to the first one. Therefore, the effective interaction between adatoms is repulsive. If two adatoms are placed on the solid surface a distance r apart, their repulsive interaction energy is proportional to r^{-3}.

There are two classes of kinetic roughening process, which are characterized by nonconserved and conserved surface relaxation processes, respectively. Non-conserved processes are typically described by the KPZ equation, which was the subject of the previous chapter. Vapor deposition processes that are carried out at either low or high temperatures where defect formation and desorption become important fall into the KPZ class. In this chapter we consider the conserved case, which corresponds to films grown by deposition processes such as MBE. For such technologically relevant deposition methods, both defect formation and desorption are negligible, especially at the early stage of growth. For coherent epitaxial films, surface diffusion is the dominant relaxation process. Three basic growth modes have been seen in experiments on heteroepitaxial films. The Frank–van der Merwe mode corresponds to layer-by-layer growth. The Volmer–Weber mode involves island formation and occurs in systems where a large amount of energy is required to form the substrate/film interface. For the Stranski–Kranstanov mode, the first few monolayers are almost flat and are followed by an increasingly rough surface, which exhibits an island-like morphology as the film thickens. The type of growth that occurs in a specific system depends on the relative importance of elastic strain and interfacial energies. Based on these considerations, we use linear continuum elasticity theory to describe the growth of epitaxial films.

During growth, a partially formed monolayer necessarily has step edges. Some systems have potential barriers near step edges, and these barriers suppress the diffusion of adatoms to lower terraces. The presence of these barriers, known as the Ehrlich–Schwoebel barriers (Ehrlich and Huda, 1966; Schwoebel and Shipsey, 1966), give rise to an instability (Villain, 1991; Pimpinelli and Villain, 1998) that results in pyramidal structures on the growing surface during homoepitaxy. The growth of such structures occurs as a result of a nonequilibrium surface diffusion current which is not derivable from a free energy functional (Krug *et al.*, 1993). The surface diffusion coefficient D_s and the deposition rate F play important roles (Krug, 1997) in the film growth. For large values of D_s/F the growing film develops wedding-cake–like structures. For certain step orientations, it is found that the Ehrlich–Schwoebel barrier vanishes, since the potential that an adatom sees has a maximum at the center of the terrace instead of the edge. We consider such systems below.

Another ubiquitous instability occurs as a result of a lattice mismatch between the substrate and the growing film. It is this instability in heteroepitaxial films that is the central theme of this chapter.

17.1 Lattice misfit

Consider a heteroepitaxial film growing on a semi-infinite substrate as shown schematically in Fig. 17.1(a). The substrate, located in the region $z \leq 0$, has a lattice constant a_s. The coherently grown solid thin film within the region $0 < z < h(x, y; t)$ tends to have its normal bulk-state lattice constant a_f (Fig. 17.1(b)). However,

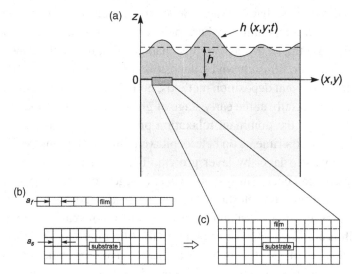

Fig. 17.1. Schematic diagram showing the coherent growth of film on a substrate

coherency requires that the lattice constant of the film be the same as that of the substrate (Fig. 17.1(c)). This leads to a lattice misfit and, within the film, gives rise to a strain $\epsilon_o = (a_f - a_s)/a_s$ in directions x and y parallel to the interface at $z = 0$. There is also a strain in the perpendicular z direction. The stress associated with this strain is relieved in the perpendicular direction. This is called Poisson relaxation. The stress in the xy plane remains. A planar surface of such a uniaxially stressed solid is unstable to a range of wavelengths, bounded from below by a length given by the ratio of the surface and strain energies (Asaro and Tiller, 1972; Grinfeld, 1986; Srolovitz, 1989).

This morphological instability, which is often referred to as Asaro–Tiller–Grinfeld instability, can be seen to originate as follows (Spencer *et al.*, 1993). A uniform uniaxial stress applied to a flat film results in a uniform strain throughout the film. This leads to a uniform strain energy density, which is a product of stress and strain. Since the applied stress and strain are always of the same sign, the strain energy density is always positive. Such a perfectly flat film is in a state of unstable equilibrium in the absence of surface fluctuations. If the same film has a sinusoidal perturbation on its surface, the applied stress leads to a nonuniform stress distribution within the film. Regardless of the sign of the stress (compressional or dilatational), there is a relaxation of the stress at the peaks and a concentration of stress at the valleys of the perturbation. The peaks are less constrained by the bulk and relax towards a stress-free state more easily than the valleys, which become stress concentrators. Thus, the strain energy density is higher in the valleys than at the relaxed peaks. This difference in the strain energy drives the instability. Just as the stabilizing surface energy drives mass transport from regions of high curvature to regions of low curvature, strain energy drives the mass transport from regions of high strain energy (valleys) to regions of low strain energy (peaks). The peaks then grow at the expense of valleys, leading to an even larger strain energy variation between them. This positive feedback leads to growing perturbations and is responsible for the morphological instability of a uniaxially stressed film.

The analysis of this instability presented below is based on continuum linear elasticity theory and the mesoscopic Langevin model. Its applicability is restricted to length scales much larger than lattice spacing. For some aspects of thin film growth, step–step interactions and step growth are important. The energy required to form a step on an infinite $2d$ surface vanishes at the roughening temperature T_r, which is lower than the melting temperature T_m. For $T_r < T < T_m$, thermal fluctuations liberate the surface from the ordering influence of the underlying lattice, and the crystal's equilibrium shape is rough. For $T < T_r$, step kinetics is an important part of thin film growth. In some situations Ostwald ripening can create smoothing below T_r (Pimpinelli and Villain, 1998). Our interest is in instabilities. If the characteristic wavelength of an instability is much longer than the lattice spacing, a quasi-continuum analysis can be used to describe the film growth.

During the epitaxial growth of a film, if the deposition of material occurs under ultrahigh-vacuum conditions, similar to growth through MBE, evaporation and recondensation are negligible. The system evolution would then follow a conserved dynamics analogous to that introduced in Chapter 6. We assume that the film completely wets the substrate, that there is no interdiffusion between the film and the substrate, and that diffusion in the bulk film is negligible compared with diffusion on the surface. The neglect of bulk diffusion implies that the layers within the film are buried metastable layers, which are frozen on account of negligible mobility of atoms in the bulk.

Consequently, the deposition rate determines the mean surface height $\bar{h} : \bar{h} = h_o + vt$. The deposition rate F is the number of atoms deposited per unit area per unit time along the $-z$ direction on the top surface of the film. The growth velocity of the film is $v = F/N_v$, where N_v is the number density of the atoms per unit volume. The results of the analysis below, in the limit of no deposition, correspond to the case of a static film of thickness h_o. In a reference frame moving with the average surface position $\bar{h}$, the surface diffusion process conserves the total amount of material. Thus, the motion of the interface satisfies a continuity equation

$$N_v \frac{\partial r_n}{\partial t} = -\nabla \cdot \mathbf{J}_h + \tilde{F}, \tag{17.1}$$

where $\partial r_n / \partial t$ is the rate of change of the interface position along its normal and $\tilde{F} = F / \sqrt{g}$. The Cartesian components of the normal unit vector $\hat{n}$, pointing away from the film, are given by $(-\partial h / \partial x, -\partial h / \partial y, 1) / \sqrt{g}$, where $g = (1 + |\nabla h|^2)$ is the determinant of the surface metric (see also Appendix to Chapter 7 and Section 16.4), with $|\nabla h|^2 = (\partial h / \partial x)^2 + (\partial h / \partial y)^2$. Then the equation for the z component of the interface motion is

$$\frac{\partial h}{\partial t} = \sqrt{g} \frac{\partial r_n}{\partial t} = -N_v^{-1} \sqrt{g} \nabla \cdot \mathbf{J}_h + v. \tag{17.2}$$

Here the surface flux of atoms $\mathbf{J}_h$ is proportional to the gradient of a diffusion potential, $\delta \mathcal{F} / \delta h$. More precisely,

$$\mathbf{J}_h = -\frac{D_s N_s}{k_b T N_v} \nabla \frac{\delta \mathcal{F}}{\delta h}, \tag{17.3}$$

where N_s is the number density of atoms per unit surface area. From the above two equations, the conserved dynamics evolution equation for $h(x, y; t)$ can be deduced and is

$$\frac{\partial h}{\partial t} = \Gamma_h \sqrt{g} \nabla_s^2 \frac{\delta \mathcal{F}}{\delta h} + v, \tag{17.4}$$

where ∇_s^2 is the surface Laplacian and $\Gamma_h = D_s N_s / k_b T N_v^2$ is the kinetic coefficient, which depends on temperature T, surface diffusion coefficient D_s, the number density of atoms per unit surface area N_s, and per unit volume N_v. The free energy functional $\mathcal{F}$ consists of surface energy and elastic energy parts,

$$\mathcal{F} = \mathcal{F}_s + \mathcal{F}_{el}. \tag{17.5}$$

The surface free energy functional is given by

$$\mathcal{F}_s = \gamma \int d^2 r \sqrt{g}, \tag{17.6}$$

where the integral is over the top surface of the film and γ is the surface tension at the top surface. For simplicity, we assume γ to be isotropic. For real systems, the surface tension anisotropy is often important. The surface free energy clearly plays a central role in determining equilibrium crystal shape (Schukin and Bimberg, 1999). The elastic free energy functional will be deduced from linear elasticity theory in the next section.

Many films of interest are alloys: for example, an alloy of silicon and germanium grown on a silicon substrate. For alloys, an alloy composition field $\phi(x, y, z; t)$ must be included in the description (Guyer and Voorhees, 1995; Léonard and Desai, 1998; Huang and Desai, 2002). The free energy functional now includes an additional term, $\mathcal{F} = \mathcal{F}_s + \mathcal{F}_{el} + \mathcal{F}_{GLW}$. The form of $\mathcal{F}_{GLW}$ is given in Eqs. (3.2) and (3.15) within the film (this implies that the limit of the z-integration ranges between 0 and $h(x, y; t)$), and the elastic free energy functional also depends on ϕ. For simplicity we restrict our considerations to a simpler case, such as a germanium film grown on a silicon substrate.

17.2 Elastic free energy functional

When a deformation occurs in a body, distances between various points within it change. The position vector $\mathbf{r}$ of a point is displaced to $\mathbf{r}' = \mathbf{r} + \mathbf{u}$. The vector distance $\mathbf{dr}$ between two neighboring points $\mathbf{r}$ and $\mathbf{r} + \mathbf{dr}$ changes to $\mathbf{dr}'$. If the corresponding magnitudes of $\mathbf{dr}$ and $\mathbf{dr}'$ are dl and dl', then $dl^2 = dx_i dx_i$ changes to $dl'^2 = (dx_i + du_i)(dx_i + du_i)$. Here and below, we use the Einstein summation convention, where repeated indices are summed. For small displacements (deformations), we find

$$dl'^2 \simeq dl^2 + 2\frac{\partial u_i}{\partial x_k}dx_k dx_i$$
$$= dl^2 + 2\epsilon_{ik}dx_k dx_i, \tag{17.7}$$

where we have introduced the strain tensor ϵ, which is

$$\epsilon_{ik} = \frac{1}{2}\left(\frac{\partial u_i}{\partial x_k} + \frac{\partial u_k}{\partial x_i}\right), \tag{17.8}$$

for small deformations. The misfit strain ϵ_o, introduced in Section 17.1, is related to the strain tensor: for the example in Fig. 17.1(c) the three nonzero elements of ϵ are $\epsilon_{xx}^o = \epsilon_{yy}^o = -\epsilon_o$ and $\epsilon_{zz}^o = -\epsilon_o + \bar{\epsilon}$, where $\bar{\epsilon}$ is obtained below. The superscript o indicates that the value ϵ_o is for a film with a flat top surface.

When a deformation occurs, the body is displaced from equilibrium and internal forces arise, which may then drive the body towards a new equilibrium. Let V be the volume of the body being deformed, and let it be enclosed by a surface S. The volume can be divided into a large number of small volume elements dV. The total internal force $\int_V \mathbf{F}dV$ can then be viewed as a sum of forces acting on various elements dV. The forces internal to V at the boundaries of internal volume elements dV cancel due to Newton's third law. Thus, the total internal force reduces to a sum of forces acting on various elements $\mathbf{dS}$ of the surface S. The direction of the surface element vector $\mathbf{dS}$ is taken along the outward normal. As a result one obtains

$$\int_V F_i dV = \int_V \frac{\partial \sigma_{ik}}{\partial x_k}dV = \oint_S \sigma_{ik}dS_k. \tag{17.9}$$

We have assumed that external forces are absent. The stress tensor $\boldsymbol{\sigma}$ is symmetric (Landau and Lifshitz, 1986). For a system in mechanical equilibrium the internal stresses balance in each of the volume elements dV: thus, the condition of mechanical equilibrium is

$$\frac{\partial \sigma_{ik}}{\partial x_k} \equiv (\nabla \cdot \boldsymbol{\sigma})_i = 0. \tag{17.10}$$

Within a volume element dV, the work done by internal stresses of the system per unit volume on the external agency creating a deformation is

$$\delta W = \frac{\partial \sigma_{ik}}{\partial x_k} \delta u_i.$$

Using integration by parts, the total work done can be written as

$$\int \delta W dV = \oint \sigma_{ik} \delta u_i dS_k - \int \sigma_{ik} \frac{\partial \delta u_i}{\partial x_k} dV.$$

For an infinite medium which is not deformed at infinity the first term vanishes, and the second term can be written in a symmetric form to yield

$$\int \delta W dV = -\frac{1}{2} \int \sigma_{ik} \delta \left(\frac{\partial u_i}{\partial x_k} + \frac{\partial u_k}{\partial x_i} \right) dV = - \int \sigma_{ik} \delta \epsilon_{ik} dV,$$

so that

$$\delta W = -\sigma_{ik} \delta \epsilon_{ik}. \tag{17.11}$$

Since the change in the internal energy is $\delta Q - \delta W = TdS + \sigma_{ik} d\epsilon_{ik}$, the differential $\sigma_{ik} d\epsilon_{ik}$ is an analog of the $-pdV$ term for pressure–volume work in isotropic fluids. It then follows that elastic deformations of solids lead to a differential change in Helmholtz free energy given by

$$dF_{el} = \sigma_{ik} d\epsilon_{ik}. \tag{17.12}$$

In linear elasticity theory stress is proportional to strain, as exemplified by Hooke's law. Its tensorial extension is the relation

$$\sigma_{ik} = \lambda_{ik}^{lm} \epsilon_{lm}, \tag{17.13}$$

where λ_{ik}^{lm} is the elastic modulus tensor. A useful form of the expression for F_{el} of a deformed body can be obtained by noting that it is a quadratic function of the strain tensor, $\partial F_{el}/\partial \epsilon_{ik} = \sigma_{ik}$, and using Euler's theorem, $\epsilon_{ik} \partial F_{el}/\partial \epsilon_{ik} = 2F_{el}$. The result is

$$F_{el} = \frac{1}{2}\sigma_{ik}\epsilon_{ik} = \frac{1}{2}\lambda_{ik}^{lm}\epsilon_{lm}\epsilon_{ik}. \tag{17.14}$$

For an inhomogeneous system, such as the coherent film sketched in Fig. 17.1, F_{el} can be generalized to give the part of the free energy functional that arises from elastic deformations,

$$\mathcal{F}_{el} = \int_V f_{el}(\epsilon(\mathbf{r}))d^3r, \tag{17.15}$$

where the elastic free energy density is

$$f_{el}(\epsilon(\mathbf{r})) = \frac{1}{2}\lambda_{ik}^{lm}\epsilon_{lm}(\mathbf{r})\epsilon_{ik}(\mathbf{r}). \tag{17.16}$$

Since the strain tensor is symmetric, $\epsilon_{lm}\epsilon_{ik}$ is unchanged when the indices (i,k) or (l,m) are interchanged or the pairs of indices (i,l) and (k,m) are interchanged. Due to this symmetry, one has

$$\lambda_{ik}^{lm} = \lambda_{ik}^{ml} = \lambda_{ki}^{lm}, \tag{17.17}$$

and, for a 3d system, out of the 81 elements of λ_{ik}^{lm} only 21 are independent. This number may be further reduced by using the symmetry of the crystal. For a system with triclinic symmetry, all 21 elements are nonzero. For a hexagonal system, five independent elements are nonzero and, for a system with cubic symmetry, there are only three independent nonzero elements, λ_{xx}^{xx}, λ_{xx}^{yy}, and λ_{xy}^{xy}.

The elastic properties of most crystals in nature are anisotropic. For crystals with cubic symmetry this anisotropy is characterized by the difference $\left(\lambda_{xx}^{xx} - \lambda_{xx}^{yy} - 2\lambda_{xy}^{xy}\right)$. For simplicity we assume that this difference is zero. Such an isotropic solid is characterized by an isotropic compressibility and an isotropic resistance to shear, and its free energy is invariant under a rotation of the axes. It is this extra symmetry that further reduces the number of independent elements of the elastic modulus tensor to two and leads to

$$\lambda_{ik}^{lm} = \mu(\delta_{il}\delta_{km} + \delta_{im}\delta_{kl}) + \lambda\delta_{ik}\delta_{lm}, \tag{17.18}$$

where μ and λ are called the Lamé coefficients. The elastic free energy density for an isotropic solid simplifies to

$$f_{el}(\epsilon(\mathbf{r})) = \mu\,\epsilon_{ik}\epsilon_{ik} + \frac{\lambda}{2}\,\epsilon_{ll}^2, \tag{17.19}$$

where ϵ_{ll}^2 is the squared sum of the diagonal elements and $\epsilon_{ik}\epsilon_{ik}$ is the sum of squares of *all* of the elements of the strain tensor. The stress–strain relation (17.13) reduces to

$$\sigma_{ik} = 2\mu\epsilon_{ik} + \lambda\epsilon_{ll}\delta_{ik}. \tag{17.20}$$

Any deformation can be written as the sum of a pure shear and a hydrostatic compression:

$$\epsilon_{ik} = \left(\epsilon_{ik} - \frac{1}{3}\delta_{ik}\epsilon_{ll}\right) + \frac{1}{3}\delta_{ik}\epsilon_{ll}. \tag{17.21}$$

Using this result the elastic free energy density can be written as

$$f_{el}(\epsilon) = \mu\left(\epsilon_{ik} - \frac{1}{3}\delta_{ik}\epsilon_{ll}\right)^2 + \frac{1}{2}K\epsilon_{ll}^2, \tag{17.22}$$

while the stress–strain relation (17.20) becomes

$$\sigma_{ik} = 2\mu\left(\epsilon_{ik} - \frac{1}{3}\delta_{ik}\epsilon_{ll}\right) + K\epsilon_{ll}\delta_{ik}, \tag{17.23}$$

where the Lamé coefficient μ is referred to as the shear modulus and $K = \left(\lambda + \frac{2}{3}\mu\right)$ as the bulk modulus. In thermodynamic equilibrium the free energy is a minimum, and this implies that both K and μ are positive definite. Sometimes it is convenient to use Young's modulus E and Poisson's ratio ν in place of K and μ. The equations relating these moduli are

$$E = \frac{9K\mu}{(3K + \mu)} = \frac{\mu(3\lambda + 2\mu)}{(\lambda + \mu)}, \quad \nu = \frac{1}{2}\frac{(3K - 2\mu)}{(3K + \mu)} = \frac{\lambda}{2(\lambda + \mu)}. \tag{17.24}$$

For the simple extension of a rod along the z direction, using a force per unit area F_z, Young's modulus is $E = F_z/\epsilon_{zz}$ and Poisson's ratio is the ratio of the transverse compression to the longitudinal extension, $\nu = -\epsilon_{xx}/\epsilon_{zz}$. While K and μ are always positive, ν can be negative. If $K = 0$, then $\nu = -1$, and if $\mu = 0$, then $\nu = \frac{1}{2}$. The range of ν is $-1 \le \nu \le +\frac{1}{2}$. Among the various moduli, λ, μ, K, ν, and E, only two are independent. Some useful relations are

$$\mu = \frac{E}{2(1 + \nu)},$$

$$K = \frac{E}{3(1 - 2\nu)} = \frac{2\mu(1 + \nu)}{3(1 - 2\nu)},$$

$$\lambda = \frac{E\nu}{(1 - 2\nu)(1 + \nu)} = \frac{2\mu\nu}{(1 - 2\nu)}. \tag{17.25}$$

In terms of μ and ν, the stress–strain relation, Eq. (17.20), becomes

$$\sigma_{ik} = 2\mu\left(\epsilon_{ik} + \frac{\nu}{1 - 2\nu}\epsilon_{ll}\delta_{ik}\right). \tag{17.26}$$

If we use Eqs. (17.8) and (17.26), the mechanical equilibrium condition, Eq. (17.10), reduces to

$$\nabla(\nabla \cdot \mathbf{u}) + (1 - 2\nu)\nabla^2\mathbf{u} = 0. \tag{17.27}$$

Now consider a heterogeneous system such as that shown in Fig. 17.1(a). In the expression for the elastic free energy density in Eq. (17.19), the displacements and strains must be measured from the unconstrained state. However, for the heterogeneous film where the same reference state is used for both the substrate and the film, one of these two quantities is generally constrained. In Fig. 17.1(a) the film is constrained and its strain with respect to the unconstrained state (see Fig. 17.1(b)) has a nonzero value ϵ_{ik}^o. If ϵ_{ik} is the strain measured relative to the common reference state (flat coherent film on an unconstrained substrate, as shown in Fig. 17.1(c)), ϵ_{ik} must be replaced by $\epsilon_{ik} + \epsilon_{ik}^o$ in the expressions for the elastic free energy. The nonzero elements of ϵ_{ik}^o are $\epsilon_{xx}^o = \epsilon_{yy}^o = -\epsilon_o$ and $\epsilon_{zz}^o = -\epsilon_o + \bar{\epsilon}$. The element ϵ_{zz}^o is a combination of an isotropic strain $(-\epsilon_o)$ and a Poisson relaxation $\bar{\epsilon}$ that occurs due to the stress $\bar{\sigma}$ in the lateral directions (x, y). The Poisson relaxation and the lateral stress arise on account of the isotropic strain $(-\epsilon_o\delta_{ij})$. Thus, in Eq. (17.19), $\epsilon_{xx}, \epsilon_{yy},$ and ϵ_{zz} should be replaced by $\epsilon_{xx} - \epsilon_o, \epsilon_{yy} - \epsilon_o,$ and $\epsilon_{zz} + \epsilon_{zz}^o$. Within an additive constant, the elastic free energy density within the film is

$$f_{el}(\epsilon) = \mu\left[\epsilon_{ik}^2 - 2\epsilon_o(\epsilon_{xx} + \epsilon_{yy} + \epsilon_{zz}) + 2\bar{\epsilon}\epsilon_{zz}\right] + \frac{\lambda}{2}[(\epsilon_{xx} + \epsilon_{yy} + \epsilon_{zz}) - 3\epsilon_o + \bar{\epsilon}]^2. \tag{17.28}$$

The free energy density now contains terms that are linear in the strain. They are similar to terms in the free energy arising from external forces. We now identify

$$\bar{\epsilon} = \frac{3\lambda + 2\mu}{\lambda + 2\mu}\epsilon_o = \frac{1 + \nu}{1 - \nu}\epsilon_o. \tag{17.29}$$

This result is obtained from the constraint that the coefficient of the term linear in ϵ_{zz} vanishes. This is required since the lattice misfit does not create any additional stress in the z direction. Additional stress $\bar{\sigma}$ occurs only in the x and y directions and is obtained from the coefficient of the term which is linear in $(\epsilon_{xx} + \epsilon_{yy})$. It is $[-2\mu\epsilon_o + \lambda(-3\epsilon_o + \bar{\epsilon})]$, and reduces to

$$\bar{\sigma} = -\epsilon_o\frac{E}{(1 - \nu)} = -2\mu\epsilon_o\frac{(1 + \nu)}{(1 - \nu)}, \tag{17.30}$$

where $\bar{\epsilon}$ from Eq. (17.29) was used. Since $\bar{\sigma}$ is a constant, the mechanical equilibrium condition (17.27) obtained from Eq. (17.10) remains unaltered. Using the result for $\bar{\epsilon}$, the elastic free energy density, Eq. (17.28), becomes

$$f_{el}(\epsilon) = \mu\epsilon_{ik}^2 + \frac{\mu v}{(1-2v)}(\epsilon_{xx}+\epsilon_{yy}+\epsilon_{zz})^2 - \frac{2\mu\epsilon_o(1+v)}{(1-v)}(\epsilon_{xx}+\epsilon_{yy}), \quad (17.31)$$

within an additive constant. As before, $\epsilon_{ik}^2 \equiv \epsilon_{ik}\epsilon_{ik}$ is the sum of squares of *all* of the elements of the strain tensor. Using Eqs. (17.15) and (17.31), the elastic free energy functional within the film is given by

$$\mathcal{F}_{el} = \int_0^{h(x,y;t)} dz \int dx \int dy f_{el}(\epsilon). \quad (17.32)$$

The elastic free energy functional within the substrate is obtained by setting $\epsilon_o = 0$ in Eq. (17.31) and changing the limits of the integration over z to range from $-\infty$ to 0. To simplify the analysis we assume that each of the elastic moduli of the substrate is equal to that of the film.

17.3 Evolution equations for the morphological instability

To recapitulate, the time evolution of $h(x, y; t)$ is given by

$$\frac{\partial h}{\partial t} = \Gamma_h \sqrt{g}\nabla_s^2 \frac{\delta\mathcal{F}}{\delta h} + v, \quad (17.33)$$

where $g = (1+|\nabla h|^2)$ and the free energy functional is $\mathcal{F} = \mathcal{F}_s + \mathcal{F}_{el}$. The surface energy part of the free energy functional is

$$\mathcal{F}_s = \gamma \int d^2r \sqrt{g}, \quad (17.34)$$

and the elastic energy part is

$$\mathcal{F}_{el} = \int_0^{h(x,y;t)} dz \int dx \int dy f_{el}(\epsilon). \quad (17.35)$$

Here the elastic free energy density is

$$f_{el}(\epsilon) = \mu\epsilon_{ik}^2 + \frac{\mu v}{(1-2v)}(\epsilon_{xx}+\epsilon_{yy}+\epsilon_{zz})^2 - \frac{2\mu\epsilon_o(1+v)}{(1-v)}(\epsilon_{xx}+\epsilon_{yy}), \quad (17.36)$$

where the strain tensor ϵ_{ik} is related to the displacement field $u_i(x, y, z; t)$ through Eq. (17.8) and u_i satisfies the mechanical equilibrium condition, Eq. (17.27), which applies both to the film and to the substrate. Since diffusion is a much slower process than local lattice rearrangements, the displacement vector **u** can be assumed to instantaneously satisfy the condition of mechanical equilibrium. We also assume that the growing film–substrate system is isothermal, so that its time evolution equations reflect the slower diffusive nonequilibrium features. The equation for h is nonlinear and is coupled to **u** through the strain tensor ϵ_{ik} in $\mathcal{F}_{el}$. In Eq. (17.36) the term proportional to ϵ_o is linear in the strain tensor and does not alter the equation for the mechanical equilibrium condition, Eq. (17.27). Its solution gives the displacement vector **u** in terms of h. This is obtained in the next section.

The mechanical equilibrium condition (17.27) is to be solved subject to the following boundary conditions. For MBE growth, the pressure above the film is negligible: hence the total force on a mass element on the surface is zero. This implies that

$$\sigma_{ij}^f n_j = 0 \text{ at } z = h, \tag{17.37}$$

with the unit vector $\hat{\mathbf{n}} = (-\nabla h, 1) / \sqrt{g}$ oriented towards the half space in the positive z direction. Because the plane $z = 0$ is assumed to remain flat and coherent, the displacement vector and the stress tensor must be continuous there, implying that

$$\sigma_{zi}^f = \sigma_{zi}^s \text{ and } \mathbf{u}^f = \mathbf{u}^s \text{ at } z = 0. \tag{17.38}$$

The superscripts f and s indicate the stress tensor in the film and the substrate, respectively. Also, we require that the displacement vector within the substrate vanish far from the film/substrate interface: $\mathbf{u}^s \to 0$ as $z \to -\infty$. This also implies that the strain tensor far from the film decays to zero:

$$\epsilon_{ij} \to 0 \text{ as } z \to -\infty. \tag{17.39}$$

The nonlinear equation for the top surface profile h of the film contains the morphological instability due to lattice misfit. A linear stability analysis of the equation can provide the range of unstable wavenumbers and the growth rate of perturbations to $\bar{h}$ as a function of wavenumber. The reference state around which the perturbations occur is

$$\bar{h} = h_o + vt. \tag{17.40}$$

This is appropriate for a flat film where

$$\bar{u}_x^s = \bar{u}_y^s = \bar{u}_z^s = 0, \tag{17.41}$$

due to a completely relaxed substrate for which

$$\bar{u}_x^f = \bar{u}_y^f = 0, \tag{17.42}$$

corresponding to a coherent substrate/film interface and a state of uniform epitaxial strain in the film. Furthermore,

$$\bar{u}_z^f = (\epsilon_{zz}^o + \epsilon_o)z = \bar{\epsilon}z, \tag{17.43}$$

characterizing the Poisson relaxation in the z direction. The stress state associated with this reference state is a stress-free substrate and a state of biaxial stress in the film,

$$\bar{\sigma}_{xx}^f = \bar{\sigma}_{yy}^f = -2\mu\bar{\epsilon} = \bar{\sigma}. \tag{17.44}$$

The reference state strains and stresses are uniform throughout the system and are also independent of film thickness and growth velocity v.

The stability of the film is studied by considering small perturbations around the reference state. A general variable ξ is expanded in a two-dimensional Fourier series as

$$\xi = \bar{\xi} + \sum_q \hat{\xi}\,(\mathbf{q}, z, t)\,e^{i\,(q_x x + q_y y)}. \tag{17.45}$$

The Fourier coefficients for the film height are independent of z. Computation of the functional derivative in Eq. (17.33) yields the free energy density (Eq.(17.36)) evaluated at the top surface. Hence, for the purpose of calculating $\partial\hat{h}/\partial t$, the free energy density has to be computed to $\mathcal{O}(\hat{\xi})$. The elastic energy $\hat{\mathcal{E}}$, to linear order in $\hat{\xi}$, is

$$\hat{\mathcal{E}} = \bar{\sigma}\,(\hat{\epsilon}_{xx} + \hat{\epsilon}_{yy}). \tag{17.46}$$

The second-order correction to the elastic energy, $\tilde{\mathcal{E}} \sim \mathcal{O}(\hat{\xi}^2)$, is

$$\tilde{\mathcal{E}} = \lambda_{ik}^{lm}\hat{\epsilon}_{lm}\hat{\epsilon}_{ik}, \tag{17.47}$$

where λ_{ik}^{lm} is given by Eq. (17.18).

17.4 Solution of mechanical equilibrium equations

The linear mechanical equilibrium equations (17.27) can be transformed to Fourier space by substituting

$$\mathbf{u}(x, y, z; t) = \bar{\mathbf{u}} + \sum_q \hat{\mathbf{u}}\,(\mathbf{q}, z, t)\,e^{i(q_x x + q_y y)}. \tag{17.48}$$

The resulting equations for the film are

$$(1 - 2\nu)\left(\frac{\partial^2}{\partial z^2} - q^2\right)\hat{u}_x + iq_x\left(iq_x\hat{u}_x + iq_y\hat{u}_y + \frac{\partial}{\partial z}\hat{u}_z\right) = 0, \quad (17.49)$$

$$(1 - 2\nu)\left(\frac{\partial^2}{\partial z^2} - q^2\right)\hat{u}_y + iq_y\left(iq_x\hat{u}_x + iq_y\hat{u}_y + \frac{\partial}{\partial z}\hat{u}_z\right) = 0, \quad (17.50)$$

$$(1 - 2\nu)\left(\frac{\partial^2}{\partial z^2} - q^2\right)\hat{u}_z + \frac{\partial}{\partial z}\left(iq_x\hat{u}_x + iq_y\hat{u}_y + \frac{\partial}{\partial z}\hat{u}_z\right) = 0. \quad (17.51)$$

Here $q^2 = q_x^2 + q_y^2$. This is a set of coupled homogeneous second-order differential equations first obtained by Spencer *et al.* (1993). In the substrate, the displacements also satisfy the same homogeneous equations where the form of the solution is

$$\begin{bmatrix} \hat{u}_x^s \\ \hat{u}_y^s \\ \hat{u}_z^s \end{bmatrix} = \begin{bmatrix} u_x^0 \\ u_y^0 \\ u_z^0 \end{bmatrix} e^{qz} - \begin{bmatrix} iq_x/q \\ iq_y/q \\ 1 \end{bmatrix} Bze^{qz}. \quad (17.52)$$

Here $B = [1/(3 - 4\nu)](iq_xu_x^0 + iq_yu_y^0 + qu_z^0)$, and we have used the condition that $\mathbf{u}^s$ vanishes far from the film/substrate boundary. In the film, the form of the solution is

$$\begin{bmatrix} \hat{u}_x^f \\ \hat{u}_y^f \\ \hat{u}_z^f \end{bmatrix} = \begin{bmatrix} \alpha_x \\ \alpha_y \\ \alpha_z \end{bmatrix} \cosh(qz) + \begin{bmatrix} \beta_x \\ \beta_y \\ \beta_z \end{bmatrix} \sinh(qz) \quad (17.53)$$

$$- \begin{bmatrix} Ciq_x/q \\ Ciq_y/q \\ D \end{bmatrix} z\sinh(qz) - \begin{bmatrix} Diq_x/q \\ Diq_y/q \\ C \end{bmatrix} z\cosh(qz).$$

In this last equation, C and D are defined as $C = [1/(3 - 4\nu)]\,(iq_x\alpha_x + iq_y\alpha_y + q\beta_z)$ and $D = [1/(3 - 4\nu)]\,(iq_x\beta_x + iq_y\beta_y + q\alpha_z)$. The above relations for the displacements in the film and in the substrate contain nine coefficients that must be determined using the boundary conditions. To $\mathcal{O}(\hat{\xi})$, the boundary conditions at $z = 0$ are

$$\hat{\sigma}_{zi}^f = \hat{\sigma}_{zi}^s \text{ and } \hat{\mathbf{u}}^f = \hat{\mathbf{u}}^s, \quad (17.54)$$

which lead to the relations $u_i^0 = \alpha_i = \beta_i$, and $C = D = B$. The $\mathcal{O}(\hat{\xi})$ boundary conditions at $z = h$ are

$$\hat{\sigma}_{xz} = iq_x\bar{\sigma}\hat{h}, \quad \hat{\sigma}_{yz} = iq_y\bar{\sigma}\hat{h} \text{ and } \hat{\sigma}_{zz} = 0, \quad (17.55)$$

which result in the expression for C and the relation between α_z and C:

$$C = A\left[-\frac{q\bar{\sigma}\hat{h}}{2\mu}\right], \quad q\alpha_z = C(1 - 2v + q\bar{h}), \tag{17.56}$$

where $A = \left[\cosh(q\bar{h}) + \sinh(q\bar{h})\right]^{-1}$.

Using these solutions of the mechanical equilibrium conditions, we can evaluate the free energy density to first and second orders using Eqs. (17.46) and (17.47), respectively. For the linear analysis that is used here, only the first-order result is needed. It is

$$\widehat{\mathcal{E}} = \bar{\sigma}[2\epsilon_o(1 + v)q\hat{h}] = -4\mu\epsilon_o^2\frac{(1 + v)^2}{(1 - v)}q\hat{h}. \tag{17.57}$$

The evolution equation for $\hat{h}$ in Fourier space is then obtained as

$$\frac{\partial\hat{h}}{\partial t} = -\Gamma_h q^2\left(-4\mu\epsilon_o^2\frac{(1 + v)^2}{(1 - v)}q + \gamma q^2\right)\hat{h} \equiv \Omega\hat{h}. \tag{17.58}$$

An analogous result for the more general case where μ and v are different for the substrate and the film has been given by Spencer *et al.* (1993). For this case the strain energy term, where μ and v correspond to the film values μ_f and v_f, respectively, is multiplied by a function that depends on $q\bar{h}$, μ_f/μ_s, v_f, and v_s. This function reduces to unity in the limit where the moduli are equal. For situations where the elastic moduli are equal the dynamical equation for $\hat{h}$ is independent of $\bar{h}$ (see Eq. (17.58)) and, therefore, is the same for static and growing films.

17.5 Linear stability analysis

The solution of the evolution equation (17.58) is

$$\hat{h}(q, t) = \hat{h}_o(q)\exp(\Omega(q)t), \tag{17.59}$$

where the growth rate can be written as

$$\Omega(q) = \Gamma_h\gamma q^3\left(4\mu\frac{\epsilon_o^2}{\gamma}\frac{(1 + v)^2}{(1 - v)} - q\right) \equiv \Gamma_h\gamma q^3(q_c - q). \tag{17.60}$$

Here the second equality defines a cutoff wavenumber q_c. The strain energy term is destabilizing regardless of the sign of the misfit since it is proportional to ϵ_o^2. The surface energy term is stabilizing and dominates for $q > q_c$. The misfit-related morphological instability occurs for long wavelengths $0 < q < q_c$.

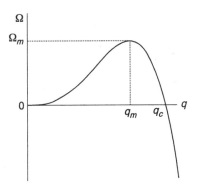

Fig. 17.2. Schematic diagram showing the growth rate of the misfit-related morphological instability.

As a function of q, the growth rate Ω has zero slope at the origin, rises to a positive maximum and then decreases monotonically, crossing the $\Omega = 0$ axis at the cutoff wavenumber q_c (Fig. 17.2). The maximum in $\Omega(q)$ occurs at $q_m = 3q_c/4$ and has a value $\Omega_m = \frac{27}{256}\Gamma_h\gamma q_c^4$. Due to exponential growth the maximal mode (q_m, Ω_m) dominates all other modes in time. The maximal wavenumber $q_m \propto \epsilon_o^2$ and the maximal growth exponent $\Omega_m \propto \epsilon_o^8$. The wavelength $\lambda_m = 2\pi/q_m \sim \epsilon_o^{-2}$ can be viewed as an approximation for the periodicity of surface patterns, such as coherent islands, which result from the morphological instability.

It is appropriate to define the characteristic length of the instability as $l_o = q_c^{-1}$ and the characteristic time as $t_o = l_o^4/(\Gamma_h\gamma)$. For a germanium film coherently grown over a silicon substrate, $\epsilon_o = 0.0418$, $\gamma = 1.927\,\mathrm{J\,m^{-2}}$, $v_f = 0.198$, and $\mu_f = 0.568 \times 10^{11}\,\mathrm{J\,m^{-3}}$, giving $l_o = 2.7\,\mathrm{nm}$ (Spencer *et al.*, 1993). The time scale t_o is very sensitive to misfit and temperature. It varies with misfit as ϵ_o^{-8} so that changing misfit from 2% to 4% decreases t_o by a factor of 256. The sensitive dependence on temperature arises from the Arrhenius temperature dependence of the surface diffusivity D_s, which is contained within Γ_h. For a germanium film on silicon, $D_s = 8.45 \times 10^{-10}\exp(-0.83eV/k_BT)\,\mathrm{m^2\,s^{-1}}$, $N_s = 1.25 \times 10^{19}\,\mathrm{atoms/m^2}$ and $N_v = 4.41 \times 10^{28}\,\mathrm{atoms/m^3}$ which gives $t_o = 236\,\mathrm{s}$ at 423 K and 0.23 s at 623 K.

From the linear analysis, the height of the film in real space is given by

$$h(x, y; t) = \overline{h}(t) + \sum_{\mathbf{q}} \hat{h}(\mathbf{q}; t)e^{i(q_x x + q_y y)}$$

$$= vt + \sum_{\mathbf{q}} \hat{h}_o(q)e^{\Omega(q)t}e^{i(q_x x + q_y y)}. \tag{17.61}$$

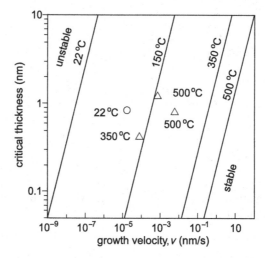

Fig. 17.3. Variation of critical thickness h_c with growth velocity v for Ge/Si film for four different temperatures. Experimental observations of planar film (circle) and coherent dislocation free islands (triangles) as shown indicate consistency with theory. Reprinted with permission from Spencer *et al.* (1993). Copyright 1993, American Institute of Physics.

For linearly unstable growth of h, at early times the mean height $\bar{h}$ grows with a rate v and the initial height perturbations $\hat{h}_o$ grow with a growth rate $\sim \Omega(q_m)$. This competition between deposition and instability can be used to define a critical thickness h_c,

$$h_c = \frac{v}{\Omega(q_m)} = \frac{256}{27}\frac{v}{\Gamma_h \gamma q_c^4} = \frac{v\gamma^3(1-v)^4}{27\Gamma_h\mu^4\epsilon_o^8(1+v)^8}. \tag{17.62}$$

As $\bar{h}(t)$ passes through h_c the instability growth rate overtakes the linear growth rate of the mean height. The critical thickness sharply decreases with increasing misfit as ϵ_o^{-8}. As the misfit increases, the development of the instability becomes more abrupt. For $\bar{h}(t) < h_c$ the film growth is kinetically stabilized so that a growing film remains flat. In Fig. 17.3 the variation of h_c with v is shown for four temperatures. These results were obtained by Spencer *et al.* (1993) for a more general case where the elastic moduli of the film are different from those of the substrate. A comparison with experimental results for germanium films on a silicon substrate (Asai *et al.*, 1985; Eaglesham and Cerullo, 1990; LeGoues *et al.*, 1990) is also shown. For a given temperature, the curve in the figure is a boundary for linear stability; instability is apparent above the curve. In experimental observations the instability leads to coherent dislocation-free islands.

18

Propagating chemical fronts

In earlier chapters we considered the structure of the interfaces separating coexisting phases in systems described by models A and B. Here we examine the nature of propagating chemical fronts and show how the front profiles and speeds may be determined. Propagating chemical fronts were observed and analyzed by Luther (1906) in an early investigation of reacting systems. Luther noted that the front velocity is given by $v \propto (kD)^{1/2}$, where k is a chemical rate constant and D is the diffusion coefficient (Showalter and Tyson, 1987). We shall show how such a result follows from the propagating-front solution of the reaction–diffusion equation, and discuss the conditions under which a unique front velocity is selected by the system.

In the analysis of the propagation of chemical fronts it is important to distinguish between two cases: a stable state propagates into another linearly stable state, and a stable state propagates into an unstable state. In the former case a unique front velocity is selected and the front structure is determined by the interior of the boundary region that separates the stable states. As we shall see, this leads to a nonlinear eigenvalue problem for the front velocity and profile. In the case where a stable state invades an unstable state, the front dynamics is determined by the dynamics of the leading edge of the front. In this case there is a continuous spectrum of velocities, and one must solve a selection problem to determine which class of initial conditions will lead to a particular front velocity.

We restrict our considerations mostly to one-dimensional systems. Phenomena in higher-dimensional systems will be considered in subsequent chapters. Suppose the concentration field $c(x,t)$ satisfies the reaction–diffusion equation

$$\frac{\partial c(x,t)}{\partial t} = \mathcal{R}(c(x,t)) + D\frac{\partial^2 c(x,t)}{\partial x^2}, \qquad (18.1)$$

where $\mathcal{R}(c(x,t))$ is the reaction rate. We are interested in solutions of this equation for a class of physically realizable initial conditions leading to a propagating front that moves with speed v and connects two different concentration regions at $x = \pm\infty$. The investigation of such front propagation has a long history (Fisher, 1937; Kolmogorov *et al.*, 1937; Aronson and Weinberger, 1975).

Traveling wave solutions have the form $c(x, t) = c(x - vt) = c(u)$ and satisfy the equation

$$D\frac{d^2c(u)}{du^2} + v\frac{dc(u)}{du} + \mathcal{R}(c(u)) = 0. \tag{18.2}$$

When examining the properties of this equation, a mechanical analogy is useful. If we define a momentum-like variable $D\,dc/du = p$, we can write Eq. (18.2) as the coupled pair of equations

$$\frac{d}{du}c(u) = \frac{p(u)}{D},$$
$$\frac{d}{du}p(u) = -\frac{v}{D}p(u) + F(c(u)) = 0. \tag{18.3}$$

Interpreting u as a time-like variable, these are the equations of motion for a particle of mass D acted on by a force $F(c(u)) = -\mathcal{R}(c(u)) = -dV(c)/dc$, derived from a potential $V(c)$ and a frictional force $-(v/D)p$, where v plays the role of the friction coefficient.

18.1 Propagation into a metastable state

First we consider the propagation of a stable state into a metastable (linearly stable) state where the mechanical analogy may be used. Suppose that $\mathcal{R}(c) = 0$ possesses three fixed points, $c^{\pm}$ and c^u. The fixed points $c^{\pm}$ are linearly stable so that $R'(c^{\pm}) < 0$, while c^u is unstable so that $\mathcal{R}'(c^u) > 0$. We also assume that $c(x, t) \to c^{\pm}$ for $x \to \pm\infty$. An example of a chemical system with these characteristics is the Schlögl model discussed earlier in Chapter 5 in connection with model A of critical phenomena. For this model the reaction rate is $\mathcal{R}(c) = k_1a - k_{-1}c + k_2bc^2 - k_{-2}c^3$. Cubic models of this type can be written in the simpler form $\mathcal{R}(c) = -c^3 + c + a$. The potential function for this model is $V(c) = -\frac{1}{4}c^4 + \frac{1}{2}c^2 + ac$, corresponding to the force $F(c) = -\mathcal{R}(c)$. This potential function is sketched in Fig. 18.1 for $a > 0$. In the mechanical analogy, we seek solutions that start at the stable state c^+ near $c = 1$ with zero kinetic energy and end at the metastable state c^- near $c = -1$. Since both states are local maxima, now there is a unique "friction" $v = v^*$ that will accomplish this. Also, if $a = 0$ so that the maxima in the potential corresponding to the two linearly stable states have equal values, then the friction needed for the particle to start at c^+ with zero kinetic energy and end at c^- is $v = 0$. So, the front does not propagate. This is the case considered earlier in Chapter 5 for critical quenches in model A.

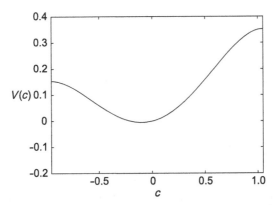

Fig. 18.1. Sketch of the potential $V(c) = -\frac{1}{4}c^4 + \frac{1}{2}c^2 + ac$ in the mechanical analogy used in the analysis of the cubic propagating front with $a = 0.1$.

It is not difficult to solve the nonlinear eigenvalue problem (18.2) for the unique front velocity and profile. The profile of the propagating front is given by

$$c_0(u) = \frac{1}{2}(c^+ - c^-) \tanh\left(\frac{(c^+ - c^-)}{2\sqrt{2D}}u\right) + \frac{1}{2}(c^+ + c^-), \qquad (18.4)$$

while the front velocity is $v = (c^+ + c^- - 2c^u)\sqrt{D/2}$. This is the unique solution to this problem.

The stability of the one-dimensional front can be studied by linearizing the reaction–diffusion equation (18.1) about the solution $c_0(u)$. We let

$$c(x, t) = c_0(u) + \delta c(u, t). \qquad (18.5)$$

Substitution into the reaction–diffusion equation (18.1) gives

$$\frac{\partial \delta c(u, t)}{\partial t} = \left(D\frac{\partial^2}{\partial u^2} + v\frac{\partial}{\partial u} + \frac{\delta R}{\delta c_0}\right) \delta c(u, t)$$
$$\equiv \Omega_R(u)\delta c(u, t), \qquad (18.6)$$

which defines the linear operator Ω_R. We use an abstract notation similar to that in quantum mechanics to simplify the equations. We let $\langle u|c\rangle = c(u)$ with an analogous representation for operators: $\langle u|\hat{\Omega}_R|u'\rangle = \Omega_R(u)\delta(u - u')$. The eigenvalue problem for the operator $\hat{\Omega}_R$ may then be written as

$$\hat{\Omega}_R|\zeta_i\rangle = \lambda_i|\zeta_i\rangle. \qquad (18.7)$$

From Eq. (18.6) we see that $\zeta_0(u) = dc_0/du$ is an eigenfunction of $\hat{\Omega}_R$ with eigenvalue $\lambda_0 = 0$. This is a consequence of the broken translational invariance

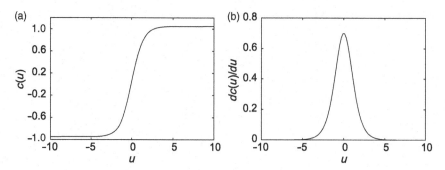

Fig. 18.2. (a) The profile of a propagating front for $a = 0.1$ and $D = 1$; (b) its derivative, the eigenfunction corresponding to the zero eigenvalue of the linearized eigenvalue problem.

along u and is the analog of the Goldstone mode discussed in connection with the interface dynamics in models A and B. The front profile and the eigenfunction $\zeta_0(u) = dc_0/du$ are plotted in Fig. 18.2. The figure shows that $\zeta_0(u)$ is localized in the vicinity of the front. The operator $\hat{\Omega}_R$ is not self-adjoint so we must also consider the adjoint eigenvalue problem,

$$\langle \zeta_i | \hat{\Omega}_R^\dagger = \lambda_i \langle \zeta_i |. \tag{18.8}$$

Here the adjoint operator is

$$\Omega_R^\dagger(u) = D\frac{\partial^2}{\partial u^2} - v\frac{\partial}{\partial u} + \frac{\delta \mathcal{R}}{\delta c_0}. \tag{18.9}$$

The eigenfunctions satisfy the orthonormality condition $\langle \zeta_i | \zeta_j \rangle = \delta_{ij}$. The stability of the front is determined by the sign of the nonzero eigenvalues. In the next chapter we will make use of this formalism to determine the stability of interfacial dynamics to transverse perturbations in two spatial dimensions.

18.2 Propagation into an unstable state

Next, we consider propagation of a linearly stable state into an unstable state, which we term Fisher-type fronts. Suppose the reaction rate has the following properties: $\mathcal{R}(0) = \mathcal{R}(1) = 0$ and $\mathcal{R}(c) > 0$ for all $c \in [0, 1]$. A specific example of a reaction rate with these properties is $\mathcal{R}(c) \equiv \mathcal{R}_F(c) = kc(1-c)$ with $k > 0$. Since $\mathcal{R}_F'(0) = k$ and $\mathcal{R}_F'(1) = -k$, the steady state $c = 0$ is linearly unstable and the state $c = 1$ is linearly stable. Then there is a stable traveling front that connects $c = 0$ and $c = 1$ with a propagation speed $v \geq v^*$. The minimum speed v^* has a lower bound: $v^* \geq v_0 = 2(\mathcal{R}_F'(0))^{1/2}$, where v_0 is the speed with which small linear perturbations

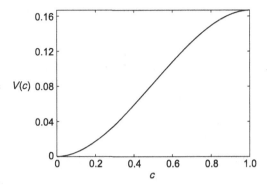

Fig. 18.3. Sketch of the potential $V(c) = \frac{k}{2}c^2(1 - \frac{2}{3}c)$ in the mechanical analogy used in the analysis of the Fisher propagating front with $k = 1$.

about the unstable state spread. For initial conditions such that $c(x,0) \in [0,1]$ vanishes beyond some finite value x_0, Aronson and Weinberger (1975) showed that $c(x,t)$ converges to a traveling wave solution of the reaction–diffusion equation with speed v^*. Thus, for experimentally realizable initial conditions with compact support, this result shows that the slowest stable-traveling wave solution is selected.

The fronts corresponding to Fisher-type equations where a stable state propagates into an unstable state are usually classified as being *pulled* or *pushed* (van Saarloos, 2003). If the front speed $v^* = v_0$, the fronts are called pulled because, effectively, they are pulled along by the spreading of small perturbations about the unstable state. Pushed fronts have $v^* > v_0$. For such fronts the dynamics in the bulk behind the front drives the front propagation and pushes the front from behind. The speed of such fronts is higher than that of pulled fronts because the "force" exerted from behind the front leads to velocities greater than that due to the natural speed that results from small perturbations about the unstable state.

For the Fisher model, the mechanical analogy is also useful. The associated potential $V(c) = \frac{k}{2}c^2(1 - \frac{2}{3}c)$ is sketched in Fig. 18.3. This potential increases monotonically from $c = 0$ to $c = 1$. The trajectory of the "particle" starts at $c = 1$ with zero kinetic energy and falls to $c = 0$. Let v^* be the critical value of the friction that allows the particle to come to rest there. If $v < v^*$ the particle will overshoot $c = 0$ and lead to negative values of the concentration. Since the force $F(c) = -\mathcal{R}(c) \leq 0$, for any $v \geq v^*$ the particle will also come to rest at $c = 0$. From this mechanical analogy we can understand the existence of a critical speed and the origin of the problem of the selected velocity.

The front-velocity selection problem for propagation of a stable state into an unstable state has been a topic of considerable interest (Dee and Langer, 1983; Paquette and Oono, 1994; Brunet and Derrida, 1997; Kessler *et al.*, 1998;

van Saarloos, 2003). Starting from an initial state, the tail of the front penetrates farther and farther ahead of the front into the bulk unstable phase, leading to a very slow approach to the steady-state propagating front. In addition, if a cutoff is applied to the reaction term ahead of the front, any value of the cutoff, regardless of how small, will lead to the selection of a unique velocity, which approaches a finite value as the cutoff is removed (Brunet and Derrida, 1997). These features serve to indicate the sensitivity of these fronts to the behavior at the leading edge of the front. Stochastic models of such front propagation show that fluctuations in the particle number in the leading edge have strong effects on the front dynamics. Discrete stochastic models effectively introduce a cutoff in the reaction dynamics, and indeed the scaling predictions of the discrete stochastic and mean field reaction–diffusion dynamics with a cutoff are similar, although there are nontrivial differences in prefactors (Kessler *et al.*, 1998).

The one-dimensional equation (18.1) for the potential shown in Fig. 18.3 can be written in a dimensionless form ($x/l \rightarrow x, kt \rightarrow t, l = \sqrt{(D/k)}$):

$$\frac{\partial c(x,t)}{\partial t} = \frac{\partial^2 c(x,t)}{\partial x^2} + c(x,t)(1 - c(x,t)). \qquad (18.10)$$

This is the one-dimensional Fisher equation. Its d-dimensional version has been solved by Puri *et al.* (1989) using the KYG singular perturbation theory described in Chapter 9. The solution is

$$c(\mathbf{r},t) = \frac{c^o(\mathbf{r},t)}{1 + c^o(\mathbf{r},t)} \qquad (18.11)$$

where $c^o(\mathbf{r},t) = \exp[t(1 + \nabla^2)]c(\mathbf{r},0)$. For an initial condition $c(\mathbf{r},0) = c_o \delta(\mathbf{r} - \mathbf{r_0})$, it reduces to

$$c(\mathbf{r},t) = \left[1 + c_o^{-1}(4\pi t)^{d/2} \exp\left(\frac{(\mathbf{r} - \mathbf{r_0})^2 - 4t^2}{4t} \right) \right]^{-1}. \qquad (18.12)$$

Let $\mathbf{r_0} = 0$. The center of the moving front is located at $c(r(t),t) = \frac{1}{2}$ and is given by

$$r(t) = 2t \left(1 - \frac{d \ln(4\pi t)}{2t} \right)^{\frac{1}{2}}. \qquad (18.13)$$

From this it follows that the front speed is

$$v(t) = 2 \left(1 - \frac{d \ln(4\pi t)}{2t} \right)^{\frac{1}{2}} + \frac{d}{2} \left(\frac{\ln(4\pi t)}{t} - \frac{1}{t} \right) \left(1 - \frac{d \ln(4\pi t)}{2t} \right)^{-\frac{1}{2}}, \qquad (18.14)$$

which has the following long time-expansion ($\ln(4\pi t)/t$ denoted by $1/\tau$):

$$v(t) = 2 - \frac{d}{2t} + \frac{d^2}{8\tau}\left(\frac{1}{2\tau} - \frac{1}{t}\right) + \frac{d^3}{16\tau^2}\left(\frac{1}{2\tau} - \frac{3}{4t}\right) + \cdots \quad (18.15)$$

The asymptotic velocity is $v(\infty) = 2$ in all dimensions and is approached from below at long times. Comparison with the numerical solution of the nonlinear Eq. (18.10) in one dimension shows that the analytic result obtained using the KYG approximation gives an overestimate at all times.

Whether the reaction–diffusion equation is discretized, a stochastic term is added to it, or it is solved using the KYG approximation, a unique velocity, independent of the initial condition, is found. This is a characteristic of a pushed front. The Fisher equation itself is structurally unstable (Brunet and Derrida, 1997; Kessler *et al.*, 1998), and a small perturbation will slow the front so that the velocity is less than v_0. The slow approach to v_0 has been known for a long time (Bramson, 1983; Bramson *et al.*, 1986).

19

Transverse front instabilities

The description of propagating chemical fronts in the preceding chapter was restricted to one-dimensional systems. In higher dimensions a front may develop structure transverse to its propagation direction. We now consider propagating chemical fronts in two-dimensional systems and examine the circumstances under which such transverse instabilities develop.

In reaction–diffusion media, front instabilities typically occur in systems whose dynamics is described by several species variables. The mechanism of the instability relies on the fact that the chemical species may diffuse at different rates. Thus, we are led to the study of propagating fronts in a system described by the reaction–diffusion equation

$$\frac{\partial}{\partial t}\mathbf{c}(\mathbf{r}, t) = \mathcal{R}(\mathbf{c}) + \mathbf{D}\nabla^2\mathbf{c}. \tag{19.1}$$

Here $\mathbf{c}$ is a vector of local chemical concentrations, $\mathcal{R}$ stands for the vector of reaction rates, and $\mathbf{D}$ is the matrix of diffusion coefficients, which is assumed to be diagonal. A general framework for treating transverse front instabilities was developed by Kuramoto and Sivashinsky (Kuramoto and Tsuzuki, 1976; Sivashinsky, 1977, 1983; Kuramoto, 1984). Transverse front instabilities can be analyzed by reducing the reaction–diffusion equation for the chemical concentrations to the Kuramoto–Sivashinsky equation for the front dynamics.

19.1 Planar traveling fronts

We begin by studying a planar front which travels with constant speed v along the x direction and is stable to transverse perturbations along its length in the y direction. When a system parameter is varied, the planar front will lose stability at some critical value of the parameter. We investigate the nature of this bifurcation and show how the critical value of the bifurcation parameter can be calculated.

An essential element in this analysis is the structure of the initially stable planar front. Since the front is planar, it is homogeneous in the y direction, and the problem reduces to finding the front profile as a function of x. Thus, as in the previous chapter, we may focus on the solution of the one-dimensional reaction–diffusion equation in a system where a front moves with speed v and connects two different concentration regions at $x = \pm\infty$. Let $c_0(u)$ be the concentration field in a system where such a front exists, where $u = x - vt$ is the coordinate in a frame moving with speed v. This concentration field is the solution of the equation

$$\left(\mathbf{D}\frac{d^2}{du^2} + v\frac{d}{du} \right) c_0(u) + \mathcal{R}(c_0(u)) = 0, \tag{19.2}$$

which can be solved to yield both the front speed v and the front profile. In what follows we assume such a solution has been determined.

The stability of the front can be studied by linearizing the one-dimensional version of the reaction–diffusion equation (19.1) about the solution $c_0(u)$. To this end, we write

$$\mathbf{c}(x, t) = \mathbf{c}_0(u) + \delta\mathbf{c}(u, t). \tag{19.3}$$

Substitution into Eq. (19.1) yields

$$\frac{\partial \delta\mathbf{c}(u, t)}{\partial t} = \left(\mathbf{D}\frac{d^2}{du^2} + \mathbf{I}v\frac{d}{du} + \frac{\delta\mathcal{R}}{\delta\mathbf{c}_0} \right) \delta\mathbf{c}(u, t),$$
$$\equiv \Omega_R(u)\delta\mathbf{c}(u, t), \tag{19.4}$$

which defines the linear operator Ω_R. Here $\mathbf{I}$ is the unit matrix. As discussed earlier, it is convenient to introduce an abstract notation similar to that in quantum mechanics to simplify the presentation. We let $\langle u|\mathbf{c}\rangle = \mathbf{c}(u)$, with an analogous representation for operators: $\langle u|\hat{\Omega}_R|u'\rangle = \Omega_R(u)\delta(u - u')$. The eigenvalue problem for the operator $\hat{\Omega}_R$ may then be written as

$$\hat{\Omega}_R|\boldsymbol{\zeta}_i\rangle = \lambda_i|\boldsymbol{\zeta}_i\rangle. \tag{19.5}$$

From Eq. (19.2) we see that $\boldsymbol{\zeta}_0(u) = d\mathbf{c}_0/du$ is an eigenfunction of $\hat{\Omega}_R$ with eigenvalue $\lambda_0 = 0$. This is a consequence of the broken translational invariance along u, and is the analog of the Goldstone mode discussed in connection with the interface dynamics in models A and B.

The operator $\hat{\Omega}_R$ is not self-adjoint, so we must also consider the adjoint eigenvalue problem,

$$\langle \boldsymbol{\zeta}_i|\hat{\Omega}_R^\dagger = \lambda_i\langle \boldsymbol{\zeta}_i|. \tag{19.6}$$

Here the adjoint operator is

$$\Omega_R^\dagger(u) = \mathbf{D}\frac{d^2}{du^2} - \mathbf{I}v\frac{d}{du} + \frac{\delta\mathcal{R}}{\delta c_0}. \tag{19.7}$$

The eigenfunctions satisfy the orthonormality condition $\langle \zeta_i | \zeta_j \rangle = \delta_{ij}$. We shall assume that this eigenvalue problem has been solved and will make use of the solutions to determine the interfacial dynamics in two spatial dimensions.

19.2 Kuramoto–Sivashinsky equation

Given this background, we return to the situation where a propagating front exists in a two-dimensional system. While the entire front propagates with speed v, we no longer assume the front is planar and instead suppose that it has transverse structure as shown in Fig. 19.1. The position of the interfacial profile may be determined by choosing a value of u corresponding to some marker value of the concentration field for each value of y. The interfacial profile will be denoted by $\chi_0(y, t)$. Provided the transverse variations in the front are not too large, at any point y on the front the solution at any time instant may be written as the planar front profile plus a correction term: thus, we write the solution to the reaction–diffusion equation in the form

$$\mathbf{c}(\mathbf{r}, t) = \mathbf{c}_0(u + \chi_0(y, t)) + \delta\mathbf{c}(u, y, t). \tag{19.8}$$

The concentration field corresponding to the one-dimensional front solution is evaluated at $u + \chi_0(y, t)$ to account for the displacement of the location of the front from u arising from the transverse variations in the front position. This displacement will vary with time since the front profile need not be stationary. The deviation from the planar front solution $\delta\mathbf{c}(u, y, t)$ depends on both u and y. Since we have assumed $\delta\mathbf{c}$ is small, the reaction–diffusion equation (19.1) may be linearized about $\mathbf{c}_0$ using

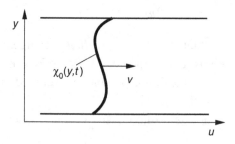

Fig. 19.1. Variation of the interfacial profile $\chi_0(y, t)$ as a function of y at a given time instant.

Eq. (19.8) to give

$$
\frac{\partial \delta c(u, y, t)}{\partial t} + \frac{\partial \chi_0(y, t)}{\partial t} \zeta_0(u) = \left\{ \Omega_R(u) + D \frac{\partial^2}{\partial y^2} \right\} \delta c(u, y, t)
$$
$$
+ D \left\{ \frac{\partial^2 \chi_0(y, t)}{\partial y^2} \zeta_0(u) + \left(\frac{\partial \chi_0(y, t)}{\partial y} \right)^2 \zeta_0'(u) \right\}, \qquad (19.9)
$$

where $\zeta_0'(u) = d\zeta_0(u)/du$.

The eigenfunctions of $\hat{\Omega}_R$ form a complete set in u space, so we may expand the deviation in these eigenfunctions to obtain

$$
\delta c(x, y, t) = \sum_{i>0} \chi_i(y, t) \zeta_i(u), \qquad (19.10)
$$

where the expansion coefficients are denoted by $\chi_i(y, t)$. Substituting the expansion of $\delta c(u, y, t)$ in terms of the eigenfunctions of the one-dimensional problem and using the abstract notation for variables involving u, we find,

$$
\sum_{i>0} \frac{\partial \chi_i(y, t)}{\partial t} |\zeta_i\rangle + \frac{\partial \chi_0(y, t)}{\partial t} |\zeta_0\rangle = \sum_{i>0} \left\{ \lambda_i + D \frac{\partial^2}{\partial y^2} \right\} \chi_i(y, t) |\zeta_i\rangle
$$
$$
+ D \left\{ \frac{\partial^2 \chi_0(y, t)}{\partial y^2} |\zeta_0\rangle + \left(\frac{\partial \chi_0(y, t)}{\partial y} \right)^2 |\zeta_0'\rangle \right\}. \qquad (19.11)
$$

The goal of this calculation is to extract an equation of motion for the front profile, $\chi_0(y, t)$. In Eq. (19.11) the time variation of $\chi_0(y, t)$ is multiplied by the eigenket $|\zeta_0\rangle$, so to extract the interface dynamics we must project this equation onto the interfacial mode $|\zeta_0\rangle$. The analysis is similar to that presented earlier in Chapter 7 for model B. There is one technical difference: since the eigenvalue problem for $\hat{\Omega}_R$ is not self-adjoint we need both the left and right eigenfunctions corresponding to the zero eigenvalue to carry out this projection. Projecting Eq. (19.11) onto $|\zeta_0\rangle$ we find

$$
\frac{\partial \chi_0(y, t)}{\partial t} = \sum_{i=0}^{\infty} \langle \zeta_0 | D | \zeta_i \rangle \frac{\partial^2 \chi_i}{\partial y^2} + \langle \zeta_0 | D | \zeta_0' \rangle \left(\frac{\partial \chi_0(y, t)}{\partial y} \right)^2. \qquad (19.12)
$$

This equation expresses the time variation of $\chi_0(y, t)$ in terms of $\chi_0(y, t)$ as well as all other $\chi_i(y, t)$, $(i > 0)$ terms arising from the first term on the right-hand side of Eq. (19.11). The $i = 0$ terms on the right-hand side of Eq. (19.12) come from the second term in Eq. (19.11). Consequently, to obtain a closed solution we must also construct evolution equations for the $\chi_i(y,t)$. These equations may be found by projecting Eq. (19.11) onto $|\boldsymbol{\zeta}_i\rangle$ $(i \neq 0)$ to obtain

$$
\begin{aligned}
\frac{\partial \chi_i(y,t)}{\partial t} &= \lambda_i \chi_i + \sum_{j=0}^{\infty} \langle \boldsymbol{\zeta}_i | \mathbf{D} | \boldsymbol{\zeta}_j \rangle \frac{\partial^2 \chi_j}{\partial y^2} + \langle \boldsymbol{\zeta}_i | \mathbf{D} | \boldsymbol{\zeta}_0' \rangle \left(\frac{\partial \chi_0(y,t)}{\partial y} \right)^2 \\
&= \sum_{j>0}^{\infty} \mathcal{W}_{ij} \chi_j + \langle \boldsymbol{\zeta}_i | \mathbf{D} | \boldsymbol{\zeta}_0 \rangle \frac{\partial^2 \chi_0(y,t)}{\partial y^2} \\
&\quad + \langle \boldsymbol{\zeta}_i | \mathbf{D} | \boldsymbol{\zeta}_0' \rangle \left(\frac{\partial \chi_0(y,t)}{\partial y} \right)^2.
\end{aligned}
\tag{19.13}
$$

In the second line of this equation we introduced the matrix operator $\mathcal{W}$, whose elements are

$$
\mathcal{W}_{ij} = \lambda_i \delta_{ij} + \langle \boldsymbol{\zeta}_i | \mathbf{D} | \boldsymbol{\zeta}_j \rangle \frac{\partial^2}{\partial y^2}.
\tag{19.14}
$$

Equation (19.13) for χ_i, $(i > 0)$ may be solved formally treating χ_0 as an independent function and the result substituted into Eq. (19.12) to obtain a closed equation for $\chi_0(y,t)$. This full solution is complicated and does not admit a simple analysis (Malevanets *et al.*, 1995). It is possible to obtain an approximate solution for weakly curved interfaces whose structure is much simpler. For weakly curved interfaces we can neglect the spatial gradients along y in $\mathcal{W}$ and approximate $\mathcal{W}_{ij} \approx \lambda_i \delta_{ij}$ for $j \neq 0$. Furthermore, since the λ_i are negative for $i > 0$ and are assumed to be well separated from zero, we neglect the time dependence of χ_i in the computation of χ_0. Using these approximations we obtain

$$
\chi_i \approx -\lambda_i^{-1} \left[\langle \boldsymbol{\zeta}_i | \mathbf{D} | \boldsymbol{\zeta}_0 \rangle \frac{\partial^2 \chi_0(y,t)}{\partial y^2} + \langle \boldsymbol{\zeta}_i | \mathbf{D} | \boldsymbol{\zeta}_0' \rangle \left(\frac{\partial \chi_0(y,t)}{\partial y} \right)^2 \right].
\tag{19.15}
$$

Substitution into Eq. (19.12) and neglect of the higher-order terms which are quadratic in χ_0 and involve up to four derivatives with respect to y, $(\partial^2 \chi_0/\partial y^2)^2$, and $(\partial \chi_0/\partial y)(\partial^3 \chi_0/\partial y^3)$ yields

$$
\begin{aligned}
\frac{\partial \chi_0(y,t)}{\partial t} &= \langle \boldsymbol{\zeta}_0 | \mathbf{D} | \boldsymbol{\zeta}_0 \rangle \frac{\partial^2 \chi_0}{\partial y^2} + \langle \boldsymbol{\zeta}_0 | \mathbf{D} | \boldsymbol{\zeta}_0' \rangle \left(\frac{\partial \chi_0(y,t)}{\partial y} \right)^2 \\
&\quad - \left[\sum_{i>0} \frac{\langle \boldsymbol{\zeta}_0 | \mathbf{D} | \boldsymbol{\zeta}_i \rangle \langle \boldsymbol{\zeta}_i | \mathbf{D} | \boldsymbol{\zeta}_0 \rangle}{\lambda_i} \right] \frac{\partial^4 \chi_0}{\partial y^4}.
\end{aligned}
\tag{19.16}
$$

In order to simplify the appearance of this equation it is convenient to introduce the following definitions:

$$\nu = \langle \zeta_0 | \mathbf{D} | \zeta_0 \rangle, \tag{19.17}$$

$$\kappa = \sum_{i>0} \frac{\langle \zeta_0 | \mathbf{D} | \zeta_i \rangle \langle \zeta_i | \mathbf{D} | \zeta_0 \rangle}{\lambda_i}. \tag{19.18}$$

The values of these coefficients must be computed for the problem of interest but, typically, one finds that the sign of ν depends on the values of various diffusion coefficient ratios, while κ is positive. The coefficient of the nonlinear term in $\chi_0(y, t)$ may be evaluated explicitly in the following way. Using the fact that $\langle \zeta_0 | \hat{\Omega}_R | \zeta_0 \rangle = 0$, it follows that

$$\langle \zeta_0 | u \rangle \langle u | \hat{\Omega}_R | \zeta_0 \rangle - \langle \zeta_0 | \hat{\Omega}_R | u \rangle \langle u | \zeta_0 \rangle = 0, \tag{19.19}$$

for any u. Then, by using the definition of Ω_R in Eq. (19.4) and integrating over u from $-\infty$ to x, one may show that

$$\begin{aligned}
0 &= \int_{-\infty}^{x} du \left[\langle \zeta_0 | \hat{\Omega}_R | u \rangle \langle u | \zeta_0 \rangle - \langle \zeta_0 | u \rangle \langle u | \hat{\Omega}_R | | \zeta_0 \rangle \right] \\
&= \left[2\langle \zeta_0 | u \rangle \langle u | \mathbf{D} \frac{\partial}{\partial u} | \zeta_0 \rangle + \left(\nu - \mathbf{D} \frac{\partial}{\partial u} \right) \langle \zeta_0 | u \rangle \langle u | \zeta_0 \rangle \right] \bigg|_x.
\end{aligned} \tag{19.20}$$

Finally, integrating this result over the entire domain we find $\langle \zeta_0 | \mathbf{D} | \zeta_0' \rangle = -\nu/2$. Substituting these results into Eq. (19.16) we obtain the Kuramoto–Sivashinsky (KS) equation,

$$\frac{\partial \chi_0(y, t)}{\partial t} = \nu \frac{\partial^2 \chi_0}{\partial y^2} - \frac{\nu}{2} \left(\frac{\partial \chi_0}{\partial y} \right)^2 - \kappa \frac{\partial^4 \chi_0}{\partial y^4}, \tag{19.21}$$

which describes the evolution of the front profile.

19.3 Linear stability analysis and front dynamics

The stability of the planar interfacial profile may be investigated by carrying out a linear stability analysis of the KS equation. Linearizing Eq. (19.21) and Fourier transforming in y we obtain

$$\frac{\partial \chi_0(k, t)}{\partial t} = -(\nu k^2 + \kappa k^4) \chi_0(k, t) \equiv \lambda(k) \chi_0(k, t). \tag{19.22}$$

The decay rate $\lambda(k) = -\nu k^2 - \kappa k^4$ is sketched in Fig. 19.2 for values of the parameters above the instability threshold. In making this plot we have assumed

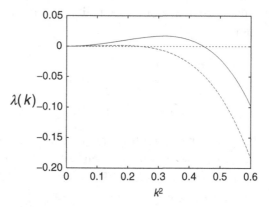

Fig. 19.2. Sketch of $\lambda(k)$ versus k^2 for parameters near to (dotted line) and well above (solid line) the front instability threshold.

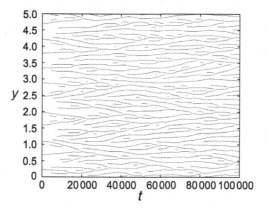

Fig. 19.3. Space–time plot of the Kuramoto–Sivashinsky chaotic dynamics. The positions of the minima of χ_0 are shown as a function of y and t. Parameters in scaled units are $-\nu = \kappa = \upsilon/2 = 1$.

that $\nu < 0$ so that the diffusion term has a destabilizing influence on the front dynamics and $\kappa > 0$ so that the fourth-order derivative term provides stability at high wavenumbers. The plot of $\lambda(k)$ versus k has the form characteristic of a phase instability: a band of unstable wavenumbers extends from zero to some maximum value. The maximum in the dispersion relation lies at $k_m = (-\nu/2\kappa)^{1/2}$ and the maximum unstable wavenumber is $k_{max} = (-\nu/\kappa)^{1/2}$. One expects a long wavelength instability with dissipation at short wavelengths. The characteristic dissipation length in the problem is $\ell_c \approx 2\pi/k_m$.

The results of a direct simulation of the KS equation are plotted in Fig. 19.3. Rather than plotting the entire $\chi_0(y, t)$ profile as a space–time plot, we have chosen to track only the minima in this profile along the y direction as a function of time. This representation of the front dynamics allows one to focus on the essential features of the spatiotemporal state that exists beyond the bifurcation. The existence of minima signals that the planar front has undergone a transverse instability and developed structure along y. If the resulting state were an inhomogeneous stationary state one would simply see a set of parallel lines in the space–time plot. Instead one sees that the locations of the minima evolve in time. These minima execute reactive particle-like dynamics. When the trajectories of two neighboring minima collide, both minima (particles) are destroyed. If the gap between two minima becomes too large a new minimum (particle) is created. As a result of these destruction and creation processes, the mean number of minima is fixed but their instantaneous value fluctuates. The resulting front dynamics exhibits spatiotemporal chaos.

From a quantitative analysis of the data that underlie this figure one can see that there is an average characteristic distance ℓ_c between the minima and measurement of its value shows that this distance is given by $\ell_c \approx 2\sqrt{2}\pi$ in accord with the above estimate of the dissipation length. When the distance between extrema exceeds ℓ_c, new extrema are formed to preserve the average characteristic length. Dissipation occurs at short wavelengths where extrema in fronts collide and disappear.

20

Cubic autocatalytic fronts

Transverse front instabilities occur in autocatalytic chemical reactions. A system that illustrates the interplay between the nonlinear kinetics and diffusion is the cubic autocatalytic reaction,

$$A + 2B \xrightarrow{k} 3B, \tag{20.1}$$

where the autocatalyst B consumes the fuel A. Imagine a two-dimensional system infinitely extended along x and of width L along y. Initially, let the domain $x < 0$ of the system contain B and the domain $x \geq 0$ contain A, with a sharp planar interface separating them at $x = 0$. Then, as time evolves, the autocatalyst will consume the fuel and the front will move to the right (increasing x) with velocity v. We would like to know if the front remains planar or develops structure along y. The same question can be posed for any propagating front in a system described by the reaction–diffusion equation (19.1).

A simple physical argument can be given to indicate the nature of the profile one might expect to see in different parameter regimes. Suppose the interface is initially slightly nonplanar. We wish to determine whether the reaction–diffusion dynamics will tend to eliminate this nonplanarity or accentuate it. First, let the diffusion coefficient of B be much greater than that of A, $D_B \gg D_A$. The situation is depicted schematically in Fig. 20.1, where large diffusion fluxes are indicated by heavy arrows and smaller diffusion fluxes by dotted arrows. For the part of the B front that protrudes into the A region, fast diffusion of B will lead to dispersal of B and suppress the autocatalytic reaction that requires two molecules of B. The front will have difficulty advancing in these regions. In the portion of the front where A protrudes into B, A will react with B, leading to advancement of the front. The net effect is to remove any initial nonplanarity and produce a planar front.

Next, suppose $D_B \ll D_A$. Now, in regions where B protrudes into A, rapid A diffusion will lead to conversion of A to B, leading to front advance. In regions

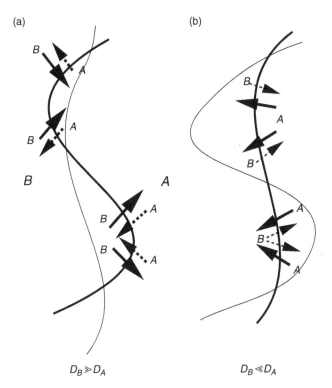

Fig. 20.1. Schematic picture of how the front instability (a) is suppressed for $D_B \gg D_A$ and (b) occurs for $D_B \ll D_A$. The heavy lines represent the front at some time instant while the light lines show its position at a later time.

where A protrudes into B, small diffusion of B into the A region will lead to slow autocatalytic conversion of A to B, so the front will not advance rapidly. Hence, any small nonplanarity will grow to make the front even more nonplanar.

As a result of these considerations, for some ratio of diffusion coefficients $D_A/D_B > 1$, we expect that the planar front will lose its stability. Of course, this argument provides no estimate of the precise value of D_A/D_B where this instability will occur or how the dynamics of the nonplanar interface may be described.

One can see that even this qualitative argument relies on the nature of the autocatalytic reaction: for example, if cubic autocatalysis is replaced by quadratic autocatalysis the above conclusions no longer apply, nor are the properties of the planar front the same (Billingham and Needham, 1991). There have been numerous studies, both theoretical (Scott and Showalter, 1992; Horváth *et al.*, 1993; Zhang and Falle, 1994; Malevanets *et al.*, 1995; Milton and Scott, 1995) and experimental (Horváth and Showalter, 1995), of cubic autocatalysis.

20.1 Analysis of front instability

The analysis of the mechanism of a transverse instability discussed in Chapter 19 was general in character, and the application of the method hinges on the reduction of the reaction–diffusion equation of interest to the KS equation. Such a reduction requires that one solve the eigenvalue problem posed in Eq. (19.5) and, using the results of this solution, find the coefficients v and κ that determine the location of the instability. Such a procedure must often be carried out numerically but in some cases one may obtain exact or approximate analytical solutions. We illustrate this procedure by carrying out such a reduction of the reaction–diffusion equation for the cubic autocatalysis reaction.

On the basis of the physical considerations discussed above, for sufficiently large D_A/D_B we expect that the planar chemical front separating the fuel A and the autocatalyst B will lose its stability and develop transverse structure. This is borne out by simulations of the reaction–diffusion equation for this system. Given the mechanism in Eq. (20.1) the reaction–diffusion equation (19.1) can be written explicitly as

$$\frac{\partial c_A(\mathbf{r},t)}{\partial t} = -kc_A(\mathbf{r},t)c_B(\mathbf{r},t)^2 + D_A\nabla^2 c_A(\mathbf{r},t)$$

$$\frac{\partial c_B(\mathbf{r},t)}{\partial t} = kc_A(\mathbf{r},t)c_B(\mathbf{r},t)^2 + D_B\nabla^2 c_B(\mathbf{r},t). \tag{20.2}$$

With initial conditions to produce an initially planar front separating the two species, the reaction–diffusion equation may be solved for various values of D_A/D_B to determine whether the planar front will remain stable to perturbations transverse to its propagation direction. For small values of D_A/D_B, simulations show that the front remains planar. However, for sufficiently large D_A/D_B, propagating fronts of the type shown in Fig. 20.2 are found. The front has a complex cellular structure.

We may examine the front dynamics of cubic autocatalysis in terms of the same space–time representation used in Fig. 19.3. Such a plot of the front profile minima is shown in Fig. 20.3 (see Malevanets *et al.*, 1995). The space–time dynamics has the same appearance as that seen in the KS equation simulations.

To determine the precise value of D_A/D_B where the instability occurs and quantitatively characterize the nature of the resulting dynamics, we must extract the front evolution equation from the full reaction–diffusion equation. This program can be carried out only numerically for the general case of arbitrary diffusion coefficients. However, it is possible to give an analytical estimate of the KS parameters using perturbation theory about the equal-diffusion case where the eigenvalue problems can be solved for the quantities needed for the calculation.

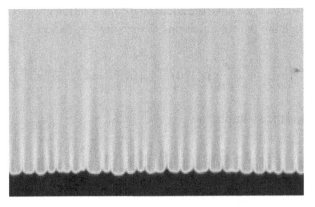

Fig. 20.2. Cubic autocatalysis front profile at one time instant for $D_A/D_B = 5$.

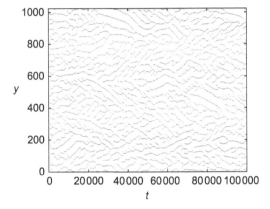

Fig. 20.3. Space–time plot of the cubic autocatalysis reaction–diffusion equation for $D_A/D_B = 5$. The positions of the minima of χ_0 are shown as a function of y and t.

We first require the planar front profiles. For cubic autocatalysis, in a frame moving with the front velocity, Eq. (20.2) takes the form

$$\left(\delta^{-1}\frac{d^2}{du^2} + v\frac{d}{du}\right)c_A(u) - c_A(u)c_B^2(u) = 0,$$

$$\left(\frac{d^2}{du^2} + v\frac{d}{du}\right)c_B(u) + c_A(u)c_B^2(u) = 0, \qquad (20.3)$$

where $\delta^{-1} = D_A/D_B$. We have scaled time as $t \to kt$ and space so that $\mathbf{r} \to (D_B/k)^{-1/2}\mathbf{r}$. While these equations are difficult to solve for arbitrary D_A and D_B, they can be solved for the equal-diffusion case, $D_A = D_B = D$, where $\delta = 1$. Let the boundary conditions be $c_A(\infty) = c_B(-\infty) = 1$ and $c_A(-\infty) = c_B(\infty) = 0$. For the equal-diffusion case we have the additional conservation law $c_A(u) + c_B(u) = 1$, and we can reduce the pair of equations to the single equation

$$\left(\frac{d^2}{du^2} + v\frac{d}{du}\right) c_A(u) - c_A(u)(1 - c_A(u))^2 = 0. \tag{20.4}$$

This equation has the solution

$$c_A(u) = (1 + \exp(-vu))^{-1}, \tag{20.5}$$

with the front speed $v = 1/\sqrt{2}$. The eigenvalues and right and left eigenvectors of the linearized eigenvalue problem can be computed for $D_A = D_B$, and the KS parameters can be evaluated using perturbation theory for small deviations from this value (Malevanets *et al.*, 1995). This calculation yields the following approximate value for the v parameter:

$$v = \langle \zeta_0 | \mathbf{D} | \zeta_0 \rangle \approx 1 - \frac{7}{3}\frac{1 - \delta}{1 + \delta}. \tag{20.6}$$

A more lengthy calculation for κ to the same order yields

$$\kappa = \sum_{i > 0} \frac{\langle \zeta_0 | \mathbf{D} | \mathbf{u}_i \rangle \langle \zeta_i | \mathbf{D} | \zeta_0 \rangle}{\lambda_i} \approx \frac{264 + 40\pi^2}{27}\left(\frac{1 - \delta}{1 + \delta}\right)^2, \tag{20.7}$$

where the eigenfunctions are evaluated at $\delta = 1$. Instability occurs when v passes through zero, and this occurs at $\delta = \delta_c = 2/5$. Direct solution of the equations to obtain v gives the numerically exact critical value $\delta_c = 1/2.300$, which is close to the analytical value and agrees with numerical estimates of the critical value from simulations of the front evolution. If one computes the characteristic length ℓ_c from the data in Fig. 20.3 one finds it is well approximated by the KS estimate. These results confirm the KS mechanism of the front instability.

20.2 Experimental observation of front instability

The iodate–arsenous acid reaction has a long history (Roebuck, 1902; Dushman, 1904), and has been a favorite example for studies of isothermal chemical front propagation (Ganapathisubramanian and Showalter, 1985; Saul and Showalter, 1985; Horváth and Showalter, 1995). When carried out in excess arsenous acid, the iodate oxidation of arsenous acid reaction has the net stoichiometry

$$3H_3AsO_3 + IO_3^- \rightleftharpoons 3H_3AsO_4 + I^-. \tag{20.8}$$

Taking into account the mechanism that underlies this overall reaction, the reaction rates for the $IO_3^- = A$ and $I^- = B$ species are

$$\mathcal{R}_A(c_A, c_B) = -\mathcal{R}_B(c_A, c_B) = -(k_a + k_b c_B) c_B c_A c_{H^+}^2. \qquad (20.9)$$

The reaction may be carried out in buffered solutions where the H^+ concentration c_{H^+} is essentially constant. The reaction–diffusion equations read

$$\frac{\partial c_A(\mathbf{r}, t)}{\partial t} = \mathcal{R}_A(c_A, c_B) + D_A \nabla^2 c_A,$$

$$\frac{\partial c_B(\mathbf{r}, t)}{\partial t} = \mathcal{R}_B(c_A, c_B) + D_B \nabla^2 c_B. \qquad (20.10)$$

In the experiments on front instabilities, the reaction is carried out in a thin (α-cyclodextrin=C) gel film. Not only does the gel suppress convective effects so that the reaction–diffusion equation provides an appropriate description of the dynamics but, equally important, it complexes with the autocatalyst $I^- = B$,

$$B + C \rightleftharpoons C \cdot B, \qquad (20.11)$$

with equilibrium constant $K_C = c_{C \cdot B}/(c_C c_B)$. The species B now exists in two forms: as free B and as the complex $C \cdot B$. The chemical rate equations may be written taking into account the B, C and $C \cdot B$ species. The rate equations for the concentrations of the B and $C \cdot B$ species may be added to obtain the rate equation for the total concentration of B in both forms, c_B^T. Using the fact that we may write c_B^T as $c_B^T = c_B + c_{C \cdot B} = c_B(1 + K_C c_C) \equiv c_B \sigma$, the reaction–diffusion equation for the total concentration of B takes the form

$$\frac{\partial c_B(\mathbf{r}, t)}{\partial t} = \sigma^{-1} \mathcal{R}_B(c_A, c_B) + \sigma^{-1} D_B \nabla^2 c_B. \qquad (20.12)$$

Here we assumed that $C \cdot B$ does not diffuse, since it is a gel-bound species. Consequently, by varying the concentration of gel one may change the diffusion coefficient of the autocatalyst (and its reaction rate) and trigger a front instability. We shall discuss such mechanisms in more detail when Turing bifurcations are studied in Chapter 23.

The mechanism for the instability is like that described earlier in this chapter: the planar front will be stable or unstable to transverse perturbations depending on the relative values of the diffusion coefficients of the iodine and iodate species. Rapid diffusion of the autocatalyst iodine (B) causes the decay of perturbations, stabilizing a planar front, while the diffusion of iodate (A) tends to be destabilizing. Thus, as in the cubic autocatalysis case, there is a critical value of the diffusion

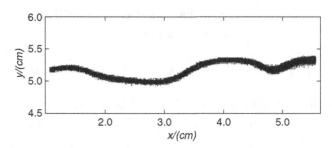

Fig. 20.4. Plot of the front in the iodate–arsenous acid system. Reprinted with permission from Horváth and Showalter (1995). Copyright 1995, American Institute
of Physics.

coefficient ratio where the planar front will lose its stability. Experiments on front
propagation in this system confirm this scenario (Horváth and Showalter, 1995).
For low values of the concentration of the gel, planar propagating fronts were
observed. For gel concentrations beyond a certain value, the planar front lost its
stability and the front developed a transverse structure which varied in time. Such
an unstable front is shown in Fig. 20.4. This figure shows the interface between the
reactant (iodate and arsenous acid species) and product (iodide) zones, which is
measured by monitoring the iodine species utilizing its strong optical absorbance.
Experiments on this system in other conditions that do not suppress convection have
shown that the reaction exothermicity in combination with small density changes
can give rise to chemical fronts in the form of complex buoyant plumes (Rogers
and Morris, 2005).

21

Competing interactions and front repulsion

Front instabilities driven by differences in the diffusion coefficients of chemical species can lead to the formation of new types of chemical pattern. In some instances the differences in diffusion coefficients can give rise to long-range repulsive coupling between local chemical concentration fields that causes chemical fronts to repel each other. In one-dimensional bistable systems, front repulsion is responsible for the existence of stable domains of the less stable phase in a sea of the more stable phase. In the absence of such front repulsion the less stable phase would be consumed by the more stable phase, as discussed earlier in Chapter 5 for model A. Long-range repulsive interactions in nonreacting systems were the topic of Chapters 13, 14, and 15.

21.1 Competing interactions in bistable media

To illustrate how long-range repulsive interactions arise in reaction–diffusion systems where species diffuse at very different rates, we consider the FitzHugh–Nagumo (FHN) model (FitzHugh, 1961; Nagumo *et al.*, 1962),

$$\frac{\partial c_A}{\partial t} = -c_A^3 + c_A - c_B + D_A \nabla^2 c_A,$$

$$\frac{\partial c_B}{\partial t} = \epsilon(c_A - \alpha c_B - \beta) + D_B \nabla^2 c_B. \tag{21.1}$$

Here c_A and c_B are the local concentrations of species A and B, respectively. This system of equations was originally constructed as a simple model for the excitable behavior of nerve tissue. It mimics the behavior of the more realistic Hodgkin–Huxley equations (Hodgkin and Huxley, 1952; Hodgkin *et al.*, 1952). While the c_A and c_B variables may be roughly associated with the membrane voltage and ion currents, respectively, their connection with these physiological variables is not direct. Although the antecedents of this model lie in nerve impulse physiology, it has seen widespread use as a generic model that exhibits many phenomena seen in

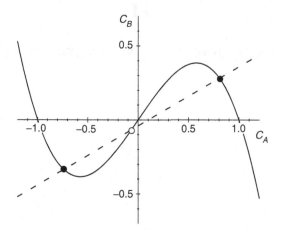

Fig. 21.1. Nullclines of the FitzHugh–Nagumo equation for parameters in the bistable regime. The linear B nullcline intersects the cubic A nullcline at three fixed points: the outer two fixed points denoted by filled circles are stable, while the central fixed point denoted by an open circle is unstable.

bistable, excitable, or oscillatory chemical media. It is in this context that we study this system. While the FHN model can display excitability associated with a single stable steady state, or oscillations arising when a steady state loses its stability in a Hopf bifurcation, we consider the parameter regime where bistability exists and domains of the stable states compete in a spatially distributed medium. Excitability and oscillations will be studied in Chapters 24 and 25.

The FitzHugh–Nagumo equation is an example of an activator–inhibitor system. The FHN equations can be derived from mass action kinetics for a model chemical reaction scheme (Malevanets and Kapral, 1997). An increase in the concentration of the activator A leads to an increase in its own production and that of B; an increase in the concentration of the inhibitor B suppresses the production of both A and B species. The interplay between these two "species" as a result of reaction and diffusion is responsible for the effects we shall describe.

The steady-state solutions of the FHN equation can be analyzed by consider-ing the structure of the nullclines of the underlying ordinary differential equation obtained from Eq. (21.1) by dropping the diffusion terms. The c_A and c_B nullclines, determined from the equations $\dot{c}_A = 0$ and $\dot{c}_B = 0$, are shown in Fig. 21.1. We have chosen parameters so that the FHN system possesses bistable steady states. There are three intersections between the c_A and c_B nullclines. Linear stability analysis shows that the outer two fixed points (solid circles) are stable while the central fixed point (open circle) is unstable. In this bistable regime, depending on the initial values of c_A and c_B, the system will evolve to either of the two stable fixed points. The existence of S-shaped and linear nullclines is a feature common to many systems and allows for the possibility of a variety of dynamical states.

If one considers the spatially distributed bistable system (21.1) with the diffusion terms present, inhomogeneous initial states will evolve to form domains of the two stable phases. The structure and dynamics of these domains depends on the interplay between the activator–inhibitor kinetics and the relative values of the diffusion coefficients of the species.

It is instructive to cast the FHN equation into a form that resembles the free energy functional models in Chapters 5 and 13 where domain coarsening was studied. While this is not possible in general, such a reduction can be carried out in certain limits. The first of Eqs. (21.1) has a structure similar to that of model A but contains the variable c_B that couples the two equations. To explore the relation to model A in more detail and understand how the long-range coupling arises, we may write the FHN equation in variational form as (Goldstein *et al.*, 1996)

$$\frac{\partial c_A}{\partial t} = -\frac{\delta \mathcal{F}_A[c_A]}{\delta c_A} - \frac{\delta \mathcal{F}_{AB}[c_A, c_B]}{\delta c_A},$$
$$\frac{\partial c_B}{\partial t} = -\frac{\delta \mathcal{F}_B[c_B]}{\delta c_B} + \epsilon \frac{\delta \mathcal{F}_{AB}[c_A, c_B]}{\delta c_B},$$

(21.2)

where the functionals $\mathcal{F}_A$, $\mathcal{F}_B$ and $\mathcal{F}_{AB}$ are defined as

$$\mathcal{F}_A[c_A] = \int d^d r \left\{ \frac{c_A^4}{4} - \frac{c_A^2}{2} + \frac{1}{2} D_A |\nabla c_A|^2 \right\},$$ (21.3)

$$\mathcal{F}_B[c_B] = \int d^d r \left\{ \alpha \epsilon \frac{c_B^2}{2} + \frac{1}{2} D_B |\nabla c_B|^2 \right\},$$ (21.4)

$$\mathcal{F}_{AB}[c_A, c_B] = \int d^d r (c_A - \beta) c_B.$$ (21.5)

The functional $\mathcal{F}_A$ has the same form as that for model A, while the quadratic free energy functional $\mathcal{F}_B$ describes a monostable system. Because of the nature of the cross terms in Eq. (21.2) this is not a pure gradient flow. The coupling between the A and B systems through $\mathcal{F}_{AB}$ gives rise to modulated phases similar to those seen in gradient systems with long-range repulsive interactions.

Although the FHN reaction–diffusion equation is not of gradient form, and there is no free energy functional that controls the evolution to the final attracting states, it is possible to make a reduction to dynamics determined by a nonlocal free energy functional when the inhibitor dynamics is fast ($\epsilon \to \infty$). To carry out this reduction of Eq. (21.2), we first formally solve the linear equation for the inhibitor B and write the equation for the activator c_A as

$$\frac{\partial c_A(\mathbf{r},t)}{\partial t} = -c_A^3 + c_A + D_A \nabla^2 c_A \tag{21.6}$$

$$- \int_V d^d r' \int^t dt' G(\mathbf{r}-\mathbf{r}',t-t')\epsilon(c_A(\mathbf{r}',t') - \beta),$$

where the Green function $G(\mathbf{r},t)$ is given by the solution of the equation

$$\left(\frac{\partial}{\partial t} + \epsilon\alpha - D_B \nabla^2\right) G(\mathbf{r}-\mathbf{r}',t-t') = \delta(\mathbf{r}-\mathbf{r}')\delta(t-t'). \tag{21.7}$$

From the form of Eq. (21.6) we can see that the activator field at space–time point $(\mathbf{r},t)$ is influenced by its value at point $(\mathbf{r}',t')$ due to coupling with the inhibitor field. In this general form the evolution equation cannot be written as the gradient of a free energy functional. However, in the *fast inhibitor limit* where ϵ is large, so that B varies on a much faster time scale than that of A, we may let $c_A(\mathbf{r}',t') \approx c_A(\mathbf{r}',t)$ in the integral in Eq. (21.6) and, after integration, obtain

$$\frac{\partial c_A(\mathbf{r},t)}{\partial t} = -c_A^3 + c_A + \frac{\beta}{\alpha} + D_A \nabla^2 c_A - \int d^d r' G(\mathbf{r}-\mathbf{r}')c_A(\mathbf{r}',t), \tag{21.8}$$

with $G(\mathbf{r})$ the time-independent Green function determined from the solution of the equation

$$\left(1 - \frac{\tilde{D}_B}{\alpha}\nabla^2\right)G(\mathbf{r}) = \frac{1}{\alpha}\delta(\mathbf{r}). \tag{21.9}$$

The scaled diffusion coefficient of the inhibitor is $\tilde{D}_B = D_B/\epsilon$. Now Eq. (21.8) can be written in gradient form,

$$\frac{\partial c_A(\mathbf{r},t)}{\partial t} = -\frac{\delta\mathcal{F}_{nl}[c_A]}{\delta c_A}, \tag{21.10}$$

where the nonlocal free energy functional $\mathcal{F}_{nl}$ is

$$\mathcal{F}_{nl}[c_A] = \int d^d r \left(\mathcal{V}(c_A) + \frac{D_A}{2}|\nabla c_A|^2\right) \tag{21.11}$$

$$+ \frac{1}{2}\int d^d r \int d^d r' c_A(\mathbf{r})G(\mathbf{r}-\mathbf{r}')c_A(\mathbf{r}').$$

The potential $\mathcal{V}$ is defined as $\mathcal{V}(c_A) = c_A^4/4 - c_A^2/2 - \beta c_A/\alpha$. Since the Green function is the solution of Eq. (21.9), we may substitute $G(\mathbf{r}) = \alpha^{-1}(\tilde{D}_B \nabla^2 G(\mathbf{r}) +$

$\delta(\mathbf{r})$) into Eq. (21.11) and integrate by parts to obtain an alternate expression for the nonlocal free energy functional,

$$\mathcal{F}_{nl}[c_A] = \int d^d r \left(V(c_A) + \frac{D_A}{2}|\nabla c_A|^2 \right) \tag{21.12}$$

$$- \frac{1}{2\kappa^2} \int d^d r \int d^d r' (\nabla c_A(\mathbf{r})) \cdot G(\mathbf{r} - \mathbf{r}')(\nabla' c_A(\mathbf{r}')),$$

where $\kappa = (\alpha/\tilde{D}_B)^{1/2}$. The modified potential energy $V(c_A)$ is defined by $V(c_A) = \mathcal{V}(c_A) + c_A^2/(2\alpha) = c_A^4/4 - (1 - \alpha^{-1})c_A^2/2 - \beta c_A/\alpha$. For a certain range of parameters the quartic potential $V(c_A)$ has two minima, denoted by $c_A^{\pm}$, corresponding to the stable homogeneous states of the system, and a maximum corresponding to the unstable state.

It is useful at this point to review some of the features of activator–inhibitor kinetics described by Eq. (21.1). If the scaled diffusion coefficient of the inhibitor $\tilde{D}_B$ tends to zero, Eq. (21.8) reduces to the time-dependent Ginzburg–Landau equation (model A for a nonconserved order parameter without noise). The domain coarsening described by this model was discussed in Chapters 5, 7, and 9. Recall that if the system is prepared in the unstable state, domains of the two stable phases will form and coarsen. The growth law and the nature of the domains depend on whether the potential V is symmetric or asymmetric. For an asymmetric free energy potential with $\beta \neq 0$, if two phases are separated by a planar interface the more stable phase will consume the less stable phase at a rate determined by the planar front velocity. For more complicated domain geometries the front velocity and domain curvature determine the coarsening rate (Allen and Cahn, 1979).

The situation changes if $D_B \gg D_A$. As indicated above, in the fast inhibitor limit the additional nonlocal term in the free energy functional gives rise to a competing interaction that leads to the possibility of modulated phases or localized patterns. These localized patterns are observed both in the fast inhibitor limit (Eq. (21.8)) and in the general case or slow inhibitor limit (Eq. (21.6)) (Ohta *et al.*, 1989, 1990; Petrich and Goldstein, 1994; Goldstein *et al.*, 1996; Hagberg and Meron, 1993, 1994; Malevanets and Kapral, 1997). When the inhibitor diffuses rapidly compared with the activator, the long-ranged B field influences the interactions among the A-field fronts, leading to front repulsion in certain parameter regimes. Such front repulsion can give rise to stationary domains of the less stable phase in a sea of the more stable phase (Ohta *et al.*, 1989, 1990). These phenomena are not restricted to the FHN equations, and for a different model with a nonlocal energy functional it was demonstrated (Gorshkov *et al.*, 1996; Aranson *et al.*, 1990) that compact stable patterns exist.

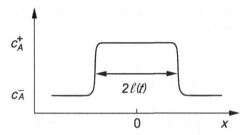

Fig. 21.2. Schematic diagram of a domain of length $2\ell(t)$ of the c_A^+ stable phase in the c_A^- stable phase.

In the next section, we restrict our considerations to the fast inhibitor limit where a description in terms of a nonlocal free energy functional applies. For this case we may examine the origin of the long-range repulsive interaction in some detail for one- and two-dimensional systems (Petrich and Goldstein, 1994; Goldstein et al., 1996).

21.2 Front repulsion

In one dimension, curvature plays no role in the dynamics of the fronts, and we study the conditions under which a domain of the less stable phase persists in a sea of the more stable phase. Since the FHN equation does not have a conservation law, such stable patterns would not be possible without front repulsion, since the more stable phase would consume the less stable phase.

Suppose the length of the c_A^+ domain is $2\ell(t)$ and it resides in a sea of the c_A^- phase (see Fig. 21.2). To begin the analysis, we write the $1d$ form of the free energy functional given in Eq. (21.12) as

$$\mathcal{F}_{nl}[c_A] = \int dx \left(V(c_A) + \frac{D_A}{2} \left(\frac{dc_A}{dx} \right)^2 \right) \tag{21.13}$$
$$- \frac{1}{2\kappa^2} \int dx \int dx' \left(\frac{dc_A(x)}{dx} \right) G(x - x') \left(\frac{dc_A(x')}{dx'} \right).$$

At this point it is useful to refer to Chapter 3, where the free energy functional was discussed and the surface tension was introduced. In parallel with the discussion leading to Eq. (3.12), we introduce the field $\Delta V(c_A)$ as

$$\Delta V(c_A) = \left(V(c_A) - V(c_A^-) \right) (1 - \theta_B(x)) + \left(V(c_A) - V(c_A^+) \right) \theta_B(x), \tag{21.14}$$

where $\theta_B(x) = (1 - \theta(-x - \ell(t)) - \theta(x - \ell(t))$ is a characteristic function that is nonzero for $-\ell(t) \le x \le \ell(t)$ and $\theta(x)$ is the Heaviside function. From its definition, one can see that $\Delta V(c_A)$ is nonzero only in the interfacial regions.

Using the properties of the Heaviside function and rearranging terms, Eq. (21.14) may be rewritten in the form

$$V(c_A) - V(c_A^-) = \Delta V(c_A) + \Delta V_{\pm}\theta_B(x), \tag{21.15}$$

where $\Delta V_{\pm} = V(c_A^+) - V(c_A^-)$. Since the $c_A(x)$ field is constant for $x \to \pm\infty$, the free energy is unbounded and it is convenient instead to consider $\Delta\mathcal{F}_{nl}[c_A] = \mathcal{F}_{nl}[c_A] - \mathcal{F}_{nl}[c_A^-]$, which takes the form

$$\Delta\mathcal{F}_{nl}[c_A] = \Delta V_{\pm} \int dx\theta_B(x) + \int dx \left(\Delta V(c_A(x)) + \frac{D_A}{2}\left(\frac{dc_A}{dx}\right)^2 \right)$$

$$- \frac{1}{2\kappa^2} \int dx \int dx' \left(\frac{dc_A(x)}{dx}\right) G(x - x') \left(\frac{dc_A(x')}{dx'}\right). \tag{21.16}$$

Using Eq. (3.8) we may identify the second term on the right-hand side of this equation with twice the surface tension since there are two interfacial zones on either side of the c_A^+ domain,

$$2\sigma = \int dx \left(\Delta V(c_A(x)) + \frac{D_A}{2}\left(\frac{dc_A}{dx}\right)^2 \right) = D_A \int dx \left(\frac{dc_A}{dx}\right)^2. \tag{21.17}$$

To reduce the expression for the nonlocal free energy functional further, we assume that the interfaces of the c_A^+ domain are very sharp so that $c_A(x) \approx c_A^- + \Delta c_A\theta(\ell(t) - x)\theta(x + \ell(t))$, with $\Delta c_A = (c_A^+ - c_A^-)$. We may approximate the free energy functional by

$$\Delta\mathcal{F}_{nl}[c_A] = \Delta V_{\pm}2\ell(t) + 2\sigma \tag{21.18}$$

$$- \frac{1}{2\kappa^2} \int dx \int dx' \Delta c_A[\delta(x + \ell(t)) - \delta(x - \ell(t))]$$

$$\times G(x - x')\Delta c_A[\delta(x' + \ell(t)) - \delta(x' - \ell(t))].$$

In writing this equation we have used the fact that $dc_A(x)/dx \approx \Delta c_A[\delta(x+\ell(t)) - \delta(x - \ell(t))]$. Finally, performing the integrals over the delta functions we obtain

$$\Delta\mathcal{F}_{nl}[c_A] = \Delta V_{\pm}2\ell(t) + 2\sigma - \kappa^{-2}(\Delta c_A)^2[G(0) - G(2\ell(t))]. \tag{21.19}$$

The last step in the reduction of the free energy functional is to insert the explicit form of the $1d$ Green function. The $1d$ Green function is determined from the solution of the equation $(1 - \kappa^{-2}d^2/dx^2)G(x - x') = \alpha^{-1}\delta(x - x')$, and is given by

$$G(x) = \frac{\kappa}{2\alpha}e^{-\kappa x}. \tag{21.20}$$

Inserting this expression for the Green function in Eq. (21.19), we obtain

$$\Delta \mathcal{F}_{nl}[c_A] = \Delta V_{\pm} 2\ell(t) + 2\sigma - \frac{(\Delta c_A)^2}{2\alpha\kappa}\left(1 - e^{-2\kappa\ell(t)}\right), \qquad (21.21)$$

as the approximate expression for the nonlocal free energy functional for a domain of length 2ℓ of the c_A^+ phase in the c_A^- phase. From this free energy functional an equation of motion for the domain size can be derived.

We may construct the evolution equation for the domain boundaries by starting with the Lagrangian equation of motion for a dissipative system (Goldstein, 1950),

$$\frac{d}{dt}\frac{\delta \mathcal{L}}{\delta \dot{c}_A} - \frac{\delta \mathcal{L}}{\delta c_A} = -\frac{\delta \mathcal{R}_d}{\delta \dot{c}_A}. \qquad (21.22)$$

Here $\mathcal{L}$ is the Lagrangian and $\mathcal{R}_d$ is the Rayleigh dissipation function, which is defined by

$$\mathcal{R}_d = \frac{1}{2}\int dx \, \dot{c}_A^2. \qquad (21.23)$$

For the interfacial dynamics under consideration, we may neglect the inertial term in the Lagrangian equation and write

$$\frac{\delta \Delta \mathcal{F}_{nl}}{\delta c_A} = -\frac{\delta \mathcal{R}_d}{\delta \dot{c}_A}, \qquad (21.24)$$

where we have used the definition of the Lagrangian, $\mathcal{L} = -\Delta \mathcal{F}_{nl}$.

The concentration field $c_A(x,t)$ depends on t through the front positions $\pm\ell(t)$ and $c_A(x) = c_A(x - \ell(t))\theta(x) + c_A(-x - \ell(t))\theta(-x)$. As a result we may write the Rayleigh dissipation function approximately as

$$\mathcal{R}_d \approx \frac{1}{2}\left(\frac{d\ell(t)}{dt}\right)^2 \int dx \left(\frac{dc_A(x)}{dx}\right)^2 = \left(\frac{d\ell(t)}{dt}\right)^2 \frac{\sigma}{D_A}. \qquad (21.25)$$

Using the expressions for the Lagrangian and Rayleigh dissipation function in terms of $\ell(t)$, Eqs. (21.21) and (21.25), respectively, the Lagrangian equation of motion reads

$$\frac{\partial \Delta \mathcal{F}_{nl}}{\partial \ell} = -\frac{\partial \mathcal{R}_d}{\partial \dot{\ell}}. \qquad (21.26)$$

This yields the following equation of motion for the domain front positions:

$$\frac{d\ell(t)}{dt} = -\frac{D_A}{\sigma}\left(\Delta V_{\pm} - \frac{(\Delta c_A)^2}{2\alpha}e^{-2\kappa\ell(t)}\right). \qquad (21.27)$$

In order to determine whether it is possible to have a stable finite domain of the c_A^+ phase in a sea of the c_A^- phase, we seek a stationary solution $d\ell(t)/dt = 0$ of Eq. (21.27). Such a solution exists if $\Delta V_\pm > 0$ or $V(c_A^+) > V(c_A^-)$: i.e. the c_A^+ phase is the less stable phase for localized solutions. The stationary solution is $\ell = \ell^*$, which satisfies

$$\ell^* = \frac{1}{2\kappa} \ln \frac{(\Delta c_A)^2}{2\alpha \, \Delta V_\pm}. \tag{21.28}$$

Consequently, a stationary localized solution is possible in one-dimensional systems.

To determine the stability of these localized stationary solutions we may linearize Eq. (21.27) about ℓ^*, $\delta\ell(t) = \ell(t) - \ell^*$, to obtain

$$\frac{d\delta\ell(t)}{dt} = -\left(\frac{D_A(\Delta c_A)^2 \kappa}{\sigma\alpha} e^{-2\kappa\ell^*}\right)\delta\ell(t), \tag{21.29}$$

which shows that the pulse solution is stable. An example of the evolution of the FHN model in the bistable region in one dimension is shown in Fig. 21.3. For the

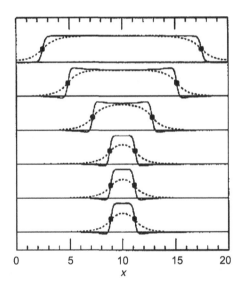

Fig. 21.3. Simulation of the FHN equation showing evolution to a stable stationary domain of finite length of the less stable phase in the more stable phase. The A concentration field is shown as a solid line while the B field is depicted by a dashed line. Filled circles indicate locations where $c_A = \frac{1}{2}$. Note that the profile of the rapidly diffusing B field spans the stable domain and provides a coupling between the two sharp A-field fronts. Reprinted Figure 5 with permission from Goldstein *et al.* (1996). Copyright 1996 by the American Physical Society.

parameters chosen for this plot, one sees that a domain with large spatial extent shrinks under the dynamics and evolves to a stable inhomogeneous solution with fixed $\ell = \ell^*$.

From this analysis we see that it is possible to stabilize a domain of the less stable phase in the more stable phase as a result of the long-range repulsive coupling that arises from the fast diffusion of the B species. This behavior should be contrasted with that of model A without long-range repulsive interaction. We now turn to the consequences of such front repulsion in two dimensions, where transverse front instabilities and front curvature play a role in determining the nature of the patterns.

22

Labyrinthine patterns in chemical systems

In two and higher dimensions front repulsion arising from competing interactions can lead to the formation of new types of chemical pattern. Since front repulsion can prevent simple curvature-driven coarsening from proceeding to completion, more complex labyrinthine patterns can form in the system. Figure 22.1 shows the development of a labyrinthine pattern in a chemically reacting system. The experiment on the iodine–ferrocyanide–sulfite chemical reaction was carried out in a gel reactor. This chemically reacting system is known to possess bistable steady states and oscillations (Edblom *et al.*, 1986; Gáspár and Showalter, 1990). The evolution and bifurcations of labyrinthine patterns in this chemical system have been studied extensively in experiments (Lee *et al.*, 1993; Lee and Swinney, 1995; Li *et al.*, 1996). We now consider the origin of such patterns.

22.1 Interface dynamics in two dimensions

Consider the $2d$ domain $\mathcal{B}$ of the c_A^+ phase in the c_A^- phase shown in Fig. 22.2. The boundary separating these two phases is denoted by $\mathbf{R}(s)$ and is parameterized by the arc length s. We use the same local interface-based coordinate system as in Chapter 7, where any point $\mathbf{r}$ in the system is given by $\mathbf{r} = \mathbf{R}(s) + u\hat{\mathbf{n}}(s)$, with $\hat{\mathbf{n}}(s)$ a unit normal to the interface at s. The tangent vector at s is $\hat{\mathbf{t}}(s)$. It is then convenient to make the change of variables $\mathbf{r} = (x, y) \rightarrow (u, s)$ and work in the (u, s) coordinate system. In this coordinate system a unit vector along $\mathbf{r}$ may be written as $\hat{\mathbf{r}} = (\hat{x}, \hat{y}) = (\hat{\mathbf{t}}, \hat{\mathbf{n}})$.

We may now follow a procedure that is analogous to that used in $1d$ to approximate the free energy functional in Eq. (21.12). Again, letting $\Delta \mathcal{F}_{nl}[c_A] = \mathcal{F}_{nl}[c_A] - \mathcal{F}_{nl}[c_A^-]$, we may write

$$\Delta \mathcal{F}_{nl}[c_A] = \int d^2r \left(\Delta V_{\pm}\theta_{\mathcal{B}}(\mathbf{r}) + \Delta V(c_A(\mathbf{r})) + \frac{D_A}{2}|\nabla c_A|^2 \right)$$

$$- \frac{1}{2\kappa^2} \int d^2r \int d^2r' (\nabla c_A(\mathbf{r})) \cdot G(\mathbf{r} - \mathbf{r}')(\nabla' c_A(\mathbf{r}')), \quad (22.1)$$

189

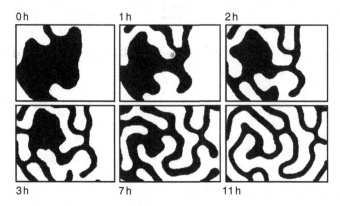

Fig. 22.1. Formation and evolution of a labyrinthine pattern in the iodine–ferrocyanide–sulfite chemically reacting system. White regions correspond to low pH and black regions to high pH. The evolution to the stationary pattern takes place in several hours. Reprinted Figure 1 with permission from Lee and Swinney (1995). Copyright 1995 by the American Physical Society.

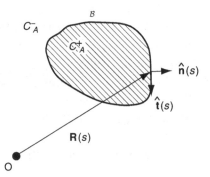

Fig. 22.2. A two-dimensional domain $\mathcal{B}$ of the c_A^+ phase in the c_A^- phase. The diagram shows the interface-based coordinate system with normal $\hat{\mathbf{n}}(s)$ and tangent $\hat{\mathbf{t}}(s)$.

where $\theta_{\mathcal{B}}(\mathbf{r})$ is a characteristic function that is unity within domain $\mathcal{B}$ and zero otherwise.

The first term in Eq. (22.1) involving the potential difference $\Delta V_{\pm}$ may be written as

$$\Delta V_{\pm} \int d^2 r \theta_{\mathcal{B}}(\mathbf{r}) = \Delta V_{\pm} A, \qquad (22.2)$$

where A is the area occupied by the c_A^+ domain. The reduction of the remaining terms in $\Delta \mathcal{F}_{nl}[c_A]$ to tractable forms may be accomplished by first making a coordinate transformation to the local coordinate system. In (u, s) coordinates we have

$$\nabla c_A = \hat{\mathbf{n}}\left(\frac{\partial c_A}{\partial u}\right) + \hat{\mathbf{t}}\left(\frac{\partial c_A}{\partial s}\right), \tag{22.3}$$

from which it follows that

$$|\nabla c_A|^2 = \left(\frac{\partial c_A}{\partial u}\right)^2 + \left(\frac{\partial c_A}{\partial s}\right)^2 \approx \left(\frac{\partial c_A}{\partial u}\right)^2. \tag{22.4}$$

In the last approximate equality we have used the small-curvature approximation to neglect the derivative of c_A with respect to the arc length compared with the derivative of c_A normal to the interface. Consequently, we may write

$$\int d^2 r \left(\Delta V(c_A(\mathbf{r})) + \frac{D_A}{2}|\nabla c_A|^2\right) \approx D_A \int ds\, du \left(\frac{\partial c_A}{\partial u}\right)^2$$

$$= \int ds\, \sigma = \sigma L, \tag{22.5}$$

where L is the length of the interface.

Finally, the nonlocal contribution in Eq. (22.1) may be evaluated using the sharp-interface approximation for the $c_A(\mathbf{r})$ concentration field,

$$c_A(\mathbf{r}) \approx c_A^+ \theta_B(\mathbf{r}) + c_A^-(1 - \theta_B(\mathbf{r})). \tag{22.6}$$

Given this form for $c_A(\mathbf{r})$, its gradient may be written as

$$\nabla c_A(\mathbf{r}) = (\Delta c_A)\hat{\mathbf{n}}(s)\delta(u), \tag{22.7}$$

where $\Delta c_A = c_A^+ - c_A^-$, as defined earlier for the $1d$ system. Using this expression we may now write the nonlocal contribution to the free energy in the simpler form

$$-\int d^2 r d^2 r' \, (\nabla c_A(\mathbf{r})) \cdot G(\mathbf{r} - \mathbf{r}')(\nabla' c_A(\mathbf{r}')) \tag{22.8}$$

$$\approx -(\Delta c_A)^2 \int ds ds' \, [\hat{\mathbf{n}}(s) \cdot \hat{\mathbf{n}}(s')]G(\mathbf{R}(s) - \mathbf{R}(s'))$$

$$= -(\Delta c_A)^2 \int ds ds' \, [\hat{\mathbf{t}}(s) \cdot \hat{\mathbf{t}}(s')]G(\mathbf{R}(s) - \mathbf{R}(s')).$$

Finally, collecting the results in Eqs. (22.2), (22.5), and (22.8), we may express the free energy functional approximately as

$$\Delta \mathcal{F}_{nl}[c_A] \approx \sigma L + A\Delta V_\pm \tag{22.9}$$

$$- \frac{(\Delta c_A)^2}{2\kappa^2} \int ds ds' \, [\hat{\mathbf{t}}(s) \cdot \hat{\mathbf{t}}(s')]G(\mathbf{R}(s) - \mathbf{R}(s')).$$

Once again, the equation of motion for the interface can be constructed using the Lagrangian equations with the inertial terms neglected,

$$\frac{\delta \Delta \mathcal{F}}{\delta c_A} = -\frac{\delta \mathcal{R}_d}{\delta \dot{c}_A}, \tag{22.10}$$

where the Rayleigh dissipation function in two dimensions is a simple extension of that for the one-dimensional case,

$$\mathcal{R}_d = \frac{1}{2} \int d^2r \; \dot{c}_A^2. \tag{22.11}$$

Using the fact that $\dot{c}_A = \dot{\mathbf{R}}(s) \cdot \nabla c_A$, together with the expression for ∇c_A in Eq. (22.3) and the small curvature approximation, we obtain

$$\dot{c}_A^2 = (\dot{\mathbf{R}}(s) \cdot \hat{\mathbf{n}}(s))^2 \left(\frac{\partial c_A}{\partial u}\right)^2. \tag{22.12}$$

Substituting this expression into Eq. (22.11) for the Rayleigh dissipation function, and using the definition of the surface tension, we are able to write $\mathcal{R}_d$ in the form

$$\mathcal{R}_d = \frac{\sigma}{2D_A} \int ds \; (\dot{\mathbf{R}}(s) \cdot \hat{\mathbf{n}}(s))^2. \tag{22.13}$$

The functional derivative of $\mathcal{R}_d$ is now easily evaluated to give $\delta \mathcal{R}_d / \delta \dot{\mathbf{R}}(s) = \frac{\sigma}{D_A}(\dot{\mathbf{R}}(s) \cdot \hat{\mathbf{n}}(s))\hat{\mathbf{n}}(s)$.

To complete the calculation we need to evaluate the functional derivatives of the free energy functional in Eq. (22.9). Using the expression for the length of the interface of the c_A^+ domain as $L = \int_{\partial B} |d\mathbf{R}(s)|$, its functional derivative may be computed to be $\delta L / \delta \mathbf{R}(s) = K(s)\hat{\mathbf{n}}(s)$, where $K(s)$ is the curvature. Likewise, using $A = \int d^2r \theta_B(\mathbf{r})$ we find $\delta A / \delta \mathbf{R}(s) = \hat{\mathbf{n}}(s)$. The final quantity that must be computed is the functional derivative of the nonlocal term in the free energy functional (22.9). Let

$$\mathbf{N} \equiv \frac{\delta}{\delta \mathbf{R}(s)} \int ds ds' \; [\hat{\mathbf{n}}(s) \cdot \hat{\mathbf{n}}(s')] G(R(s,s')), \tag{22.14}$$

where we have used the fact that $\hat{\mathbf{t}}(s) \cdot \hat{\mathbf{t}}(s') = \hat{\mathbf{n}}(s) \cdot \hat{\mathbf{n}}(s')$ and defined $\mathbf{R}(s,s') = \mathbf{R}(s) - \mathbf{R}(s')$ whose magnitude is $|\mathbf{R}(s,s')| = R(s,s')$. Performing the functional derivative, we obtain

$$\mathbf{N} = \int ds' \; (\hat{\mathbf{t}}(s')\hat{\mathbf{t}}(s) - \hat{\mathbf{t}}(s) \cdot \hat{\mathbf{t}}(s')\mathbf{I}) \cdot \hat{\mathbf{R}}(s,s') G'(R(s,s')), \tag{22.15}$$

with G' the derivative of G with respect to its argument. The x and y components of N are given by

$$N_x = -\hat{t}_y(s) \int ds' \; (\hat{t}_x(s')\hat{R}_y(s,s') - \hat{t}_y(s')\hat{R}_x(s,s')) \cdot G'(R(s,s')),$$

$$\equiv -\hat{t}_y(s) \int ds' \; [\hat{t}(s') \times \hat{R}(s,s')]G'(R(s,s')),$$

$$N_y = \hat{t}_x(s) \int ds' \; (\hat{t}_x(s')\hat{R}_y(s,s') - \hat{t}_y(s')\hat{R}_x(s,s')) \cdot G'(R(s,s')),$$

$$\equiv \hat{t}_x(s) \int ds' \; [\hat{t}(s') \times \hat{R}(s,s')]G'(R(s,s')). \tag{22.16}$$

The second equalities in these equations introduce the two-dimensional cross product of vectors $\mathbf{a}$ and $\mathbf{b}$ as $\mathbf{a} \times \mathbf{b} = a_x b_y - a_y b_x$. Note that the two-dimensional cross product, even though denoted in boldface, is a scalar quantity. Using the fact that (see Eq. (7.10)) $-\hat{t}_y(s) = \hat{n}_x(s)$ and $\hat{t}_x(s) = \hat{n}_y(s)$, we obtain a compact expression for $\mathbf{N}$ in the form

$$\mathbf{N} = \hat{n}(s) \int ds' \; [\hat{t}(s') \times \hat{R}(s,s')]G'(R(s,s')). \tag{22.17}$$

The equation of motion for the interface is obtained by using Eq. (22.9) and the results for the functional derivatives computed above. We get

$$\dot{\mathbf{R}}(s) \cdot \hat{n}(s) = -\frac{D_A}{\sigma}\Big(K(s)\sigma + \Delta V_{\pm} \tag{22.18}$$

$$-\frac{(\Delta c_A)^2}{2\kappa^2} \int ds' \; [\hat{t}(s') \times \hat{R}(s,s')]G'(R(s,s'))\Big).$$

This equation shows that the curvature enters the evolution equation of the domain boundary. In addition, the nonlocal term is responsible for the repulsive interaction between two arc length positions with anti-parallel tangent vectors.

One may apply this equation to determine the evolution of the radius $R(t)$ of a small disk-shaped domain of c_A^+ in the c_A^- phase. The $2d$ Green function is given by

$$G(R) = \frac{\kappa}{2\pi\alpha}K_0(\kappa R), \tag{22.19}$$

whose derivative is

$$G'(R) = -\frac{\kappa^2}{2\pi\alpha}K_1(\kappa R), \tag{22.20}$$

where K_0 and K_1 are modified Bessel functions and $\kappa = (\alpha/\tilde{D}_B)^{1/2}$. Making use of the fact that $R(t)$ is small to evaluate the Green functions, and the fact that the curvature for a disk-shaped domain is $K(s) = 1/R$, we find

$$\frac{dR(t)}{dt} = \left(-D_A + \frac{D_A(\Delta c_A)^2}{4\alpha\kappa\sigma}\right)\frac{1}{R} - \frac{\Delta V_\pm D_A}{\sigma}. \qquad (22.21)$$

Note that $D_A/(\alpha\kappa\sigma) = \tilde{D}_B^{1/2}/\alpha^{3/2}$ is independent of D_A. The second term in the coefficient of R^{-1} on the right-hand side arises from nonlocal repulsive interactions. If there are no long-range interactions, the disk will always shrink, since $\Delta V_\pm > 0$. On the other hand, in the presence of long-range interactions, it is possible to stabilize a disk of the less stable phase in the more stable phase because of the nonlocal coupling from the B field.

22.2 Transverse instabilities of planar fronts

The boundary of the domain $\mathcal{B}$ may not be stable to inhomogeneous perturbations which are transverse to its length. In order to address this issue in a simple form, we consider whether a planar front in a $2d$ system is stable to inhomogeneous perturbations along its length. This is the analog of the front instability problem considered in Chapters 19 and 20. To begin this analysis we consider a system with an initially planar front along the x direction and propagating along the y direction. Such a planar front will move with constant speed given by $D_A\Delta V_\pm/\sigma$. This follows from Eq. (22.18), since the terms on the right-hand side that depend on the curvature and nonlocal interactions vanish for a planar front. As in earlier discussions of propagating fronts of this type, in a coordinate system moving with this velocity, the front appears stationary. We shall assume that such a change of coordinates has been made. We then suppose the front is subject to a time-dependent perturbation at each point x in the y direction and determine whether the perturbation decays or grows.

Given this situation we may write the location of the interface as

$$\mathbf{R}(x) = x\hat{\mathbf{x}} + \chi(x, t)\hat{\mathbf{y}}, \qquad (22.22)$$

where $\chi(x, t)$ is the deviation from the planar profile at x, and $\hat{\mathbf{x}}$ and $\hat{\mathbf{y}}$ are unit vectors along x and y, respectively. In writing this form we have used x to parameterize the interface curve. The interfacial curve has tangent vector,

$$\hat{\mathbf{t}}(x) = \frac{d\mathbf{R}(x)}{dx} = \hat{\mathbf{x}} + \frac{d\chi(x, t)}{dx}\hat{\mathbf{y}}, \qquad (22.23)$$

and the curvature $K(x)$ can be determined from

$$-K(x)\hat{\mathbf{n}}(x) = \frac{d\mathbf{t}(x)}{dx} = \frac{d^2\chi(x,t)}{dx^2}\hat{\mathbf{y}}. \tag{22.24}$$

The vector distance between any two points on the interface is given by

$$\mathbf{R}(x,x') = \mathbf{R}(x) - \mathbf{R}(x') = (x - x')\hat{\mathbf{x}} + (\chi(x,t) - \chi(x',t))\hat{\mathbf{y}}. \tag{22.25}$$

For small deviations from planarity the magnitude of this difference is simply given by $R(x,x') = |\mathbf{R}(x,x')| \approx |x - x'|$.

We may specialize Eq. (22.18) to this case to find the following evolution equation for the profile:

$$\dot{\mathbf{R}}(x) \cdot \hat{\mathbf{n}}(x) = -D_A K(x) \tag{22.26}$$
$$+ \frac{D_A(\Delta c_A)^2}{2\sigma\kappa^2} \int dx'\, [\hat{\mathbf{t}}(x') \times \hat{\mathbf{R}}(x,x')]G'(R(x,x')).$$

We investigate the stability of an interface satisfying this equation by considering the Fourier components of the deviation from planarity, and take $\chi(x,t) = \chi_k(t)\cos(kx)$. In this case $\hat{\mathbf{t}}(x) = \hat{\mathbf{x}} - k\chi_k(t)\sin(kx)\hat{\mathbf{y}}$ and $K(x) = k^2\chi_k(t)\cos(kx)$. Using these expressions, the two-dimensional cross product appearing in the evolution equation may be calculated to give

$$\hat{\mathbf{t}}(x') \times \hat{\mathbf{R}}(x,x') = \hat{\mathbf{t}}_x(x')\hat{\mathbf{R}}_y(x,x') - \hat{\mathbf{t}}_y(x')\hat{\mathbf{R}}_x(x,x')$$
$$\approx |x - x'|^{-1}\chi_k(t)\Big(\cos(kx) - \cos(kx')$$
$$+ k(x - x')\sin(kx')\Big). \tag{22.27}$$

We now have all the ingredients to evaluate Eq. (22.26), which takes the form

$$\frac{d\chi_k(t)}{dt} = \left(-D_A k^2 + \frac{D_A(\Delta c_A)^2}{2\sigma\kappa^2}\int_{-\infty}^{\infty} dx'\,|x - x'|^{-1}\right. \tag{22.28}$$
$$\left.\times \left[1 - \frac{\cos(kx')}{\cos(kx)} + k(x - x')\frac{\sin(kx')}{\cos(kx)}\right]G'(|x - x'|)\right)\chi_k(t).$$

The integral on the right-hand side of this equation may be evaluated by converting the x' integral into an integral with $-\infty \leq x' \leq x$ where $x > x'$, and letting $z = x - x'$; and an integral with $x \leq x' \leq \infty$ where $x' \geq x$, and letting $z = x' - x$. We then find

$$\int_{-\infty}^{\infty} dx'\,|x - x'|^{-1}\left[1 - \frac{\cos(kx')}{\cos(kx)} + k(x - x')\frac{\sin(kx')}{\cos(kx)}\right]G'(|x - x'|)$$
$$= 2\int_0^{\infty} dz\,\frac{1}{z}(kz\sin(kz) + \cos(kz) - 1)G'(z). \tag{22.29}$$

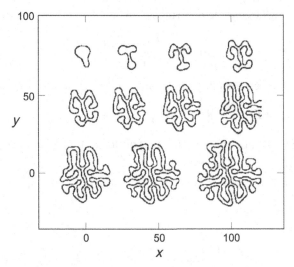

Fig. 22.3. Growth of a labyrinthine pattern from a distorted disk as a result of a transverse front instability. Reprinted Figure 19 with permission from Goldstein *et al.* (1996). Copyright 1996 by the American Physical Society.

Finally, we use Eqs. (22.19) and (22.20) for $G(z)$ and $G'(z)$. With these results, the linearized equation of motion for $\chi_k(t)$ may be written as

$$
\frac{d\chi_k(t)}{dt} = \left(-D_A k^2 + \frac{D_A(\Delta c_A)^2}{4\pi\alpha\sigma} \int_{-\infty}^{\infty} dz \, [kz\sin(kz) \right.
$$
$$
\left. + \cos(kz) - 1]\frac{1}{z}K_1(\kappa z) \right) \chi_k(t)
$$
$$
\equiv \lambda_k \chi_k(t).
\tag{22.30}
$$

The stability of the planar front can be determined by analyzing the growth rate λ_k. If one evaluates the growth rate to lowest order in k, one finds $\lambda_k = D_A(-1 + (\Delta c_A)^2/(4\alpha\sigma\kappa^2))k^2 + \mathcal{O}(k^4)$. Thus, the nonlocal force can lead to a transverse destabilization of a planar front.

If one starts with a stripe of the c_A^+ phase in the c_A^- phase, or other initial domain shapes, transverse instabilities can lead to labyrinthine patterns. For example, Fig. 22.3 shows the growth of a labyrinthine pattern from a distorted disk initial condition as a result of a transverse front instability.

For certain values of the parameters where the compact disk is stable, instead of growth of a labyrinthine pattern from an arbitrary initial condition, one may observe shrinkage of a labyrinthine pattern to a stable disk. The labyrinth shrinks by contracting its free ends along its arms until a compact structure is obtained. Such evolution is shown in Fig. 22.4.

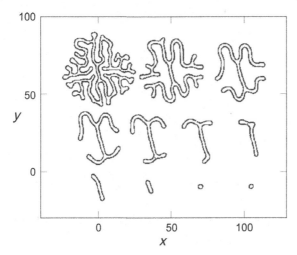

Fig. 22.4. Shrinkage of a labyrinthine pattern to a disk using the front equations. Reprinted Figure 20 with permission from Goldstein *et al.* (1996). Copyright 1996 by the American Physical Society.

22.3 Three-dimensional patterns

In three-dimensional geometries front repulsion and front instabilities can lead to new types of pattern formation. Some of the patterns are simple three-dimensional generalizations of those already discussed in one and two dimensions. For example, a ball of the less stable c_A^+ phase in a sea of the more stable c_A^- phase is the $3d$ analog of a stable $1d$ "bump" or a $2d$ disk. The existence of a stable $3d$ ball can be understood from a generalization to $3d$ of the results presented in the previous sections for one and two dimensions. In the sharp interface approximation, the concentration field for a ball with radius R has the form $c_A(\mathbf{r}) = c_A^+ \theta(R - r) + c_A^- \theta(r - R)$. From this expression it follows that $\nabla c_A(\mathbf{r}) \approx -(\Delta c_A)\hat{\mathbf{n}}\delta(r - R)$. Using these results in Eq. (21.12), we may write the free energy functional as

$$\Delta \mathcal{F}_{nl}[c_A] = \Delta V_{\pm} V + \sigma A \tag{22.31}$$
$$- \frac{(\Delta c_A)^2}{2\kappa^2} R^4 \int d\hat{S} \int d\hat{S}' \, [\hat{\mathbf{n}}(S) \cdot \hat{\mathbf{n}}(S')] G(|\mathbf{R}(S) - \mathbf{R}(S')|),$$

where S and S' are points on the surface of the ball and $d\hat{S}$ and $d\hat{S}'$ are differential surface elements. The free energy functional may be evaluated by inserting the expression for the $3d$ Green function,

$$G(\mathbf{r}) = \frac{\kappa^2}{4\pi\alpha r} e^{-\kappa r}, \tag{22.32}$$

and computing the integral by letting S define the polar axis. We find

$$\Delta \mathcal{F}_{nl}[c_A] = \Delta V_{\pm} V + \sigma A - \frac{2\pi (\Delta c_A)^2 R^3}{\alpha} \tag{22.33}$$
$$\times \left[\frac{1}{2\kappa R} - \frac{4}{(2\kappa R)^3} + e^{-2\kappa R} \left(\frac{4}{(2\kappa R)^3} + \frac{4}{(2\kappa R)^2} + \frac{1}{2\kappa R} \right) \right].$$

For small R this simplifies to

$$\Delta \mathcal{F}_{nl}[c_A] = \left(\Delta V_{\pm} - \frac{(\Delta c_A)^2}{2\alpha} \right) V + \sigma A. \tag{22.34}$$

The equation of motion for the ball radius follows from the 3d Lagrangian equations $d\mathcal{F}_{nl}/dR = -d\mathcal{R}_d/d\dot{R}$, and we obtain,

$$\frac{dR(t)}{dt} = -\frac{D_A}{\sigma} \left(\Delta V_{\pm} - \frac{(\Delta c_A)^2}{2\alpha} \right) - \frac{2D_A}{R}. \tag{22.35}$$

Since $\Delta V_{\pm} > 0$ for the less stable phase c_A^+ in the more stable c_A^- sea, a stable ball is only possible due to the existence of the term $(\Delta c_A)^2/(2\alpha)$, which arises from the repulsive interaction. The radius R^* of the stationary ball is given by

$$R^* = \frac{4\sigma\alpha}{(\Delta c_A)^2 - 2\alpha \Delta V_{\pm}}, \tag{22.36}$$

and its stability may be determined from a linear stability analysis of Eq. (22.35).

Thus far we have restricted the examples of pattern formation in systems with competing interactions arising from rapid inhibitor diffusion to cases where the inhibitor kinetics is fast (ϵ large). As discussed earlier, similar patterns can be found in the slow inhibitor limit (ϵ small), although the analysis can no longer be carried out in terms of a nonlocal free energy functional. For example, a stable 3d ball obtained under these conditions is shown in Fig. 22.5.

The basic building block for many 3d chemical patterns is a tubular domain of the less stable phase in the more stable phase, the analog of a "stripe" domain in two dimensions. Depending on the system parameters, the tubular domain may either grow in length or shrink to a ball, as shown in Fig. 22.5(a). If the parameters are such that transverse instabilities can occur, the instability can lead to fingering which results in 3d labyrinthine patterns, as shown in Fig. 22.5(b).

All of these patterns are 3d versions of their lower-dimensional analogs; however, in three-dimensional systems new stable patterns can appear that have no such analogs. A new class of chemical bistable states can be found in these systems as a result of stabilization due to the topological structure of the pattern.

(a)

(b)

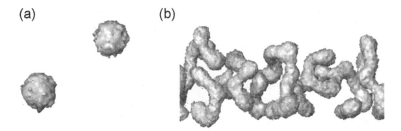

Fig. 22.5. (a) Stable balls of the c_A^+ phase in the c_A^- sea in the slow inhibitor limit of the FHN equation. (b) Growth of a tubular domain accompanied by fingering instabilities leading to the formation of a labyrinthine pattern. Reprinted with permission from Malevanets and Kapral (1998). Copyright 1998 World Scientific Publishing Company.

(a)   (b)

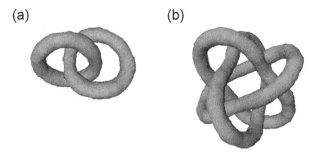

Fig. 22.6. (a) Hopf and (b) Borromean links. This and the following figure show isoconcentration surfaces corresponding to the activator.

(a)   (b)

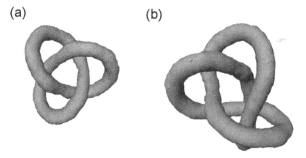

Fig. 22.7. (a) Trefoil (3_1) and (b) figure-8 (4_1) knots.

Suppose the system parameters are such that a tubular domain shrinks by contracting at its free ends while maintaining a fixed diameter. This is the $3d$ analog of the contraction shown in Fig. 22.4. This process will lead to a stable ball, as described above. However, if the tubular domain has no free ends but is a completely connected domain in three-dimensional space, the entire continuous tubular domain will contract. Such contraction may again yield a ball, but it is possible that the local repulsion between different segments of the tubular domain that arises from the rapidly diffusing inhibitor species will be sufficiently strong to prevent collapse: thus, the topology of the domain structure will be preserved.

Examples of topologically stabilized patterns are given in Fig. 22.6, which shows stable Hopf and Borromean links, while Fig. 22.7 shows stable trefoil (3_1) and figure-eight (4_1) knots. The results were obtained from reactive lattice-gas simulations of the FHN model starting from initial conditions composed of straight segments of tubular domains joined to form structures with these topologies (Malevanets and Kapral, 1996). Thus, topology in conjunction with long-range repulsive interactions can lead to the formation of new types of chemical pattern.

23

Turing patterns

A Turing pattern forms when a spatially homogeneous steady state, which is stable to small *spatially homogeneous* perturbations, loses its stability to small *spatially inhomogeneous* perturbations. The mechanism responsible for such instabilities was first described by Turing (1952), in his paper *The chemical basis of morphogenesis*, as a model for pattern formation in biology. The appearance of Turing patterns relies on the interplay between autocatalytic chemical kinetics and diffusion. The basic Turing mechanism can be described in terms of the kinetics of two chemical species termed the activator and the inhibitor. The activator tends to increase the production of chemical species while the inhibitor tends to inhibit such concentration growth. A Turing pattern can form if the diffusion coefficient of the inhibitor is much greater than that of the activator. While there is still controversy over the role of Turing patterns in morphogenesis, these patterns have been unambiguously identified in chemically reacting media.

The formation of a chemical Turing pattern in a continuously fed unstirred reactor was reported by Castets *et al.* (1990). The chlorite–iodide–malonic acid system was studied in a thin gel reactor schematically depicted in Fig. 23.1. The top and bottom sides of the thin hydrogel in which the reaction takes place are in contact with reservoirs containing chemical reagents. The chemical species diffuse into the gel, and reaction takes place in a thin layer within the gel shown in the center panel of the figure. Within this reaction zone a stationary inhomogeneous periodic pattern of chemical concentrations develops, as seen in the right panel of the figure. The chemical pattern was visualized by loading the gel with a soluble starch-like indicator, which changes from yellow to blue as a function of the concentration ratio $[I_3^-]/[I_2]$ in the course of the CIMA reaction.

Lengyel and Epstein (1991) were able to model Turing pattern formation in this reaction by a simple two-variable model involving $[I^-]$ and $[ClO_2^-]$ chemical species, which act as the activator and inhibitor, respectively. An important feature of this model was the fact that the $[I^-]$ species can complex with the

201

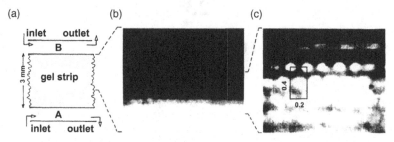

Fig. 23.1. Turing pattern formation in the CIMA reaction. (a) Schematic picture of the gel reactor; (b) the reaction zone and (c) Turing pattern. Reprinted Figure 1 with permission from Castets *et al.* (1990). Copyright 1990 by the American Physical Society.

starch in the gel, and this has the effect of significantly reducing the diffusion coefficient of the [I$^-$] species so that Turing pattern formation is favored. Turing patterns have been extensively studied in the CIMA reaction (Ouyang and Swinney, 1991). Figure 23.2 shows a variety of Turing patterns obtained under different experimental conditions. More recently, Turing patterns have been observed in the Belousov–Zhabotinsky reaction (Belousov, 1958; Zhabotinsky, 1964) carried out in a microemulsion (Vanag and Epstein, 2001). In this case water-soluble species confined to small reverse micelles diffuse more slowly than species that reside in the organic phase. This diffusion coefficient difference is probably responsible for the appearance of Turing patterns in this system. Liesegang precipitation patterns have been interpreted in terms of a Turing bifurcation by Flicker and Ross (1974).

23.1 Turing bifurcation conditions

In order to analyze how Turing patterns form, we study systems described by a set of reaction–diffusion equations of the form

$$\frac{\partial c_A}{\partial t} = \mathcal{R}_A(c_A, c_B) + D_A \nabla^2 c_A,$$

$$\frac{\partial c_B}{\partial t} = \mathcal{R}_B(c_A, c_B) + D_B \nabla^2 c_B. \tag{23.1}$$

We first consider a spatially homogeneous system where the diffusion terms do not appear, and study the homogeneous steady states of Eq. (23.1) given by the solutions of $\mathcal{R}_A(c_A^s, c_B^s) = \mathcal{R}_B(c_A^s, c_B^s) = 0$. A linear stability analysis can be performed to determine the stability of these steady states to *homogeneous* perturbations. Letting $\delta c_A(t) = c_A(t) - c_A^s$ and $\delta c_B(t) = c_B(t) - c_B^s$, and linearizing the kinetic equation,

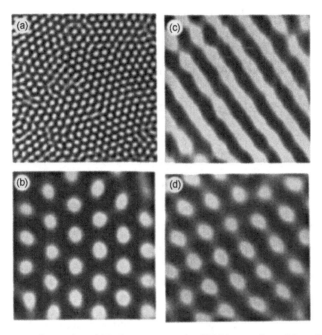

Fig. 23.2. Two-dimensional Turing patterns in CIMA reaction. (a) and (b) show hexagonal patterns, (c) is a striped pattern with evidence of a transverse instability, while the pattern in panel (d) is a mixed pattern composed of spots arranged in stripes. Reprinted by permission from Ouyang and Swinney (1991). Copyright 1991 Macmillan Publishers Ltd.

we find

$$\frac{d\delta\mathbf{c}(t)}{dt} = \mathbf{M}\delta\mathbf{c}(t), \tag{23.2}$$

where

$$\mathbf{M} = \begin{pmatrix} \mathcal{R}_{AA} & \mathcal{R}_{AB} \\ \mathcal{R}_{BA} & \mathcal{R}_{BB} \end{pmatrix}, \tag{23.3}$$

and the derivatives are evaluated at the steady state: e.g. $\mathcal{R}_{AA} = (\partial \mathcal{R}_A/\partial c_A)|_{c_A^s, c_B^s}$. The roots of $\mathbf{M}$ are given by

$$\omega = \frac{1}{2}\text{Tr } \mathbf{M} \pm \frac{1}{2}\left\{(\text{Tr } \mathbf{M})^2 - 4\det \mathbf{M}\right\}^{1/2}. \tag{23.4}$$

For linear stability of the steady state we must have $\Re\omega < 0$, which, in turn, implies that

$$\text{Tr } \mathbf{M} < 0, \quad \text{and} \quad \det \mathbf{M} > 0. \tag{23.5}$$

This imposes conditions on the reaction kinetics.

To determine the stability of the homogeneous steady states to *inhomogeneous* perturbations, we linearize the full reaction–diffusion equation. Letting $\delta c_A(\mathbf{r}, t) = c_A(\mathbf{r}, t) - c_A^s$ and $\delta c_B(\mathbf{r}, t) = c_B(\mathbf{r}, t) - c_B^s$, linearization yields

$$\frac{d\delta\mathbf{c}(\mathbf{r}, t)}{dt} = \mathbf{L}\delta\mathbf{c}(\mathbf{r}, t), \tag{23.6}$$

where $\mathbf{L} = \mathbf{M} + \mathbf{D}\nabla^2$. It is convenient to Fourier-transform this equation, letting, for example, $\hat{\mathbf{c}}(\mathbf{k}) = \int_V d\mathbf{r} e^{-i\mathbf{k}\cdot\mathbf{r}}\mathbf{c}(\mathbf{r})$. The (k-dependent) roots are given by Eq. (23.4) with $\mathbf{M}$ replaced by $\mathbf{L}(k) = \mathbf{M} - \mathbf{D}k^2$. For instability to inhomogeneous ($k \neq 0$) perturbations we must have $\Re\omega(k) > 0$. This implies that Tr $\mathbf{L}(k) > 0$ or $\det \mathbf{L}(k) < 0$ or both. Since Tr $\mathbf{M} < 0$ from the condition that the homogeneous state is stable, we have Tr $\mathbf{L}(k) < 0$. Hence, a necessary condition for the Turing instability is $\det \mathbf{L}(k) < 0$. Using this condition and the fact that $\det \mathbf{M} > 0$ we have

$$\mathcal{R}_{AA}D_B + \mathcal{R}_{BB}D_A > 0. \tag{23.7}$$

Since Tr $\mathbf{M} = \mathcal{R}_{AA} + \mathcal{R}_{BB} < 0$, we have $D_B/D_A \neq 1$, and the bifurcation occurs when the diffusion coefficients of the two species are unequal.

The inequality, Eq. (23.7), does not completely determine the bifurcation. A Turing bifurcation occurs when $\Re\omega(k)$ passes through zero for some finite value of $k = k_c$ and $\det \mathbf{L}(k) = 0$. Since the bifurcation must occur at finite k, we must also have $d\det \mathbf{L}(k)/dk^2 = 0$ at $k = k_c$, yielding

$$k_c^2 = \frac{(\mathcal{R}_{AA}D_B + \mathcal{R}_{BB}D_A)}{2D_A D_B}, \tag{23.8}$$

for the critical wavenumber. The diffusion coefficient ratio at the bifurcation point is determined from the condition $\det \mathbf{L}(k_c) = 0$, giving

$$4D_A D_B\det \mathbf{M} - (\mathcal{R}_{AA}D_B + \mathcal{R}_{BB}D_A)^2 = 0. \tag{23.9}$$

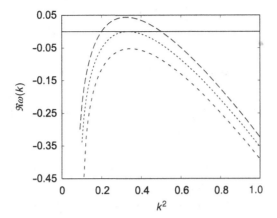

Fig. 23.3. Plot of $\Re\omega(k)$ versus k^2 for $\delta = D_B/D_A = \delta_c$, the critical diffusion ratio (middle line) and for two other values of δ lying above and below δ_c. The upper line is for $\delta > \delta_c$ while the lower line is for $\delta < \delta_c$.

With these results in hand we may sketch the picture of the bifurcation in Fig. 23.3. This figure shows $\Re\omega(k)$ versus k^2 for several values of the ratio of diffusion coefficients, with the kinetic parameters held fixed. One can see that at the bifurcation point a spatial mode with wavenumber k_c becomes unstable, corresponding to an intrinsic spatial scale $\ell_T = 2\pi/k_c$, which is determined by the kinetic parameters and diffusion coefficients of the species.

Consideration of a specific case will help to illustrate the nature of the bifurcation. The Turing bifurcation conditions imply that $\mathcal{R}_{AA}$ and $\mathcal{R}_{BB}$ must have opposite signs. Suppose that $\mathcal{R}_{AA} > 0$ and $\mathcal{R}_{BB} < 0$, and the system is of activator–inhibitor type with $\mathcal{R}_{AB} < 0$ and $\mathcal{R}_{BA} > 0$: A is the activator, since an increase in A will lead to production of chemical species; B is the inhibitor, since an increase in B will lead to suppression of chemical species growth. In this case, from Eq. (23.7) and the homogeneous stability condition Tr $M < 0$, we see that $D_B > D_A$, and the instability is driven by a large diffusion of the inhibitor compared with that of the activator. One may imagine growth in a local region of space triggered by an inhomogeneity in the activator concentration. This will produce a growth in the inhibitor which will diffuse rapidly out of the region and suppress concentration growth in nearby spatial regions. In this way one can see how the stable homogeneous state can become unstable in the presence of fast inhibitor diffusion.

From Fig. 23.3, we see that a band of unstable wavenumbers exists for $D_B/D_A > \delta_c$. In order to determine the nature of the chemical pattern that forms we need to go beyond the linear stability regime and carry out a nonlinear analysis to find the wavevectors that are selected and determine the stationary pattern.

23.2 Pattern selection

While the preceding analysis predicts the existence of a Turing bifurcation for parameters satisfying the Turing conditions, it does not provide information on the nature of the chemical pattern that forms. In order to determine the actual form of the Turing pattern, we must keep the nonlinear terms which were dropped in the above analysis (Pismen, 1980; Manneville, 1990; Cross and Hohenberg, 1993).

The unstable modes are responsible for the instability of the homogeneous steady state to inhomogeneous perturbations since they grow exponentially. We shall now derive equations for the amplitudes of these unstable modes near the bifurcation point. These amplitudes will constitute the order parameters for this problem.

We write the reaction–diffusion equation in the general form

$$\frac{\partial \mathbf{c}}{\partial t} = \mathcal{R}(\mathbf{c}, \mu) + \mathbf{D}\nabla^2 \mathbf{c}, \tag{23.10}$$

where, as earlier, $\mathbf{c}$ is a column vector of concentrations, $\mathbf{c}^T = (c_A, c_B)$. Here μ is a control parameter that may be varied to tune the system through the Turing bifurcation. We want to extract the behavior of the amplitudes of the unstable modes. Let $\mathbf{c}_0(\mu)$ be the homogeneous steady solution, $\mathcal{R}(\mathbf{c}_0(\mu); \mu) = 0$, for any μ. At the Turing bifurcation point $\mu = \mu_0$, an eigenvalue of the linearized problem vanishes: $\omega(k_c) = 0$ for $k_c \neq 0$ and $\omega(k) < 0$ for $k \neq k_c$. More specifically, the linearized reaction–diffusion equation is

$$\frac{\partial \delta \mathbf{c}(\mathbf{r}, t)}{\partial t} = \left(\frac{\delta \mathcal{R}}{\delta \mathbf{c}_0} + \mathbf{D}\nabla^2 \right) \delta \mathbf{c}(\mathbf{r}, t) = \mathbf{L}\delta \mathbf{c}(\mathbf{r}, t), \tag{23.11}$$

with $\delta \mathbf{c} = \mathbf{c} - \mathbf{c}_0$. In Fourier space we have

$$\frac{\partial \delta \mathbf{c}(\mathbf{k}, t)}{\partial t} = \left(\frac{\delta \mathcal{R}}{\delta \mathbf{c}_0} - \mathbf{D}k^2 \right) \delta \mathbf{c}(\mathbf{k}, t) = \mathbf{L}(k)\delta \mathbf{c}(\mathbf{k}, t). \tag{23.12}$$

The right eigenvalue problem is

$$\mathbf{L}(k)\mathbf{e}(k) = \omega(k)\mathbf{e}(k), \tag{23.13}$$

while the corresponding adjoint eigenvalue problem is

$$\mathbf{L}^\dagger(k)\mathbf{e}^\dagger(k) = \omega(k)\mathbf{e}^\dagger(k). \tag{23.14}$$

To carry out the analysis we introduce a small parameter λ that gauges the distance from the bifurcation point and expand both the concentrations and bifurcation parameter in this parameter:

$$\mathbf{c} = \mathbf{c}_0 + \lambda \mathbf{c}_1 + \lambda^2 \mathbf{c}_2 + \cdots = \sum_{i=0}^{\infty} \lambda^i \mathbf{c}_i, \qquad (23.15)$$

and

$$\mu = \mu_0 + \lambda \mu_1 + \lambda^2 \mu_2 + \cdots = \sum_{i=0}^{\infty} \lambda^i \mu_i. \qquad (23.16)$$

We must also introduce multiple time scales to account for slowing down near the bifurcation point:

$$\frac{\partial}{\partial t} = \lambda \frac{\partial}{\partial \tau_1} + \lambda^2 \frac{\partial}{\partial \tau_2} + \cdots \qquad (23.17)$$

We expand the reaction–diffusion equation about $\mathbf{c}_0(\mu)$ and about $\mu = \mu_0$. To do this we may use the formal Taylor expansion,

$$\mathcal{R}(\mathbf{c}, \mu) = \sum_{m,n} (m!n!)^{-1} (\mu - \mu_0)^m \mathbf{R}_{mn} \odot (\mathbf{c} - \mathbf{c}_0)^n. \qquad (23.18)$$

In this equation $\odot$ stands for tensor contraction, where, for nth rank tensors $\mathbf{A}$ and $\mathbf{B}$, $\mathbf{A} \odot \mathbf{B} = \sum_{i_1, i_2, \ldots, i_n} A_{i_1 i_2 \ldots i_n} B_{i_n i_{n-1} \ldots i_1}$. The tensor $\mathbf{R}_{mn}$ is defined as

$$\mathbf{R}_{mn} = \left(\frac{d^m}{d\mu^m} \left(\frac{\delta^n \mathcal{R}}{\delta \mathbf{c}^n} \right)_{\mathbf{c}_0(\mu)} \right)_{\mu_0}. \qquad (23.19)$$

Now we can collect terms in λ. The equations up to order λ^3 are

$$\lambda : \mathbf{Lc}_1 \equiv \left(\mathbf{R}_{01} + \mathbf{D}\nabla^2 \right) \cdot \mathbf{c}_1 = \partial_{\tau_1} \mathbf{c}_0 = 0, \qquad (23.20)$$

$$\lambda^2 : \mathbf{Lc}_2 = \partial_{\tau_1} \mathbf{c}_1 - \mu_1 \mathbf{R}_{11} \cdot \mathbf{c}_1 - \frac{1}{2} \mathbf{R}_{02} \odot \mathbf{c}_1 \mathbf{c}_1,$$

$$\lambda^3 : \mathbf{Lc}_3 = \partial_{\tau_1} \mathbf{c}_2 + \partial_{\tau_2} \mathbf{c}_1 - \frac{1}{2} \mu_1 \mathbf{R}_{12} \odot \mathbf{c}_1 \mathbf{c}_1 - \frac{1}{6} \mathbf{R}_{03} \odot \mathbf{c}_1 \mathbf{c}_1 \mathbf{c}_1$$

$$- \frac{1}{2} \mu_1^2 \mathbf{R}_{21} \cdot \mathbf{c}_1 - \mu_2 \mathbf{R}_{11} \cdot \mathbf{c}_1 - \mu_1 \mathbf{R}_{11} \cdot \mathbf{c}_2 - \mathbf{R}_{02} \odot \mathbf{c}_1 \mathbf{c}_2.$$

The first equation is homogeneous and has a nontrivial kernel which is proportional to the eigenvectors of $\mathbf{L}$ with zero eigenvalue. The remaining problems are inhomogeneous. Since the left-hand sides have a nontrivial kernel then, according to the Fredholm theorem, they have no solutions unless the right-hand sides are orthogonal to the kernel of $\mathbf{L}$.

23.3 Steady-state patterns

The steady-state inhomogeneous solutions of this set of equations are obtained by setting the time derivatives equal to zero. We have

$$\lambda : L\mathbf{c}_1 = 0, \tag{23.21}$$

$$\lambda^2 : L\mathbf{c}_2 = -\mu_1 \mathbf{R}_{11} \cdot \mathbf{c}_1 - \frac{1}{2} \mathbf{R}_{02} \odot \mathbf{c}_1 \mathbf{c}_1,$$

$$\lambda^3 : L\mathbf{c}_3 = -\frac{1}{2}\mu_1 \mathbf{R}_{12} \odot \mathbf{c}_1 \mathbf{c}_1 - \frac{1}{6} \mathbf{R}_{03} \odot \mathbf{c}_1 \mathbf{c}_1 \mathbf{c}_1$$

$$- \frac{1}{2}\mu_1^2 \mathbf{R}_{21} \cdot \mathbf{c}_1 - \mu_2 \mathbf{R}_{11} \cdot \mathbf{c}_1 - \mu_1 \mathbf{R}_{11} \cdot \mathbf{c}_2 - \mathbf{R}_{02} \odot \mathbf{c}_1 \mathbf{c}_2.$$

The right eigenfunctions of L may be written as

$$\mathbf{e}(\mathbf{r}) = \mathbf{e}(k)e^{i\mathbf{k}\cdot\mathbf{r}}, \tag{23.22}$$

where $\mathbf{e}(k)$ is the right eigenfunction of the problem

$$L(k)\mathbf{e}(k) = 0, \tag{23.23}$$

and $\mathbf{e}^{\dagger}(k)$ is the corresponding left eigenfunction. The function $\mathbf{c}_1$ must be proportional to $\mathbf{e}(\mathbf{r})$ in view of the order-λ equation (23.21), and we may write

$$\lambda\mathbf{c}_1(\mathbf{r}, t) = \mathbf{e}(k_c) \left[\sum_j A_j(t)e^{i\mathbf{k}_j \mathbf{r}_j} + c.c. \right]. \tag{23.24}$$

The complex conjugate is added since $\mathbf{c}_1$ is real (a concentration). The sum on j runs over all directions of $\mathbf{r}$, and $|\mathbf{k}| = k_c$, since we evaluate the operator at the bifurcation point. The A_j are amplitudes whose equations we need to determine.

23.4 Stripe patterns

In order to complete the analysis of these equations we must account for the possible symmetries of the patterns in order to find the equations for the amplitudes. The stabilities of the patterns can then be determined using these amplitude equations. To illustrate the method in a simple context, we restrict our considerations to the case of stripe solutions where the concentration is homogeneous along y and varies only along x. In this case we have a one-dimensional problem, and the concentration field takes the form

$$\mathbf{c}_1(x) = 2\mathbf{e}(k_c) \cos(k_c x), \tag{23.25}$$

where the steady-state amplitude A_s has been identified with the parameter λ. Since $\mathbf{e}^\dagger(k)$ is a left eigenfunction of $\mathbf{L}(k)$, multiplying from the left by $\mathbf{e}^\dagger$, the order-λ^2 equation gives

$$0 = -4\mathbf{e}^\dagger \cdot \mathbf{R}_{11} \cdot \mathbf{e}\mu_1 \cos^2(k_c x) - 4\mathbf{e}^\dagger \cdot \mathbf{R}_{02} \odot \mathbf{ee} \cos^3(k_c x). \quad (23.26)$$

Integration over the spatial domain,

$$\frac{k_c}{2\pi} \int_0^{2\pi/k_c} dx \cdots, \quad (23.27)$$

yields $\mathbf{e}^\dagger \cdot \mathbf{R}_{11} \cdot \mathbf{e}\mu_1 = 0$, from which we conclude that $\mu_1 = 0$ if $\mathbf{e}^\dagger \cdot \mathbf{R}_{11} \cdot \mathbf{e} \neq 0$. We also find that $\mathbf{e}^\dagger \cdot \mathbf{R}_{02} \odot \mathbf{ee} = 0$. Using this result in the order-λ^2 equation we conclude that $\mathbf{c}_2 \propto \mathbf{c}_1$, so we may set $\mathbf{c}_2 = 0$. The order-λ^3 equation now reads

$$\mathbf{L}\mathbf{c}_3 = -\frac{1}{6}\mathbf{R}_{03} \odot \mathbf{c}_1\mathbf{c}_1\mathbf{c}_1 - \mu_2\mathbf{R}_{11} \cdot \mathbf{c}_1. \quad (23.28)$$

Substitution of the expression for $\mathbf{c}_1$ and multiplication from the left with $\mathbf{c}_1^\dagger$ yields

$$0 = -\frac{8}{3}\mathbf{e}^\dagger \cdot \mathbf{R}_{03} \odot \mathbf{eee} \cos^4(k_c x) - 4\mathbf{e}^\dagger \cdot \mathbf{R}_{11} \cdot \mathbf{e} \cos^2(k_c x)\mu_2. \quad (23.29)$$

After integration over the spatial domain it follows from this equation that

$$\mu_2 = -\frac{\mathbf{e}^\dagger \cdot \mathbf{R}_{03} \odot \mathbf{eee}}{2\mathbf{e}^\dagger \cdot \mathbf{R}_{11} \cdot \mathbf{e}} \equiv g. \quad (23.30)$$

Through this procedure we have determined the values of μ_1 and μ_2. To order λ^3 we have $\mu = \mu_0 + g\lambda^2$ and

$$\lambda \approx \pm \left(\frac{\mu - \mu_0}{g}\right)^{1/2}, \quad (23.31)$$

so that the solution for the concentration field reads

$$\mathbf{c}(\mathbf{r}) \approx \mathbf{c}_0 \pm 2\left(\frac{\mu - \mu_0}{g}\right)^{1/2} \mathbf{e}(k_c) \cos(k_c x). \quad (23.32)$$

Note that since $\mu_1 = 0$ the bifurcation is either supercritical ($g > 0$) or subcritical ($g < 0$). The steady-state amplitude is given by $A_s = \pm(\frac{\mu - \mu_0}{g})^{1/2}$, which is sketched in Fig. 23.4 as a function of the parameter μ for the supercritical case. We see that the amplitude of the stripe pattern grows parabolically with the distance from the bifurcation. An example of a stripe Turing pattern is shown in Fig. 23.5.

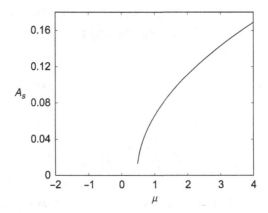

Fig. 23.4. Sketch of A_s versus μ for the stripe solution.

(a)   (b)

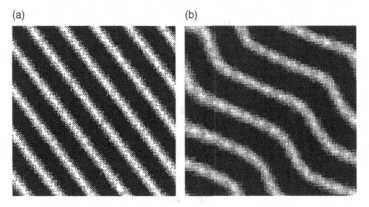

Fig. 23.5. (a) Stripe Turing pattern simulated using the Brusselator model; (b) a zigzag Turing pattern that arises from a sideband instability. Reprinted from P. Borckmans *et al.* (1992). Copyright 1992, with permission from Elsevier.

23.5 Time-dependent amplitude equation for stripe patterns

Next, we consider the time-dependent solutions with stripe symmetry. Using the results obtained above for the μ coefficient and the fact that $\mathbf{c}_2 = 0$, the evolution equations read:

$$\lambda : \mathbf{L}\mathbf{c}_1 = 0, \tag{23.33}$$
$$\lambda^2 : \mathbf{L}\mathbf{c}_2 = 0 = \partial_{\tau_1}\mathbf{c}_1,$$
$$\lambda^3 : \mathbf{L}\mathbf{c}_3 = \partial_{\tau_2}\mathbf{c}_1 - \frac{1}{6}\mathbf{R}_{03} \odot \mathbf{c}_1\mathbf{c}_1\mathbf{c}_1 - \mu_2\mathbf{R}_{11} \cdot \mathbf{c}_1.$$

Letting

$$\mathbf{c}_1(x,t) = \mathbf{e}(k_c)\left[A(t)e^{ik_cx} + A^*(t)e^{-ik_cx}\right],$$ (23.34)

and using the solvability conditions we find

$$\frac{\partial A}{\partial \tau_2} = \mu_2(\mathbf{e}^\dagger \cdot \mathbf{R}_{11} \cdot \mathbf{e})A + \frac{1}{2}(\mathbf{e}^\dagger \cdot \mathbf{R}_{03} \odot \mathbf{eee})|A|^2 A.$$ (23.35)

Now note that $\lambda^2\partial/\partial\tau_2 = \partial/\partial t + \mathcal{O}(\lambda^3)$ and $\mu_2 \equiv g$. Then we have

$$\frac{\partial A}{\partial t} = (\mathbf{e}^\dagger \cdot \mathbf{R}_{11} \cdot \mathbf{e})[(\mu - \mu_0)A(t) - g|A(t)|^2 A(t)].$$ (23.36)

By rescaling the time as $\tau = (\mathbf{e}^\dagger \cdot \mathbf{R}_{11} \cdot \mathbf{e})t$, we obtain

$$\frac{\partial A}{\partial \tau} = (\mu - \mu_0)A(\tau) - g|A(\tau)|^2 A(\tau).$$ (23.37)

This is the amplitude or order-parameter equation for the stripe solutions in a Turing bifurcation.

A linear stability analysis of this amplitude equation is easily carried out. Linearizing about the steady-state amplitude, $A_s = \pm(\frac{\mu-\mu_0}{g})^{1/2}$, $\delta A(\tau) = A(\tau) - A_s$, we find

$$\frac{d\delta A(\tau)}{d\tau} = -(\mu - \mu_0)\delta A(\tau),$$ (23.38)

so that the stripe pattern is stable if $\mu > \mu_0$.

Similar but more involved analyses can be carried out to determine pattern selection in two and higher dimensions leading to hexagonal and other patterns (Borckmans *et al.*, 1992; Dewel *et al.*, 1995). In addition, the above analysis was limited to spatial patterns constructed from critical modes with $|\mathbf{k}| = k_c$. Instabilities in the spatial patterns can arise from modes in the sidebands where $\mathbf{k} \neq \mathbf{k}_c$. For the stripe Turing patterns such sideband modes can lead to zigzag patterns, shown in the right panel of Fig. 23.5.

24

Excitable media

Excitable media are spatially distributed systems with a stable state that responds to perturbations in a distinctive way. If the normal resting state of the medium is perturbed sufficiently strongly, the perturbation is amplified before the system returns to the resting state. Such excitable media are commonly found in nature, and self-organized wave patterns in these systems control the behavior of many physical and biological systems (Zykov, 1987; Mikhailov, 1994; Kapral and Showalter, 1995). Surface catalytic oxidation reactions often proceed through the propagation of excitable waves of oxidation that sweep across the surface of the catalyst. The oxidation of CO on Pt surfaces has been especially well studied in this context (Ertl, 2000). In biological systems waves of this type occur in the aggregation stage of the slime mould *Dictyostelium discoideum*, where the chemical signaling is through periodic waves of cAMP; also, the Ca^{+2} waves in systems such as *Xenopus laevis* oocytes and pancreatic β cells fall into this category (Goldbeter, 1996). Electrochemical waves in cardiac and nerve tissue also depend on the excitability of the medium, and the appearance and/or breakup of spiral wave patterns (Fig. 24.1) are believed to be responsible for various types of arrhythmia in the heart (Winfree, 1987; Fenton *et al.*, 2002; Clayton and Holden, 2004). Excitable waves have been extensively studied (Belmonte *et al.*, 1997) for the BZ reaction, one of the first systems in which such waves were observed (Zaikin and Zhabotinsky, 1970; Winfree, 1972). Chemical waves in excitable media often take the form of spirals, and Fig. 24.2 shows spiral waves in the Belousov–Zhabotinsky reaction under conditions where this chemical medium is excitable.

The dynamics of an excitable system can be described more precisely as follows: if the stable fixed point of an excitable system is subjected to small perturbations, it returns quickly to the fixed point; however, larger perturbations that exceed a certain threshold value cause the system to make a long excursion in concentration phase space before returning to the stable state. In many physical systems this behavior

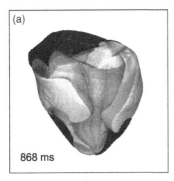
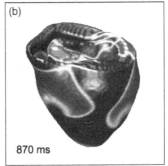

(a) 868 ms
(b) 870 ms

Fig. 24.1. Spiral wave rotors in an anatomically detailed model of the rabbit heart. (a) Isosurface of electrically active areas; (b) membrane potential across the heart surface. Figure supplied by R. Clayton, University of Sheffield.

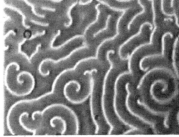

Fig. 24.2. Spiral waves in the Belousov–Zhabotinsky reaction. Experimental results provided by S. Kadar and K. Showalter, West Virginia University.

is captured by the dynamics of two concentration fields,

$$\frac{dc_A}{dt} = \mathcal{R}_A(\mathbf{c}), \quad \frac{dc_B}{dt} = \mathcal{R}_B(\mathbf{c}), \tag{24.1}$$

a fast activator or propagator variable c_A with an S-shaped nullcline, $\dot{c}_A = \mathcal{R}_A(\mathbf{c}) = 0$, and a slow inhibitor or controller variable c_B with a linear nullcline, $\dot{c}_B = \mathcal{R}_B(\mathbf{c}) = 0$ (Fife, 1984). The FitzHugh–Nagumo equation, which was discussed in Chapter 21, is one of the best known equations with such activator–inhibitor kinetics. In this case $\mathcal{R}_A(\mathbf{c}) = -c_A^3 + c_A - c_B$ and $\mathcal{R}_B(\mathbf{c}) = \varepsilon(c_B - ac_A + b)$. Figure 24.3 shows the $\dot{c}_A = 0$ and $\dot{c}_B = 0$ nullclines of this system along with trajectories corresponding to sub- and superthreshold excitations. The trajectory arising from the subthreshold perturbation quickly relaxes back to the stable fixed point. Three stages can be identified in the trajectory resulting from the superthreshold perturbation: an excited stage where the phase point quickly evolves far from the fixed point; a refractory stage where the system relaxes back to the stable state and

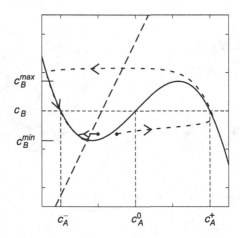

Fig. 24.3. Cubic ($\dot{c}_A = 0$) and linear ($\dot{c}_B = 0$) nullclines for the FitzHugh–Nagumo equation. The figure shows trajectories resulting from sub- and superthreshold excitations as well as c_A and c_B concentration values that enter the analysis in the text.

is not susceptible to additional perturbations; and the resting state where the system again resides at the stable fixed point.

An excitable medium is a diffusively coupled array of such local excitable elements, and is described by the set of reaction–diffusion equations

$$\frac{\partial c_A}{\partial t} = \mathcal{R}_A(\mathbf{c}) + D_A \nabla^2 c_A,$$

$$\frac{\partial c_B}{\partial t} = \mathcal{R}_B(\mathbf{c}) + D_B \nabla^2 c_B. \tag{24.2}$$

Imagine that a local superthreshold perturbation is applied to the system in its homogeneous resting state. Due to diffusive coupling, the perturbation will excite neighboring regions of the medium. The originally perturbed region will then relax to the refractory state, where it is no longer susceptible to perturbation, and finally back to the stable steady state. Consequently, a circular wave of excitation with a refractory tail will propagate outward through the medium (Fig. 24.4). If the excitable system is periodically stimulated in a local region of the medium (a pacemaker region), a target pattern comprising a set of concentric rings of excitation will be observed.

Suppose an excitable wave is broken by some means, for instance by an obstacle or inhomogeneity in the medium. Since the front velocity is smaller at the tip than the rest of the wavefront, the free ends of wavefronts will curl, leading to the formation of spiral waves in the system. An example of a spiral wave in the FHN system is shown in the right panel of Fig. 24.4.

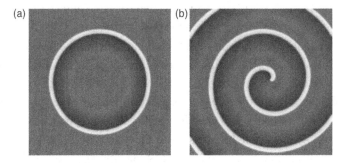

Fig. 24.4. (a) A growing ring of excitation in an excitable FitzHugh–Nagumo medium; (b) a spiral wave in the same system.

24.1 Traveling waves in excitable media

By exploiting the different time scales of the propagator and controller species, it is possible to use singular perturbation techniques to investigate the properties of chemical waves in excitable media (Tyson and Keener, 1988; Bernhoff, 1991). Scaling space and time variables by the small parameter ϵ as $\mathbf{r}\epsilon/\sqrt{D_A} = \mathbf{r}'$ and $t\epsilon = t'$, Eq. (24.2) can be written (after dropping the primes) in the form

$$\epsilon\frac{\partial c_A}{\partial t} = \overline{\mathcal{R}}_A(\mathbf{c}) + \epsilon^2\nabla^2 c_A,$$

$$\frac{\partial c_B}{\partial t} = \overline{\mathcal{R}}_B(\mathbf{c}) + \epsilon\delta\nabla^2 c_B, \tag{24.3}$$

where $\delta = D_B/D_A$, $\overline{\mathcal{R}}_A(\mathbf{c}) = \mathcal{R}_A(\mathbf{c})$, and $\overline{\mathcal{R}}_B(\mathbf{c}) = \mathcal{R}_B(\mathbf{c})/\epsilon$. As discussed earlier in Chapter 19, to describe a chemical front propagating with constant speed v_p in a one-dimensional medium, it is convenient to introduce a coordinate frame moving with the front speed, $u = x - v_p t$, and let $c_A(x, t) = c_A(u)$ and $c_B(x, t) = c_B(u)$. The subscript p is introduced to denote the velocity of a planar front. In this moving frame the reaction–diffusion equation is

$$\epsilon^2\frac{d^2 c_A}{du^2} + \epsilon v_p\frac{d c_A}{du} + \overline{\mathcal{R}}_A(\mathbf{c}) = 0,$$

$$\epsilon\delta\frac{d^2 c_B}{du^2} + v_p\frac{d c_B}{du} + \overline{\mathcal{R}}_B(\mathbf{c}) = 0. \tag{24.4}$$

A traveling front solution can be constructed for small ϵ. Setting $\epsilon = 0$ in the first of Eqs. (24.4), we find that (c_A, c_B) must satisfy $\overline{\mathcal{R}}_A(\mathbf{c}) = 0$. Referring to the nullclines in Fig. 24.3, we see that for $c_B^{min} < c_B < c_B^{max}$, where c_B^{min} and c_B^{max} are the values of c_B at the minimum and maximum of the S-shaped nullcline, there are three solutions of $\overline{\mathcal{R}}_A(\mathbf{c}) = 0$, which we call $c_A^{\pm}(c_B)$ and $c_A^0(c_B)$. The relevant solutions are $c_A^{\pm}(c_B)$, since $c_A^0(c_B)$ is an unstable solution.

A single propagating front in an infinite domain satisfies the boundary conditions $\lim_{u \to \pm\infty} c_A(u) = c_A^{\pm}(c_B)$. Starting from an initial condition of the form $c_A(u) = c_A^{\pm}$ for $u < 0$ and $u > 0$, respectively, in the vicinity of the origin the terms of $\mathcal{O}(\epsilon)$ cannot be neglected since the solution is discontinuous at that point. A boundary layer whose width is $\mathcal{O}(\epsilon)$ will develop in order to connect the asymptotic $c_A^{\pm}$ solutions smoothly. In order to investigate the structure of this boundary layer, we introduce a stretched coordinate $\xi = u/\epsilon$. In this new coordinate Eqs. (24.4) become

$$\frac{d^2 c_A}{d\xi^2} + v_p \frac{d c_A}{d\xi} + \overline{\mathcal{R}}_A(\mathbf{c}) = 0,$$

$$\delta \frac{d^2 c_B}{d\xi^2} + v_p \frac{d c_B}{d\xi} + \epsilon \overline{\mathcal{R}}_B(\mathbf{c}) = 0. \tag{24.5}$$

If we now set $\epsilon = 0$, integration of the second equation yields $\delta d c_B/d\xi + v_p c_B = \text{const.}$ If c_B is to be bounded between c_B^{min} and c_B^{max} we must have $c_B = c_B^* = \text{const.}$ Therefore, in the boundary layer connecting the solutions $c_A^{\pm}$, we have

$$\frac{d^2 c_A}{d\xi^2} + v_p \frac{d c_A}{d\xi} + \overline{\mathcal{R}}_A(c_A, c_B^*) = 0, \tag{24.6}$$

which must be solved subject to the boundary conditions $c_A(\pm\infty) = c_A^{\pm}$. This equation is a nonlinear eigenvalue problem for the front speed v_p, which depends on c_B^*. We see that while the propagator species changes rapidly across the boundary layer, the controller species variable remains nearly constant at $c_B = c_B^*$, and its value determines the wave speed.

As an example of the form of a propagating chemical front in an excitable medium we return to the FitzHugh–Nagumo model with $\overline{\mathcal{R}}_A$ written in the form $\overline{\mathcal{R}}_A(c_A, c_B^*) = -(c_A - c_A^+)(c_A - c_A^-)(c_A - c_A^0)$. The nonlinear eigenvalue problem is easily solved, and the equation for the propagating front is given by

$$c_A(u) = \frac{1}{2}(c_A^+ - c_A^-) \tanh\left(\frac{(c_A^+ - c_A^-)}{2\epsilon\sqrt{2}} u\right) + \frac{1}{2}(c_A^+ + c_A^-). \tag{24.7}$$

The front speed is $v_p = (2c_A^0 - c_A^- + c_A^-)/\sqrt{2}$. In these equations $c_A^{\pm} = c_A^{\pm}(c_B^*)$. From this explicit solution and form of the FitzHugh–Nagumo equations we see that for $c_B^* = 0$ we have $c_A^- = -c_A^+$ and $c_A^0 = 0$, so that $v_p = 0$. The sign of the front velocity depends on the sign of c_B^*. These arguments can be generalized for any reaction–diffusion system with such excitable kinetics.

24.2 Eikonal equation

In two-dimensional media, excitable chemical fronts generally take the form of curved domains whose width is approximately that of the interfacial zone separating the c_A^+ and c_A^- values obtained in the solution of the one-dimensional problem. The situation is similar to that discussed previously in Chapter 7, where interface dynamics at late times was considered. Referring to Figs. 7.1 and 7.2 we again suppose that the front position is defined by a curve $\mathbf{R}(s)$ parameterized by the arc length s lying in the interfacial zone. Provided the curvature is not too large, we may use the local coordinate system introduced in Eq. (7.8), where a point $\mathbf{x}(s, u)$ in the medium is given by $\mathbf{x}(s, u) = \mathbf{R}(s) + u\hat{\mathbf{n}}(s)$, with u measuring the distance from the front along the normal to the curve.

Following the discussion in Chapters 7 and 19, we let $\mathbf{c}(\mathbf{r}, t) = \mathbf{c}_0(u(\mathbf{R}, t)) + \delta\mathbf{c}$, where $\mathbf{c}_0(u(\mathbf{R}, t))$ is the solution to the planar (one-dimensional) front problem discussed above. Substituting this expression into the reaction–diffusion equation (24.2), making use of Eq. (7.18) for the two-dimensional Laplacian, and the same approximations that led to Eq. (7.17), we have

$$\left(\mathbf{D}\frac{d^2}{du^2} + (\mathbf{D}K(s) + v_N)\frac{d}{du}\right)\mathbf{c}_0 + \mathcal{R}(\mathbf{c}_0)$$
$$+ \left(\mathbf{D}\left(\frac{\partial^2}{\partial u^2} + \frac{\partial^2}{\partial s^2}\right) + (\mathbf{D}K(s) + v_N)\frac{\partial}{\partial u} + \frac{\delta\mathcal{R}}{\delta\mathbf{c}_0}\right)\delta\mathbf{c} \approx 0, \quad (24.8)$$

where $v_N(s) = -du(\mathbf{R}(s), t)/dt$ is the normal velocity of the front, and $K(s)$ is the curvature. The planar solution $\mathbf{c}_0$ satisfies the equation

$$\left(\mathbf{D}\frac{d^2}{du^2} + v_p\frac{d}{du}\right)\mathbf{c}_0 + \mathcal{R}(\mathbf{c}_0) = 0. \quad (24.9)$$

Following the arguments in Chapters 7 and 19, we see that $\langle u|\zeta_0\rangle = \zeta_0(u) = d\mathbf{c}_0/du$ is a right eigenfunction satisfying the eigenvalue problem

$$\left(\mathbf{D}\frac{d^2}{du^2} + v_p\frac{d}{du} + \frac{\delta\mathcal{R}}{\delta\mathbf{c}_0}\right)\zeta_0 = 0, \quad (24.10)$$

with eigenvalue zero. Using Eq. (24.9) in Eq. (24.8) and taking the scalar product with $< \zeta_0|$, the left eigenvector corresponding to eigenvalue zero, we obtain the eikonal equation for the normal velocity of the front,

$$v_N(s) = v_p - \langle\zeta_0|\mathbf{D}|\zeta_0\rangle K(s). \quad (24.11)$$

Using the approximations leading to Eq. (24.6) for the propagator–controller equations, we find $\langle \zeta_0 | D | \zeta_0 \rangle = \epsilon$, and this relation takes the form

$$v_N(s) = v_p - \epsilon K(s). \tag{24.12}$$

24.3 Spiral wave solution

The eikonal relation can be used to construct a spiral wave solution to the reaction–diffusion equation (Tyson and Keener, 1988). A rigidly rotating excitable front with the form of a spiral wave, parameterized by the radial distance r, is described by the parametric equations

$$\mathbf{R}(r) = (x, y) = (r \cos(\theta(r) - \omega t), r \sin(\theta(r) - \omega t)). \tag{24.13}$$

The change in the arc length ds is given by $ds = (1 + \psi^2(r))^{1/2} dr$, where $\psi(r) = r d\theta(r)/dr$. The tangent $\hat{\mathbf{t}}$ and normal $\hat{\mathbf{n}}$ vectors are given by

$$\hat{\mathbf{t}} = \mathbf{R}_r / |\mathbf{R}_r|, \quad \hat{\mathbf{n}} = -\hat{\mathbf{t}}_r / K. \tag{24.14}$$

Using the formula $K = |\mathbf{R}_r \times \mathbf{R}_{rr}| / |\mathbf{R}_r|^3$ for the curvature that is independent of parametrization (Tabor and Klapper, 1994), we obtain

$$K = \frac{\psi(r)}{r(1 + \psi^2(r))^{1/2}} + \frac{\psi_r(r)}{(1 + \psi^2(r))^{3/2}}. \tag{24.15}$$

The velocity of the spiral space curve can be decomposed into normal and tangential components,

$$\dot{\mathbf{R}}(r) = (\dot{x}, \dot{y}) = \omega r (\sin(\Theta(r, t)), -\cos(\Theta(r, t))) = v_N \hat{\mathbf{n}} + v_T \hat{\mathbf{t}}, \tag{24.16}$$

where $\Theta(r, t) = \theta(r) - \omega t$, to give the following results for the normal and tangential velocity components:

$$v_N = \frac{\omega r}{(1 + \psi^2(r))^{1/2}}, \quad v_T = -\frac{\omega r \psi(r)}{(1 + \psi^2(r))^{1/2}}. \tag{24.17}$$

The expressions for the curvature and normal velocity may be substituted into the eikonal equation to obtain the differential equation for the unknown function $\psi(r)$,

$$\frac{d\psi(r)}{dr} = (1 + \psi^2) \left[-\frac{\omega r}{\epsilon} + \frac{v_p}{\epsilon}(1 + \psi^2)^{1/2} - \frac{\psi}{r} \right]. \tag{24.18}$$

Notice that this equation is independent of the specific form of the reaction terms in the reaction–diffusion equation. This equation must be solved subject to boundary conditions. For example, consider a spiral wave in an infinite domain rotating around a hole with radius r_c, and assume that the hole radius is sufficiently large that v_p is independent of r. In this case we may take no-flux boundary conditions at $r = r_c$, $\psi(r_c) = 0$, and assume $\psi(r) \sim \omega r/v_p$ for large r. Here ω/v_p is the wave number of the planar fronts far from the origin. It is not difficult to verify that solutions of Eq. (24.18) for $r \approx r_c$ and large r are given by

$$\psi(r) = \begin{cases} \frac{(v_p - \omega r_c)}{\epsilon}(r - r_c) + \frac{v_p}{\epsilon}\frac{(r - r_c)^2}{r_c} + \mathcal{O}((r - r_c)^3), & (r \approx r_c) \\ \frac{\omega}{v_p}r + \frac{\omega \epsilon}{v_p^2} + \mathcal{O}(r^{-1}), & (r \to \infty). \end{cases} \quad (24.19)$$

The full solution may be obtained by matching these asymptotic results. The solution of Eq. (24.18), along with its boundary conditions, gives a relation between ω and v_p. Once the full solution for $\psi(r)$ is known, the spiral wave solution can be constructed by calculating the function $\theta(r)$ that appears in the parametric equation for the spiral wave as $\theta(r) = \int^r dr'\, \psi(r')/r'$. For large distances from the origin we have $\theta(r) \sim (\omega/v_p)r + (\omega\epsilon/v_p^2)\ln r$, so that the spirals are Archimedean, while near the radius of the hole we find $\theta(r) \sim ((v_p - \omega r_c)/\epsilon)(r - r_c) + (v_p/2r_c\epsilon)(r - r_c)^2$.

24.4 Kinematic theory

In the limit of weakly excitable kinetics, where the pitch of the spiral is much larger than the length of the refractory zone following the excitation, it is possible to construct kinematic models that capture essential features of spiral wave dynamics (Mikhailov et al., 1994). For weakly excitable media, the excitation and recovery zones of the propagating fronts are narrow, and the spiral wave can be considered to be a space curve with a free end. The asymptotic reduction of the reaction–diffusion equation to the kinematic model for weakly excitable media has been carried out by Hakim and Karma (1999). The dynamics of space curves was discussed earlier in Chapter 7, and we employ similar considerations here to see how spiral waves arise in excitable media.

We consider a space curve $\mathbf{R}(s, t)$ now parameterized by the arc length s. The tangent to the curve at s is defined by $\hat{\mathbf{t}}(s, t) = \partial \mathbf{R}(s, t)/\partial s$, while the variations of the tangent and normal with arc length are given by $\partial \hat{\mathbf{t}}/\partial s = -K\hat{\mathbf{n}}$ and $\partial \hat{\mathbf{n}}/\partial s = K\hat{\mathbf{t}}$, where again K is the curvature. We have chosen the normal so that it will point in the direction of the propagating spiral wave. Since the dot products of the tangent and normal vectors are constant, it follows that $\hat{\mathbf{t}} \cdot \hat{\mathbf{t}}_t = 0$ and $\hat{\mathbf{n}} \cdot \hat{\mathbf{n}}_t = 0$, where the subscript refers to partial differentiation with respect to time t. From these results it may be shown that these vectors satisfy the two-dimensional equations

$$\frac{\partial \hat{\mathbf{t}}(s, t)}{\partial t} = -\alpha \hat{\mathbf{n}}(s, t), \qquad \frac{\partial \hat{\mathbf{n}}(s, t)}{\partial t} = \alpha \hat{\mathbf{t}}(s, t), \quad (24.20)$$

where α is a proportionality constant.

The velocity $\mathbf{R}_t$ of the space curve can again be resolved into normal and tangential components,

$$\mathbf{R}_t(s,t) = v_N(s,t)\hat{\mathbf{n}}(s,t) + v_T(s,t)\hat{\mathbf{t}}(s,t). \tag{24.21}$$

Differentiating this equation with respect to the arc length s and using the fact that $\partial\hat{\mathbf{t}}(s,t)/\partial t = \partial\mathbf{R}_t(s,t)/\partial s$, we obtain

$$\frac{\partial\hat{\mathbf{t}}}{\partial t} = \frac{\partial v_N}{\partial s}\hat{\mathbf{n}} + v_N\frac{\partial\hat{\mathbf{n}}}{\partial s} + \frac{\partial v_T}{\partial s}\hat{\mathbf{t}} + v_T\frac{\partial\hat{\mathbf{t}}}{\partial s},$$

$$-\alpha\hat{\mathbf{n}} = \frac{\partial v_N}{\partial s}\hat{\mathbf{n}} + v_N K\hat{\mathbf{t}} + \frac{\partial v_T}{\partial s}\hat{\mathbf{t}} - v_T K\hat{\mathbf{n}}. \tag{24.22}$$

Taking the dot products with $\hat{\mathbf{n}}$ and $\hat{\mathbf{t}}$, respectively, we find

$$\frac{\partial v_N}{\partial s} = -\alpha + K v_T, \qquad \frac{\partial v_T}{\partial s} = -K v_N. \tag{24.23}$$

Next, we differentiate the first of Eqs. (24.23) with respect to s and get

$$\frac{\partial^2 v_N}{\partial s^2} = -\frac{\partial\alpha}{\partial s} + K_s v_T + K\frac{\partial v_T}{\partial s}. \tag{24.24}$$

The solution for $v_T(s,t)$ in terms of $v_N(s,t)$ can be obtained by integrating the second of Eqs. (24.23) from $s = 0$ at its free end $(v_T(0) \equiv v_{Tf})$ to s, to give

$$v_T(s,t) = v_{Tf}(t) - \int_0^s ds' K(s',t)v_N(s',t). \tag{24.25}$$

Differentiating $K(s) = \hat{\mathbf{t}} \cdot \partial\hat{\mathbf{n}}/\partial s$ with respect to time, we have

$$K_t = \frac{\partial\hat{\mathbf{t}}}{\partial t} \cdot \frac{\partial\hat{\mathbf{n}}}{\partial s} + \hat{\mathbf{t}} \cdot \frac{\partial}{\partial s}\frac{\partial\hat{\mathbf{n}}}{\partial t} = \hat{\mathbf{t}} \cdot \frac{\partial\alpha\hat{\mathbf{t}}}{\partial s} = \frac{\partial\alpha}{\partial s}. \tag{24.26}$$

In writing this equation we used the fact that $(\partial\hat{\mathbf{t}}/\partial t) \cdot (\partial\hat{\mathbf{n}}/\partial s) = -\alpha K\hat{\mathbf{n}} \cdot \hat{\mathbf{t}} = 0$. Employing these results in Eq. (24.24), we obtain the equation of motion for the curvature expressed in terms of the normal component of the velocity of the space curve and the tangential component at the free end at $s = 0$:

$$K_t = -\frac{d^2 v_N}{ds^2} - K^2 v_N - K_s \int_0^s ds' K v_N + K_s v_{Tf}(t). \tag{24.27}$$

To solve this equation we need to supply initial and boundary conditions.

Kinematic theory assumes that the normal velocity $v_N(s)$ of the space curve of an excitable wave at s is smaller than the corresponding velocity v_p if the curve were planar. The normal velocity depends linearly on the curvature through the eikonal relation, $v_N(s) = v_p - \gamma K(s)$, where we let $\gamma = \langle \zeta_0 | \mathbf{D} | \zeta_0 \rangle$. Similarly, it is assumed that the tangential velocity at the free end v_{Tf} also depends linearly on the curvature at the free end $K(0)$: i.e. the curvature extrapolated to $s = 0$, $v_{Tf} = v_{Tfp} - \gamma_f K(0)$. Depending on the magnitude of the curvature at the free end, the curve will either grow or shrink. For a critical value of the curvature at the end, $K_c = v_{Tfp}/\gamma_f$, the curve will neither grow nor shrink. In this case we can derive the form for a spiral wave rotating around a circle of a given radius, the core of the spiral wave.

When $K(0) = K_c$ we can find a steady solution of Eq. (24.27) where the spiral wave rotates with frequency ω around the core. Equation (24.27) may be simplified by using $v_N(s) = v_p - \gamma K(s)$, noting that $d^2 v_N/ds^2 = -\gamma K_{ss}$, and assuming that the curvature is sufficiently small that $\gamma K \ll v_p$, so that v_N may be replaced by v_p in the remaining terms. In this circumstance, we find the approximate equation of motion

$$K_t = \gamma K_{ss} - K^2 v_p - K_s \int_0^s ds' \, K v_p$$

$$= \gamma K_{ss} - \frac{\partial}{\partial s} \left(K \int_0^s ds' \, K v_p \right). \tag{24.28}$$

For steady rotation of the spiral around the core we have

$$\gamma K_{ss} - \frac{d}{ds} \left(K \int_0^s ds' \, K v_p \right) = 0. \tag{24.29}$$

Integrating over s we find

$$\gamma K_s = v_p K \int_0^s ds' \, K - \omega, \tag{24.30}$$

where ω is an integration constant. Letting $h(s) = \int^s ds' \, K(s')$ or $K(s) = dh(s)/ds$ in this equation, it takes the form,

$$\gamma \frac{d^2 h}{ds^2} = v_p h \frac{dh}{ds} - \omega. \tag{24.31}$$

The boundary conditions are $(dh(s)/ds)_{s=0} = K_c$, $h(0) = 0$ and, since far from the tip the curvature is zero, $\lim_{s \to \infty} h_s(s) = 0$. The second-order differential equation subject to these three boundary conditions has a solution only for a specific value of ω.

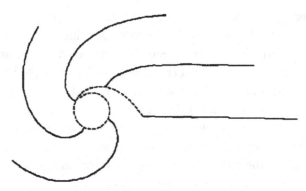

Fig. 24.5. Spiral wave determined from the solution of the kinematic equation (24.28) for $v_p = 1.5$, $\gamma_f = 0.96$, $\gamma = 1.0$, and $\gamma_f = 1.5$. Reprinted from Mikhailov *et al.* (1994). Copyright 1994, with permission from Elsevier.

An approximate solution of Eq. (24.31) can be obtained as follows. From the form of this equation we see that $\lim_{s\to 0} dK(s)/ds = -\omega/\gamma$. It then follows that near $s = 0$ the solution is $K(s) = K_c - \omega s/\gamma$. For large values of s the second derivative term in Eq. (24.31) can be neglected because the curvature tends to a constant in this region. Using this approximation we find $K(s) = (\omega/2v_p s)^{1/2}$ for large s. The unknown parameter ω can be determined by requiring that these solutions and their first derivatives match at some intermediate value of $s = s'$. The result is

$$\omega = \left(\frac{2}{3}\right)^{3/2} K_c (K_c \gamma v_p)^{1/2}. \tag{24.32}$$

To obtain a more accurate result the equation must be integrated numerically. The results of such a calculation show that prefactor $(2/3)^{3/2} \approx 0.544$ is replaced by 0.685.

Equation (24.28) may be solved numerically in a suitable laboratory frame starting from a given initial curve to determine the evolution. The result of such a calculation is shown in Fig. 24.5. It shows that an initial straight-line segment with a free end evolves to a spiral wave steadily rotating around a circular core with radius r_c. The core radius is approximately given by $r_c = v_p/\omega$.

24.5 Spiral wave meander

We have seen that spiral wave solutions, where the tip of the spiral executes a circular periodic motion about the spiral core, can be constructed. However, this is not the only type of motion the spiral tip can undergo. Shortly after spiral waves were found to exist in the excitable Belousov–Zhabotinsky reaction–diffusion system, Winfree (1973) observed that the tip could execute much more complicated "flower-pattern" dynamics of the type shown in Fig. 24.6. Subsequently, this motion was

Fig. 24.6. Spiral wave meander for the BZ reaction as a chemical concentration is changed. Reprinted from Skinner and Swinney (1991). Copyright 1991, with permission from Elsevier.

quantitatively characterized experimentally (Agladze *et al.*, 1988; Janke *et al.*, 1989; Skinner and Swinney, 1991) and the theoretical description of the origin of such complex tip dynamics was provided (Barkley, 1994; Golubitsky *et al.*, 1997; Sandstede *et al.*, 1997; Sandstede and Scheel, 2001).

If a rigidly rotating spiral wave with angular velocity ω is viewed in a rotating frame, it will appear stationary. We consider either the entire plane or a disk-shaped domain and, as in earlier chapters, let $\mathbf{c} = (c_A, c_B)$ be a vector-valued concentration field. In the rotating frame the reaction–diffusion equation is

$$\frac{\partial}{\partial t}\mathbf{c} = \omega\frac{\partial}{\partial \theta}\mathbf{c} + \mathbf{D}\nabla^2\mathbf{c} + \mathcal{R}(\mathbf{c}), \tag{24.33}$$

where the polar coordinates are (r, θ). Steadily rotating spiral waves $\mathbf{c}_s(r, \theta)$ are stationary solutions of the reaction–diffusion equation

$$\omega\frac{\partial}{\partial \theta}\mathbf{c}_s + \mathbf{D}\nabla^2\mathbf{c}_s + \mathcal{R}(\mathbf{c}_s) = 0. \tag{24.34}$$

This equation is a nonlinear eigenvalue problem, which must be solved, subject to suitable boundary conditions for the domain under consideration, to yield the field $\mathbf{c}_s$ and the angular velocity ω. Given these steady-state solutions in the rotating frame, their stability can be determined in the usual way by linearizing the reaction–diffusion system about the steady-state solutions and studying the properties of the eigenvalues. The eigenvalues λ can be determined from the solution of the linearized eigenvalue problem,

$$\left(\omega \frac{\partial}{\partial \theta} + \mathbf{D}\nabla^2 + \frac{\delta \mathcal{R}}{\delta \mathbf{c}_s} \right) \mathbf{C} = \lambda \mathbf{C}, \tag{24.35}$$

where $\delta \mathcal{R}/\delta \mathbf{c}_s$ is the functional derivative of $\mathcal{R}$ and $\mathbf{C}$ is an eigenvector. Rotating spiral waves become unstable through a Hopf bifurcation where a complex conjugate pair of eigenvalues crosses the imaginary axis (see also Chapter 25). The Hopf bifurcation introduces a second frequency ω_2 into the problem, giving rise to modulated rotating spiral waves where spiral wave tips execute the flower-pattern trajectories shown in Fig. 24.6. If $\omega_2 < \omega$ the flower petals are inward, and they are outward when $\omega_2 > \omega$. The spiral tip trajectories of modulated traveling waves are quasiperiodic in the laboratory frame but are periodic in the rotating frame since the frequency ω has been removed from the problem.

The phase diagram that shows the different types of spiral tip dynamics versus the parameters of the excitable system has been constructed for several model excitable systems, including the FitzHugh–Nagumo model and the Oregonator model for the Belousov–Zhabotinsky reaction. This diagram has also been deduced from experimental measurements. It has the general characteristics shown in Fig. 24.7, which was constructed for Barkley's variant of the FitzHugh–Nagumo model, where $\mathcal{R}_A = \varepsilon^{-1} c_A (1 - c_A)(c_A - c_A^{th}(c_B))$ and $\mathcal{R}_B = c_A - c_B$, where $c_A^{th}(c_B) = (c_B + b)/a$ is a threshold that depends on c_B. In this figure the Hopf bifurcation line separates the domain that contains periodically rotating spiral waves from the domain where modulated rotating spiral waves are found. Modulated traveling waves are found along the dashed line in this figure, which starts on the Hopf boundary where $\omega = \omega_2$. The spiral tips of these modulated traveling waves have trajectories like those shown in the figure.

In the vicinity of the point in the phase diagram where the dashed line of traveling-wave solutions meets the Hopf bifurcation line, one finds periodically rotating, modulated-rotating and traveling-wave solutions. In order to understand this structure one must consider a two-parameter bifurcation analysis. On the Hopf line there is a complex conjugate pair of eigenvalues $\lambda_H = \pm i \omega_2$ on the imaginary axis. In addition to these eigenvalues, the other relevant eigenvalues of the problem have their origins in the symmetries of the solutions. There is a zero eigenvalue that has eigenfunction $\mathbf{C}_R = \partial_\theta \mathbf{c}_s$ and is associated with the rotational symmetry of the rigidly rotating spiral state. In addition there is a complex conjugate pair of imaginary eigenvalues $\pm i \omega$ with eigenfunctions $\mathbf{C}_T^\pm = \partial_x \mathbf{c}_s \pm \partial_y \mathbf{c}_s$ arising

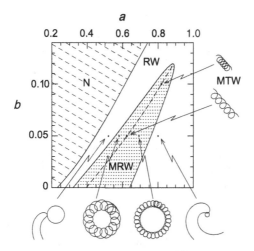

Fig. 24.7. Phase diagram showing regions where different types of spiral dynamics occur as a function of the parameters a and b. The RW region contains periodically rotating spiral waves, the MRW region has modulated rotating waves, while the region labeled N does not support spiral waves. A solid curve of supercritical Hopf bifurcation points separates the RW and MRW regions. The dashed line is the locus where $\omega = \omega_2$ separates tip trajectories with inward petals from those with outward petals. Reprinted Figure 1 with permission from Barkley (1994). Copyright 1994 by the American Physical Society.

from the translational symmetry of the system. These results can be verified by differentiation of Eq. (24.34). At the codimension-two point (intersection of the dashed and solid lines in Fig. 24.7) where the Hopf and traveling wave lines meet, the eigenvalues are no longer distinct since $\omega = \omega_2$.

To unravel the behavior at this point a bifurcation analysis must be carried out that considers the resonance between the Hopf eigenvalue $\lambda_H = i(\omega - \Delta\omega_2)$ and the translation eigenvalue. The normal form for a supercritical Hopf bifurcation near the resonance is given by (Sandstede and Scheel, 2001)

$$\frac{dp}{dt} = e^{i\Phi}(p_0 - z^*), \qquad \frac{d\Phi}{dt} = \omega$$
$$\frac{dz}{dt} = (\lambda_H + \mu - |z|^2)z, \tag{24.36}$$

where p_0 is a complex constant and μ is a bifurcation parameter. The last equation is the Hopf normal form (Nicolis, 1995). This normal form is derived in Chapter 25, which deals with oscillatory media. The spiral tip position is determined by the complex function p. This set of equations can be solved for $p(t)$ to yield

$$p(t) = \begin{cases} -i\frac{p_0}{\omega}e^{i\omega t} + i\frac{\sqrt{\mu}}{\Delta\omega_2}e^{-i\alpha}\left(e^{i\Delta\omega_2 t} - 1\right) & (\Delta\omega_2 \neq 0), \\ -i\frac{p_0}{\omega}e^{i\omega t} + \sqrt{\mu}t\,e^{i(\pi-\alpha)} & (\Delta\omega_2 = 0), \end{cases} \tag{24.37}$$

where the initial condition $z = e^{i\alpha}$ was used. Letting $p(t) = x(t) + iy(t)$, we can find the spiral tip trajectories in the plane. From the forms of the time dependence of these results we see that the spirals meander for $\Delta\omega_2 \neq 0$ and drift for $\Delta\omega_2 = 0$.

24.6 Scroll wave dynamics

In three-dimensional excitable media spiral waves take the form of scroll waves shown in Fig. 24.8. We can imagine constructing an excitable scroll wave by stacking two-dimensional spiral waves as shown in the figure. The core of the spiral wave is drawn out to form a filament, and the study of the dynamics of this filament provides information on the scroll wave motion.

The filament is a space curve $\mathbf{X}(s,t) = (x(s,t), y(s,t), z(s,t))$, where t is time and s is arc length, $0 \leq s \leq L(t)$ with $L(t)$ total length of the filament. An orthogonal coordinate system with tangent $\hat{\mathbf{t}}(s)$, normal $\hat{\mathbf{n}}(s)$, and binormal $\hat{\mathbf{b}}(s)$ can be used to describe the local geometry of the filament:

$$\hat{\mathbf{t}}(s) = \frac{d\mathbf{X}}{ds}, \quad \hat{\mathbf{n}}(s) = \frac{d\hat{\mathbf{t}}/ds}{|d\hat{\mathbf{t}}/ds|}, \quad \hat{\mathbf{b}}(s) = \hat{\mathbf{t}}(s) \times \hat{\mathbf{n}}(s). \tag{24.38}$$

The tangent, normal, and binormal satisfy the Frenet–Serret equations

$$\frac{d\hat{\mathbf{t}}}{ds} = K(s)\hat{\mathbf{n}}, \quad \frac{d\hat{\mathbf{n}}}{ds} = -K(s)\hat{\mathbf{t}} + \tau(s)\hat{\mathbf{b}}, \quad \frac{d\hat{\mathbf{b}}}{ds} = -\tau(s)\hat{\mathbf{n}}, \tag{24.39}$$

where $K(s) = |d\hat{\mathbf{t}}/ds|$ is the curvature and $\tau(s) = |d\hat{\mathbf{b}}/ds|$ is the torsion.

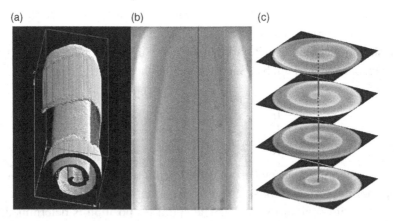

Fig. 24.8. Scroll wave for the Belousov–Zhabotinsky reaction. (a) An isoconcentration surface, (b) the projection through the cuvette, and (c) reconstructed horizontal slices through the system. The filament describing the core is also shown in the figure. Reprinted from Storb *et al.* (2003). Reproduced by permission of the PCCP Owner Societies.

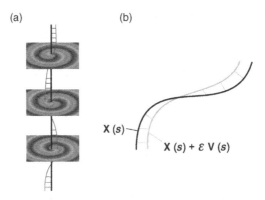

Fig. 24.9. (a) Stack of two-dimensional spiral waves to form a twisted scroll wave. The phase field of the two-dimensional spiral waves is twisted along the vertical direction leading to a ribbon curve. (b) Schematic depiction of the ribbon curve describing the scroll wave filament.

We must also account for the fact that the phase field may twist around the filament, as shown in Fig. 24.9. Thus, we are led to consider the dynamics of ribbon curves. The local phase of spiral may be defined as the angle between the unit vector $\hat{\mathbf{V}}$ and some local reference direction. We choose $\hat{\mathbf{V}} = \hat{\mathbf{b}} \cos \varphi + \hat{\mathbf{n}} \sin \varphi$, where φ is the angle that $\hat{\mathbf{V}}$ makes with the binormal $\hat{\mathbf{b}}$ (Tabor and Klapper, 1994). The edges of the ribbon curve are then defined by $\mathbf{X}(s)$ and $\mathbf{X}(s) + \epsilon \hat{\mathbf{V}}(s)$, where ϵ is a small constant. The vector $\hat{\mathbf{V}}$ twists along the filament, and the local twist rate is $w(s) = \left(\hat{\mathbf{V}} \times d\hat{\mathbf{V}}/ds \right) \cdot \hat{\mathbf{t}} = \tau(s) + d\varphi/ds$. Thus, the twist can be decomposed into the twist of the filament that measures how the binormal twists around the filament, and the ribbon twist that measures how the ribbon twists around the binormal or Frenet ribbon.

The dynamics of the concentration field in the medium is again governed by the set of reaction–diffusion equations (24.2). The motion of the filament about which the scroll wave rotates can be extracted from this equation (Keener, 1988). The reaction–diffusion equation may be expressed in the coordinate system $(s, \mathbf{u})$. Making use of Eq. (7.39) for the Laplacian in generalized coordinates, and taking $\omega^1 = s$, $\omega^2 = u_1$, and $\omega^3 = u_2$ in that equation, we find, in the limit of small-curvature K,

$$\nabla^2 \approx \left(\frac{\partial}{\partial s} - \tau \mathbf{u} \times \frac{\partial}{\partial \mathbf{u}} \right)^2 + \nabla_u^2 - K \frac{\partial}{\partial u_1}, \qquad (24.40)$$

where $\mathbf{a} \times \mathbf{b}$ stands for the two-dimensional cross product defined in Chapter 22, and ∇_u^2 is the two-dimensional Laplacian in $\mathbf{u}$ coordinates. After expressing the reaction–diffusion equation in these generalized coordinates, we transform to a frame moving with the filament, $\mathbf{X}_t = \mathbf{R}_t + u_1\hat{\mathbf{n}}_t + u_2\hat{\mathbf{b}}_t$,

$$\left(\frac{\partial}{\partial t} - \mathbf{X}_t \cdot \nabla\right)\mathbf{c}(s, \mathbf{u}, t) = D\nabla^2\mathbf{c} + \mathcal{R}(\mathbf{c}), \tag{24.41}$$

where ∇ is the gradient operator in generalized coordinates,

$$\nabla \approx \hat{\mathbf{t}}\left(\frac{\partial}{\partial s} - \tau\mathbf{u} \times \frac{\partial}{\partial \mathbf{u}}\right) + \nabla_u, \tag{24.42}$$

with ∇_u the two-dimensional gradient operator, $\nabla_u = \hat{\mathbf{n}}\partial/\partial u_1 + \hat{\mathbf{b}}\partial/\partial u_2$.

For a slowly moving filament with small curvature and torsion, we write the solution of the reaction–diffusion equation in the form $\mathbf{c}(s, \mathbf{u}, t) = \mathbf{c}_s(u, \theta + \phi(s, t) - \omega t) + \delta\mathbf{c}$. Here $\phi(s, t)$ is a slowly varying phase and $\mathbf{u} = (u_1, u_2) = (u\cos\theta, u\sin\theta)$ is expressed in polar coordinates. In this polar coordinate system $\mathbf{u} \times \partial/\partial\mathbf{u} = \partial/\partial\theta$. The field $\mathbf{c}_s(u, \theta - \omega t)$ satisfies the two-dimensional equation

$$\frac{\partial}{\partial t}\mathbf{c}_s = D\nabla_u^2\mathbf{c}_s + \mathcal{R}(\mathbf{c}_s). \tag{24.43}$$

If we substitute the expression for $\mathbf{c}(s, \mathbf{u}, t)$ into the reaction–diffusion equation, linearize assuming τ and ϕ_s are $\mathcal{O}(\epsilon)$ while $\delta\mathbf{c}, \mathbf{X}_t, \phi_t, \phi_{ss}, \tau_s$ and K are $\mathcal{O}(\epsilon^2)$, and use Eq. (24.43), we find

$$\mathcal{L}\delta\mathbf{c} = D\left(\frac{\partial}{\partial s} - \tau\frac{\partial}{\partial\theta}\right)^2\mathbf{c}_s - KD\frac{\partial\mathbf{c}_s}{\partial u_1} + (\mathbf{R}_t \cdot \hat{\mathbf{t}})\left(\frac{\partial}{\partial s} - \tau\frac{\partial}{\partial\theta}\right)\mathbf{c}_s$$
$$+ \mathbf{R}_t \cdot \nabla_u\mathbf{c}_s + (\hat{\mathbf{n}}_t \cdot \hat{\mathbf{b}} - \phi_t)\frac{\partial\mathbf{c}_s}{\partial\theta}. \tag{24.44}$$

The linear operator

$$\mathcal{L} = \omega\frac{\partial}{\partial\theta} + D\nabla_u^2 + \frac{\delta\mathcal{R}}{\delta\mathbf{c}_s}, \tag{24.45}$$

was discussed in the previous section and has three right eigenvectors corresponding to the eigenvalue zero, $\mathbf{C}_R$ and $\mathbf{C}_T^{\pm}$, which arise from the rotational and translational symmetry of the spiral. We denote these eigenvectors by $|\mathbf{C}\rangle$ and the corresponding left eigenvectors by $\langle\mathbf{C}|$. Taking the scalar product of Eq. (24.44) with $\langle\mathbf{C}|$, we have

$$0 = \langle\mathbf{C}|D\partial/\partial\theta|\mathbf{C}_R\rangle(\phi_s - \tau)^2 + \langle\mathbf{C}|D|\mathbf{C}_R\rangle(\phi_{ss} - \tau_s) - \langle\mathbf{C}|D|\mathbf{C}_+\rangle K$$
$$+ \langle\mathbf{C}|\mathbf{C}_R\rangle(\mathbf{R}_t \cdot \hat{\mathbf{t}})(\phi_s - \tau) + \langle\mathbf{C}|\mathbf{C}_-\rangle(\mathbf{R}_t \cdot \hat{\mathbf{b}})$$
$$+ \langle\mathbf{C}|\mathbf{C}+\rangle(\mathbf{R}_t \cdot \hat{\mathbf{n}}) + \langle\mathbf{C}|\mathbf{C}_R\rangle(\mathbf{n}_t \cdot \hat{\mathbf{b}} - \phi_t). \tag{24.46}$$

In writing this equation we used the fact that $\partial \mathbf{c}_s / \partial \theta = \mathbf{C}_R$, $\partial \mathbf{c}_s / \partial u_1 = \mathbf{C}_+$ with $\mathbf{C}_+ = (\mathbf{C}_T^+ + \mathbf{C}_T^-)/2$, and $\partial \mathbf{c}_s / \partial u_2 = \mathbf{C}_-$ with $\mathbf{C}_- = (\mathbf{C}_T^+ - \mathbf{C}_T^-)/2$. If we select $\langle \mathbf{C} | = \langle \mathbf{C}_R |$, we obtain the equation of motion for ϕ,

$$\phi_t = \mathbf{n}_t \cdot \hat{\mathbf{b}} + (\mathbf{R}_t \cdot \hat{\mathbf{t}})(\phi_s - \tau) + \alpha_1(\phi_{ss} - \tau_s)$$
$$+ \alpha_2(\phi_s - \tau)^2 - \alpha_3 K, \tag{24.47}$$

where $\alpha_1 = \langle \mathbf{C}_R | \mathbf{D} | \mathbf{C}_R \rangle$, $\alpha_2 = \langle \mathbf{C}_R | \mathbf{D} \partial / \partial \theta | \mathbf{C}_R \rangle$, and $\alpha_3 = \langle \mathbf{C}_R | \mathbf{D} | \mathbf{C}_+ \rangle$. The equations of motion for $\mathbf{R}_t \cdot \hat{\mathbf{n}}$ and $\mathbf{R}_t \cdot \hat{\mathbf{b}}$ can be found by letting $\langle \mathbf{C} | = \langle \mathbf{C}_\pm |$, respectively,

$$\mathbf{R}_t \cdot \hat{\mathbf{n}} = -\alpha_4(\phi_s - \tau)^2 - \alpha_5(\phi_{ss} - \tau_s) + \alpha_6 K,$$
$$\mathbf{R}_t \cdot \hat{\mathbf{b}} = -\alpha_7(\phi_s - \tau)^2 - \alpha_8(\phi_{ss} - \tau_s) + \alpha_9 K, \tag{24.48}$$

with $\alpha_{4,7} = \langle \mathbf{C}_\pm | \mathbf{D} \partial / \partial \theta | \mathbf{C}_R \rangle$, $\alpha_{5,8} = \langle \mathbf{C}_\pm | \mathbf{D} | \mathbf{C}_R \rangle$, and $\alpha_{6,9} = \langle \mathbf{C}_\pm | \mathbf{D} | \mathbf{C}_+ \rangle$. The tangential component of the velocity, $\mathbf{R}_t \cdot \hat{\mathbf{t}}$, is not specified by these equations and is arbitrary.

The coefficients in these equations are difficult to compute for general situations. The equations simplify in some cases. For example, if we consider an untwisted scroll ring the torsion is zero and ϕ is independent of the arc length. The equations of motion become

$$\mathbf{R}_t \cdot \hat{\mathbf{n}} = \alpha_6 K, \quad \mathbf{R}_t \cdot \hat{\mathbf{b}} = \alpha_9 K, \quad \phi_t = -\alpha_3 K. \tag{24.49}$$

As an example of this description consider scroll wave filaments in a three-dimensional excitable medium with a spherical shape obeying FitzHugh–Nagumo kinetics. As shown in Fig. 24.10, a scroll wave filament initiated in the interior of the sphere quickly adopts the form of a segment of a ring with ends attached to the sphere surface (see Chavez *et al.*, 2001). The filament segment (densely plotted black points) within the spherical medium with radius R lies on a circle with radius ρ. For a ring we have $\mathbf{R}_t \cdot \hat{\mathbf{n}} = \alpha_6 K = \alpha_6 / \rho$, and from this equation the length of the filament as a function of time can be computed. The result is

$$\rho(t) = R \left(e^{2\alpha_6(t_f - t)/R^2} - 1 \right)^{1/2}, \tag{24.50}$$

where t_f is the lifetime of the filament. This equation is able to describe the simulation results. The effects of twist on scroll wave filaments have also been analyzed in some detail in terms of such ribbon models for the dynamics (Echebarria *et al.*, 2006).

Scroll wave filaments can also undergo bifurcations. For example, if the coefficient α_6 is negative for some values of the parameters, a filament ring will grow

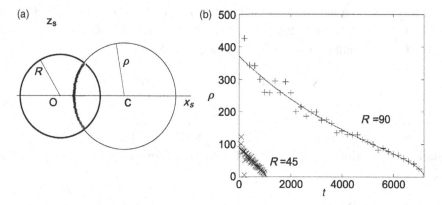

Fig. 24.10. (a) A shrinking ring segment of filament in a sphere of excitable FitzHugh–Nagumo medium (densely plotted black points); (b) the time evolution of the length of the filament obtained from a simulation of the reaction–diffusion equation compared with the theoretical prediction.

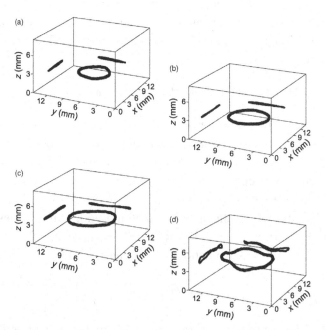

Fig. 24.11. Three-dimensional reconstruction of an expanding and buckling scroll ring with negative tension in an excitable medium. Two-dimensional projections are also shown. Reprinted Figure 3 with permission from Bánsági and Steinbock (2007). Copyright 2007 by the American Physical Society.

instead of shrinking and may develop transverse instabilities. An example of an expanding and buckling scroll ring in the Belousov–Zhabotinsky reaction is shown in Fig. 24.11. Other experimental studies have been carried out by Luengviriya *et al.* (2008). Simulations of expanding scroll rings in some excitable medium models show that they grow to fill the volume with a highly contorted filament that does not self-intersect, leading to what has been termed filament turbulence (Biktashev *et al.*, 1994; Henry and Hakim, 2000, 2002; Alonso *et al.*, 2004). Filament turbulence may play a role in processes such as cardiac fibrillation (Panfilov, 1999; Fenton *et al.*, 2002; Clayton and Holden, 2004; Clayton *et al.*, 2006).

25

Oscillatory media and complex Ginzburg–Landau equation

Thus far we have considered systems which evolve to equilibrium or are in nonequilibrium steady states. Under far-from-equilibrium conditions a number of other types of attracting state may arise. In this chapter we consider situations where stable steady states lose their stability in a Hopf bifurcation leading to stable limit cycle oscillations. In such situations a generic model for the dynamics is the complex Ginzburg–Landau (CGL) equation,

$$\frac{\partial A}{\partial \tau} = A - (1 + i\alpha)|A|^2 A + (1 + i\beta)\nabla^2 A. \tag{25.1}$$

Here $A(\mathbf{x}, \tau)$ is a complex amplitude field or order parameter that depends on position $\mathbf{x}$ and time τ and specifies the slow modulation of the envelope of the oscillatory pattern. The assumptions of slow modulation in space and time and weak nonlinearity that are used in the derivation of the CGL equation apply in many physical circumstances, giving this equation a wide domain of applicability. It has been found to describe a variety of wave and pattern formation phenomena in diverse systems, and has been the subject of frequent study. Reviews (Cross and Hohenberg, 1993; Newell *et al.*, 1993; Aranson and Kramer, 2002) and books (Kuramoto, 1984; Pismen, 1999, 2006) have been devoted to its properties and solutions.

The CGL equation has a long history, and many important contributions have been made to its development. Its origins lie in the work of Landau (1937) on phase transitions, where the free energy was expanded in terms of the order parameter, in models for superconductivity by Ginzburg and Landau (1950) and Stuart (1960) on pattern-forming instabilities (see also Chapter 5). Derivations of complex amplitude equations can be found in Newell and Whitehead (1971).

25.1 Hopf bifurcation

It is useful to review the properties of an Andronov–Hopf bifurcation before considering the derivation of the complex Ginzburg–Landau equation (Nicolis, 1995; Hoyle, 2006). Consider an ordinary differential equation $\dot{\mathbf{c}} = \mathbf{f}(\mathbf{c}, \mu)$ that depends on a parameter μ and has an equilibrium solution $\mathbf{c}_0(\mu)$ for μ near μ_0. Suppose the Jacobian matrix has one pair of complex conjugate eigenvalues that become pure imaginary when $\mu = \mu_0$. Then, as μ passes through μ_0, the equilibrium solution changes stability and a unique limit cycle bifurcates from it.

For two variables, this system of ordinary differential equations is topologically equivalent to the normal form,

$$\dot{c}_1 = (\mu - \mu_0)c_1 - c_2 + \gamma c_1 (c_1^2 + c_2^2),$$
$$\dot{c}_2 = c_1 + (\mu - \mu_0)c_2 + \gamma c_2 (c_1^2 + c_2^2). \tag{25.2}$$

Analysis of this set of equations shows that if $\gamma = -1$ the equilibrium at the origin is asymptotically stable for $(\mu - \mu_0) \le 0$ and unstable for $(\mu - \mu_0) > 0$. A unique stable limit cycle exists for $(\mu - \mu_0) > 0$ with radius $(\mu - \mu_0)^{1/2}$. The Hopf bifurcation is supercritical. If $\gamma = +1$ the equilibrium at the origin is asymptotically stable for $(\mu - \mu_0) < 0$ and unstable for $(\mu - \mu_0) > 0$. There is an unstable limit cycle for $(\mu - \mu_0) > 0$. The Hopf bifurcation is subcritical.

25.2 Complex Ginzburg–Landau equation

The reduction of a reaction–diffusion equation near a Hopf bifurcation point to the CGL equation was carried out by Kuramoto and Tsuzuki (1976) and Kuramoto (1984). In this approach, the focus is on a chemically reacting medium near a Hopf bifurcation point where a steady-state attractor (focus) loses its stability and a limit cycle is born. The system is described by a reaction–diffusion equation,

$$\frac{\partial \mathbf{c}}{\partial t} = \mathcal{R}(\mathbf{c}, \mu) + \mathbf{D}\nabla^2 \mathbf{c}. \tag{25.3}$$

Here, as usual, $\mathcal{R}$ is the reaction rate, which depends on local chemical concentrations and a set of parameters μ. The matrix $\mathbf{D}$ of diffusion coefficients is assumed to be diagonal. We suppose that Eq. (25.3) has a spatially independent steady state $\mathbf{c}_0$ determined from the solution of $\mathcal{R}(\mathbf{c}_0, \mu) = 0$.

Using the notation introduced above, the supercritical Hopf bifurcation is assumed to occur at $\mu = \mu_0$, and a stable limit cycle emerges for $\mu > \mu_0$. Its amplitude grows as $(\mu - \mu_0)^{1/2}$ while the relaxation time to the limit cycle scales as $(\mu - \mu_0)$. Thus, it is convenient to define the small parameter λ by $\lambda = (\mu - \mu_0)^{1/2}$, where $\mu - \mu_0$ gauges the distance from the Hopf bifurcation point μ_0. By introducing the scaled distance and time variables $\mathbf{x} = \lambda \mathbf{r}$ and $\tau = \lambda^2 t$, respectively, we may determine the form of the reaction–diffusion equation which is valid on the long distance and time scales that characterize the behavior near the Hopf point.

To this end, we seek a solution of the form

$$\mathbf{c}(\mathbf{x}, t) = \sum_{k=0}^{\infty} \lambda^k \mathbf{c}_k(\mathbf{x}, t, \tau), \tag{25.4}$$

where the $\mathbf{c}_k$ are bounded functions in t. The reaction term is expanded in a double Taylor series,

$$\mathcal{R}(\mathbf{c}, \lambda) = \sum_{m,n} (m!n!)^{-1} \lambda^{2m} \mathbf{R}_{mn} \odot (\mathbf{c} - \mathbf{c}_0)^n. \tag{25.5}$$

In this equation $\odot$ again stands for tensor contraction (see Chapter 23). The tensor $\mathbf{R}_{mn}$ is defined as

$$\mathbf{R}_{mn} = \left(\frac{\partial^m}{\partial \lambda^m} \left(\frac{\delta^n \mathcal{R}}{\delta \mathbf{c}^n} \right)_{\mathbf{c}_0} \right)_{\mu_0}. \tag{25.6}$$

In order to obtain the CGL equation, we insert these expressions into Eq. (25.3), along with the replacement $\mathbf{D} \cdot \nabla_{\mathbf{r}}^2 \mathbf{c}(\mathbf{r}, t) = \lambda^2 \mathbf{D} \cdot \nabla_{\mathbf{x}}^2 \mathbf{c}(\mathbf{x}, t, \tau)$. Noting that $\partial/\partial t \rightarrow \partial/\partial t + \lambda^2 \partial/\partial \tau$ and collecting terms in powers of λ, we obtain, for the first three orders in λ,

$$\lambda : \partial_t \mathbf{c}_1 = \mathbf{R}_{01} \cdot \mathbf{c}_1, \tag{25.7}$$

$$\lambda^2 : \partial_t \mathbf{c}_2 = \mathbf{R}_{01} \cdot \mathbf{c}_2 + \frac{1}{2} \mathbf{R}_{02} \odot \mathbf{c}_1 \mathbf{c}_1, \tag{25.8}$$

$$\lambda^3 : \partial_t \mathbf{c}_3 + \partial_\tau \mathbf{c}_1 = \mathbf{R}_{01} \cdot \mathbf{c}_3 + \mathbf{R}_{02} \odot \mathbf{c}_1 \mathbf{c}_2 + \frac{1}{6} \mathbf{R}_{03} \odot \mathbf{c}_1 \mathbf{c}_1 \mathbf{c}_1$$
$$+ \mathbf{R}_{11} \cdot \mathbf{c}_1 + \mathbf{D} \nabla^2 \mathbf{c}_1, \tag{25.9}$$

with similar expressions for contributions that are higher order in λ.

At a supercritical Hopf bifurcation point, $\mathbf{R}_{01} \equiv \mathbf{M}$ has a pair of purely imaginary eigenvalues $\pm i\omega_0$. Let $\mathbf{a}$ and $\mathbf{a}^\dagger$, respectively, be the right and left eigenvectors of the matrix $\mathbf{M}$ corresponding to the eigenvalue $i\omega_0$,

$$(i\omega_0 - \mathbf{M})\mathbf{a} = 0, \quad \mathbf{a}^\dagger (i\omega_0 - \mathbf{M}) = 0. \tag{25.10}$$

The solution of the order-λ equation (25.7) is

$$\mathbf{c}_1(\mathbf{x}, t, \tau) = C(\mathbf{x}, \tau) \mathbf{a} e^{i(\omega_0 t + \phi(\mathbf{x}, \tau))} + c.c., \tag{25.11}$$

plus decaying terms that have been neglected. Here *c.c.* stands for the complex conjugate of the expression that precedes it. The real functions $C(\mathbf{x}, \tau)$ and $\phi(\mathbf{x}, \tau)$ are as yet unknown. To solve the order-λ^2 equation (25.8), we substitute the solution for $\mathbf{c}_1$ into this equation and write $\mathbf{c}_2$ as

$$\mathbf{c}_2 = \bar{C}\mathbf{a}e^{i(\omega_0 t + \bar{\phi})} + \mathbf{b}_0 C^2 + \mathbf{b}_2 C^2 e^{2i(\omega_0 t + \phi)} + c.c., \tag{25.12}$$

where $\mathbf{b}_0$ and $\mathbf{b}_2$ are complex constant vectors, and $\bar{C}$ and $\bar{\phi}$, which are real functions of $\mathbf{x}$ and τ, are to be determined later. We find

$$\mathbf{b}_0 = -\frac{1}{2}\mathbf{M}^{-1}\mathbf{R}_{02} \odot \mathbf{aa}^* \tag{25.13}$$

and

$$\mathbf{b}_2 = \frac{1}{2}(2i\omega_0\mathbf{I} - \mathbf{M})^{-1}\mathbf{R}_{02} \odot \mathbf{aa}. \tag{25.14}$$

Finally, substituting the solutions for $\mathbf{c}_1$ and $\mathbf{c}_2$ into the order-λ^3 equation (25.9), we find

$$(\partial_t - \mathbf{M})\mathbf{c}_3 = \mathbf{f}, \tag{25.15}$$

where

$$\begin{aligned}
\mathbf{f} = e^{i(\omega_0 t + \phi)}\Big\{ &- (C_\tau + iC\phi_\tau)\mathbf{a} + C\mathbf{R}_{11} \cdot \mathbf{a} \\
&+ [(\nabla^2 C - C(\nabla\phi)^2 + i(2(\nabla C) \cdot (\nabla\phi) + C\nabla^2\phi)]\mathbf{D} \cdot \mathbf{a} \\
&+ C^3[\mathbf{R}_{02} \odot (\mathbf{ab}_0 + \mathbf{ab}_0^* + \mathbf{a}^*\mathbf{b}_2) + \frac{1}{2}\mathbf{R}_{03} \odot \mathbf{aaa}^*]\Big\} \\
&+ c.c. + \text{nonsecular terms.}
\end{aligned} \tag{25.16}$$

To simplify the equation we use the fact, deduced from Eq. (25.15), that

$$\mathbf{a}^\dagger(\partial_t - \mathbf{M})\mathbf{c}_3 = (\partial_t - i\omega_0)\mathbf{a}^\dagger\mathbf{c}_3 = \mathbf{a}^\dagger\mathbf{f}. \tag{25.17}$$

If we average this equation over one period of the oscillation, we find the relation

$$\int_0^{2\pi/\omega_0} dt\, e^{-i\omega_0 t}(\partial_t - i\omega_0)\mathbf{a}^\dagger\mathbf{c}_3 = \int_0^{2\pi/\omega_0} dt\, e^{-i\omega_0 t}\mathbf{a}^\dagger\mathbf{f} = 0, \tag{25.18}$$

since the integral by parts of the first term is zero. We may now multiply Eq. (25.16) by $\mathbf{a}^\dagger$ from the left, integrate over a period, and use Eq. (25.18) to obtain equations of motion for the unknown functions $C(\mathbf{x}, \tau)$ and $\phi(\mathbf{x}, \tau)$. If we define

$$\mathcal{D} = \mathbf{a}^\dagger\mathbf{D} \cdot \mathbf{a}, \quad \nu = \mathbf{a}^\dagger\mathbf{R}_{11} \cdot \mathbf{a}, \tag{25.19}$$

and

$$\sigma = -\mathbf{a}^\dagger \mathbf{R}_{02} \odot \{\mathbf{a}\mathbf{b}_0 + \mathbf{a}\mathbf{b}_0^* + \mathbf{a}^*\mathbf{b}_2\} - \frac{1}{2}\mathbf{a}^\dagger \mathbf{R}_{03} \odot \mathbf{a}\mathbf{a}\mathbf{a}^*, \qquad (25.20)$$

the equations of motion take the form

$$
\begin{aligned}
C_\tau &= (\nabla^2 C - C(\nabla\phi)^2)D_R - (2(\nabla C)\cdot(\nabla\phi) + C\nabla^2\phi)D_I \\
&\quad + Cv_R - C^3\sigma_R + \text{higher-order terms}, \qquad (25.21) \\
C\phi_\tau &= (\nabla^2 C - C(\nabla\phi)^2)D_I + (2(\nabla C)\cdot(\nabla\phi) + C\nabla^2\phi)D_R \\
&\quad + Cv_I - C^3\sigma_I + \text{higher-order terms}. \qquad (25.22)
\end{aligned}
$$

The real and imaginary parts of the complex coefficients are defined by $v_R = \Re(v)$, $v_I = \Im(v)$, $\sigma_R = \Re(\sigma)$, $\sigma_I = \Im(\sigma)$, $D_R = \Re(\mathcal{D})$, and $D_I = \Im(\mathcal{D})$.

The CGL equation may be written in the standard form given in Eq. (25.1) by scaling space, time, and the variables as: $\tau = \tau'/v_R$, $\mathbf{r} = \mathbf{r}'(D_R/v_R)^{1/2}$, $C = C'(v_R/\sigma_R)^{1/2}$, and $\phi = \phi' - v_I\tau'/v_R$. The parameters in Eq. (25.1) are given by $\alpha = \sigma_I/\sigma_R$ and $\beta = D_I/D_R$. In terms of these scaled variables we obtain

$$
\begin{aligned}
C_\tau &= (\nabla^2 C - C(\nabla\phi)^2) \qquad (25.23) \\
&\quad - (2(\nabla C)\cdot(\nabla\phi) + C\nabla^2\phi)\beta + C - C^3, \\
C\phi_\tau &= (\nabla^2 C - C(\nabla\phi)^2)\beta \qquad (25.24) \\
&\quad + (2(\nabla C)\cdot(\nabla\phi) + C\nabla^2\phi) - C^3\alpha.
\end{aligned}
$$

Finally, letting $A = Ce^{i\phi}$ be a complex amplitude, we can write this pair of equations as the complex Ginzburg–Landau equation (25.1). In general, the CGL equation cannot be written in terms of a free energy functional. However, if $\alpha = \beta = 0$, so that the coefficients are real, the CGL equation reduces to the Ginzburg–Landau equation of model A discussed earlier in Chapter 5.

In some circumstances the amplitude field is essentially constant in space and time, while the system displays substantial variations in the phase. In this case we can reduce the CGL equation to an evolution equation for the phase alone. To determine the form of the phase equation we begin with Eqs. (25.23) and (25.24) and suppose that C varies slowly in space and time in comparison with the variations that occur in the phase field. In this limit we may adiabatically eliminate C from Eq. (25.24) to obtain

$$\phi_\tau = (1 + \alpha\beta)\nabla^2\phi + (\alpha - \beta)(\nabla\phi)^2 - \frac{1}{2}\alpha^2\nabla^4\phi, \qquad (25.25)$$

which is just the Kuramoto–Sivashinsky equation discussed in Chapter 19 in connection with transverse front instabilities. This equation describes the diffusive

dynamics of the phase, provided $(1 + \alpha\beta) > 0$. If $(1 + \alpha\beta) < 0$ diffusion is destabilizing, and the fourth-order gradient term is responsible for the saturation of
the instability. As in the case of front dynamics, in this regime the Kuramoto–
Sivashinsky equation exhibits spatiotemporal chaotic dynamics associated with
phase turbulence, which will be discussed in Section 25.4.

25.3 Stability analysis of CGL equation

25.3.1 Homogeneous limit cycle

We begin the discussion of the stability of various solutions of the CGL equation
with an examination of the response of the homogeneous limit cycle to both homogeneous and inhomogeneous perturbations. In the absence of spatial gradient terms,
the CGL equation (25.1) has a limit cycle solution, $A = \exp(-i\alpha\tau)$. The stability
of this solution to radius and phase perturbations is easily carried out by considering
the linearized forms of Eqs. (25.23) and (25.24) for the deviations $\delta C = C - 1$ and
$\delta\phi = \phi + \alpha\tau$,

$$\begin{pmatrix} \delta C_\tau \\ \delta\phi_\tau \end{pmatrix} = \begin{pmatrix} \nabla^2 - 2 & -\beta\nabla^2 \\ \beta\nabla^2 - 2\alpha & \nabla^2 \end{pmatrix} \begin{pmatrix} \delta C \\ \delta\phi \end{pmatrix}. \tag{25.26}$$

Assuming $C(\mathbf{x}, \tau) \sim e^{\omega\tau} e^{i\mathbf{k}\cdot\mathbf{x}}$ in this set of linearized equations, one may compute
the dispersion relation to find

$$\omega^2 + 2\omega(1 + k^2) + k^2(2 + k^2) + \beta k^2(2\alpha + \beta k^2) = 0. \tag{25.27}$$

From an analysis of this dispersion relation it follows that the homogeneous limit
cycle is stable to inhomogeneous perturbations provided $1 + \alpha\beta > 0$. This is the
Benjamin–Feir condition.

25.3.2 Plane wave solutions

Consider a plane wave of the form

$$A_\mathbf{q}^p = \rho_q e^{i(\mathbf{q}\cdot\mathbf{r} - \omega_q\tau)}. \tag{25.28}$$

After substitution of this form into the CGL equation (25.1), one may verify that
plane waves are solutions provided the following relations are satisfied:

$$\rho_q = (1 - q^2)^{1/2}, \quad \omega_q = \alpha + q^2(\beta - \alpha). \tag{25.29}$$

To study the stability of these solutions we write

$$A = A_\mathbf{q}^p + h e^{i(\mathbf{q}\cdot\mathbf{r} - \omega_q\tau)}, \tag{25.30}$$

and linearize the CGL equation in h. The result is

$$\frac{\partial h}{\partial \tau} = \left\{ -(1+i\alpha)(1-q^2) + 2(1+i\beta)i\mathbf{q} \cdot \nabla + (1+i\beta)\nabla^2 \right\} h$$
$$- (1+i\alpha)(1-q^2)h^*. \tag{25.31}$$

A similar equation may be written for h^*, the complex conjugate of h. Fourier transformation of this pair of equations and evaluation of the secular determinant yields the dispersion relation,

$$\lambda^2 + 2\left\{1 - q^2 + k^2 - i2\beta\mathbf{q} \cdot \mathbf{k}\right\}\lambda + (1+\beta^2)\left\{k^4 - 4(\mathbf{q} \cdot \mathbf{k})^2\right\}$$
$$+ 2(1-q^2)\left\{k^2(1+\alpha\beta) + i2(\alpha-\beta)\mathbf{q} \cdot \mathbf{k}\right\} = 0. \tag{25.32}$$

This dispersion relation may be solved for λ as a power series in k to obtain

$$\lambda = i2(\beta-\alpha)(\mathbf{q} \cdot \hat{\mathbf{k}})k - d_p k^2 + \mathcal{O}(k^3), \tag{25.33}$$

where

$$d_p \equiv 1 + \alpha\beta - 2(1+\beta^2)\frac{(\mathbf{q} \cdot \hat{\mathbf{k}})^2}{(1-q^2)}. \tag{25.34}$$

The plane wave solutions will be stable to long wavelength perturbations provided $d_p > 0$. This is the Eckhaus stability criterion. The Eckhaus instability may have a *convective* character since the unstable modes have a nonzero group velocity, $v_g = 2(\beta-\alpha)(\mathbf{q} \cdot \hat{\mathbf{k}})$, if $\alpha \neq \beta$. This will be the case if the growing perturbation to the plane wave drifts sufficiently rapidly so that growth does not occur at a fixed position. The instability is *absolute* if localized perturbations grow at a fixed position.

The notions of convective and absolute instabilities can be placed in a more general context (Sandstede and Scheel, 2000). In unbounded domains, if perturbations grow in time at every point in the domain, the instability is absolute. In a convective instability, perturbations decay locally at every point in the domain and the growing perturbation is "convected" to infinity. Note that this distinction between absolute and convective instabilities depends on the choice of the coordinate system since in a moving frame a convective instability can become an absolute instability.

Usually simulations and experiments are carried out on bounded domains, and this introduces new features since waves may be reflected at the boundary. It is useful to introduce the concepts of essential and absolute spectra. Letting ω and ν be the temporal and spatial eigenvalues of the problem, the essential spectrum is the set of all ω for which one of the spatial eigenvalues ν is pure imaginary. The absolute spectrum is the set of all ω for which the spatial eigenvalues ν have the same real part. As the domain size tends to infinity all but possibly a fixed number of discrete eigenvalues will converge to the absolute spectrum. Thus, all eigenvalues of the partial differential equation on the bounded domain are close to the absolute spectrum.

Wave patterns on the unbounded domain are unstable when their essential spectrum extends into the right half-plane. Wave patterns that have unstable essential and absolute spectra are absolutely unstable. Patterns with an unstable essential spectrum but a stable absolute spectrum are only convectively unstable.

25.4 Spatiotemporal chaos

The CGL equation possesses a variety of coherent structures that often lead to complicated spatiotemporal states. In the remainder of this chapter we consider such states for one-dimensional systems. One of the simplest coherent structures that has been studied extensively is the Nozaki–Bekki hole (Nozaki and Bekki, 1984) corresponding to a sink solution for a special choice of wavenumbers. The hole solution, characterized by a sharp decrease in the amplitude $|A|$, moves with velocity v and emits plane waves (Fig. 25.1). The hole solutions are not stable to modifications of the CGL equation arising from the addition of higher-order terms. If several hole solutions exist they are separated by shocks where wave fronts collide. Complicated dynamics, which is akin to the dynamics of a collection of interacting particles, can arise from the motions of the holes and their interactions (Chaté and Manneville, 1992; Stiller *et al.*, 1995a, 1995b).

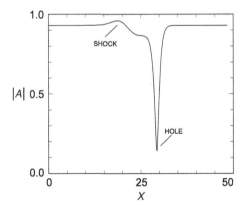

Fig. 25.1. Plot of $|A|$ versus x showing a Nozaki–Bekki hole solution of the CGL equation and its accompanying shock. Reprinted Figure 5 with permission from Aranson and Kramer (2002). Copyright 2002 by the American Physical Society.

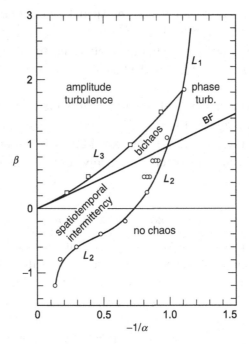

Fig. 25.2. Phase diagram of the one-dimensional CGL equation showing regions of defect/amplitude (defect-mediated) turbulence, phase turbulence, spatiotemporal intermittency and bichaos. These regions are delimited by lines L_1, L_2, and L_3. The Benjamin–Feir (BF) line is also shown. The choice of abscissa and ordinate arises from the different parametrization of the CGL equation. Reprinted from Chaté (1994), with permission of the Institute of Physics.

The approximate phase diagram of the CGL equation is shown in Fig. 25.2 and indicates where various types of spatiotemporal behavior are observed. The regions in the phase diagram include phase turbulence, defect-mediated turbulence, bichaos, and spatiotemporal intermittency.

Two prominent spatiotemporal patterns exhibited by the CGL equation are phase and defect turbulence (Sakaguchi, 1990; Shraiman *et al.*, 1992; Egolf and Greenside, 1995). As one enters the Benjamin–Feir unstable regime the CGL equation displays phase turbulence where the amplitude field $|A|$ varies weakly in space and time and remains close to its saturation value. The spatiotemporal chaos in this regime is characterized by the absence of defects where the amplitude field vanishes. A space–time plot of the dynamics in this regime is shown in Fig. 25.3. One can see that the dynamics is dominated by shocks in the $|A|$ field that propagate in space and time. Close to the BF line the full CGL dynamics may be reduced to a Kuramoto–Sivashinsky equation (25.25) for the phase variable alone, as discussed in Section 25.2.

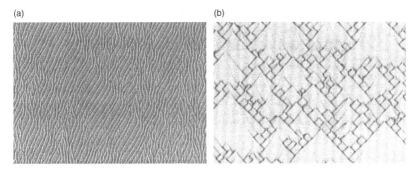

Fig. 25.3. (a) Space–time plot of $|A|$ in the phase turbulent regime. The CGL parameters are $\alpha = -4/3$ and $\beta = 1$. (b) Space–time plot of $|A|$ in the regime of spatiotemporal intermittency. The CGL parameters are $\alpha = -1.43$ and $\beta = 0.5$. In both figures the abscissa is space and the ordinate time, with time running upward in the figure. Reprinted from Chaté (1994), with permission of the Institute of Physics.

Defect-mediated turbulence is observed in the parameter domain roughly delimited by the L_1 and L_3 lines determined by Shraiman *et al.* (1992). In this regime, as the name implies, there are a finite number of defects where $|A| = 0$ and phase slips occur in the phase field. The defects are spontaneously created and destroyed, so that while the number of defects fluctuates, a constant mean number persists in the system. The nature of the transition and location of the precise boundary separating phase and defect chaos is difficult to determine since rare defect-nucleation events cannot be determined easily in simulations. This question has been addressed recently by Brusch *et al.* (2000).

The spatiotemporal dynamics in the intermittency region of the phase diagram is shown in Fig. 25.3. Intermittency is characterized by the existence of laminar domains where plane waves are found, separated by regions of localized structures where $|A|$ is very small and which move in a complicated manner. By contrast, in the bichaos regime the "laminar" domains correspond to regions of phase turbulence.

26

Spiral waves and defect turbulence

Spiral and scroll wave solutions were shown to be prominent wave patterns in two- and three-dimensional excitable media. Such solutions also exist in oscillatory media and are found in a large parameter domain of the complex Ginzburg–Landau equation, as shown in Fig. 26.1 (Rousseau *et al.*, 2008). In this chapter we discuss these spiral wave solutions and describe some of their special characteristics for oscillatory media near the Hopf bifurcation point. While explicit solutions of this type for the CGL equation can be determined only by matched asymptotic expansions or numerical methods, in polar coordinates (r, θ) these solutions take the general form

$$A_s(r, \theta, t) = F(r)e^{i[\omega_s \tau - \sigma\theta + \psi(r)]},$$ (26.1)

with $\omega_s = -\alpha + (\alpha - \beta)k_s^2$. Here $F(r)$ and $\psi(r)$ are two real functions whose asymptotic forms for large and small r may be determined by substitution into the CGL equation. The limiting forms are

$$lim_{r \to \infty} F(r) = \sqrt{1 - k_s^2}, \quad F(r) \propto r \text{ when } r \to 0,$$ (26.2)

and

$$\psi(r) \simeq k_s r \text{ when } r \to \infty, \quad \psi_r(r = 0) = 0.$$ (26.3)

The asymptotic wave number of the spiral waves is k_s. At the center or *core* of the spiral wave the amplitude vanishes, and the phase is undefined. The line integral of the gradient of the phase along a closed curve encircling the core gives the topological charge σ of the spiral (Mermin, 1979),

$$\oint \nabla\phi(\mathbf{r}, t) \cdot d\mathbf{l} = 2\pi\sigma.$$ (26.4)

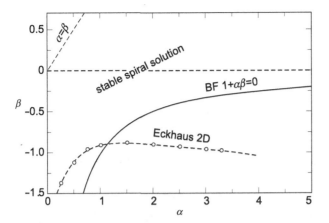

Fig. 26.1. Phase diagram for the two-dimensional complex Ginzburg–Landau equation, showing the parameter region in which stable spiral wave solutions are found. The Eckhaus (EK) and Benjamin–Feir (BF) lines are also shown in the figure.

In two space dimensions, the spiral solution with topological charge $\sigma = \pm 1$ plays a dominant role since solutions with higher topological charge are unstable (Hagan, 1982).

26.1 Core instability

Spiral waves in oscillatory media can develop instabilities that lead to complex spatiotemporal structures. In certain parameter regions the spiral core no longer remains stationary and instead accelerates. This instability occurs when β is large so that dispersion dominates over diffusion. In this regime it is convenient to rescale positions by replacing $\mathbf{r}$ by $\mathbf{r}\sqrt{|\beta|}$ so that the CGL equation takes the form (Aranson and Kramer, 2002)

$$\frac{\partial A}{\partial \tau} = A - (1 + i\alpha)|A|^2 A + (\epsilon + i)\nabla^2 A, \qquad (26.5)$$

where $\epsilon = 1/|\beta|$. For $\epsilon = 0$ this equation supports a family of spiral waves,

$$A_{\mathrm{as}}(r,\theta,t) = F(r') \exp\left(i\left[\omega'_s \tau - \sigma\theta + \psi(r') + \frac{1}{2}\mathbf{r}' \cdot \mathbf{v} \right] \right), \qquad (26.6)$$

moving with arbitrary velocity $\mathbf{v}$ where $\mathbf{r}' = \mathbf{r} + \mathbf{v}\tau$ and $\omega'_s = \omega_s - v^2/4$. For nonzero ϵ these solutions develop a slowly varying velocity. Consequently, as β increases to sufficiently large values, the core of the spiral moves. This core motion is the analog of meandering seen in excitable systems, which was discussed in

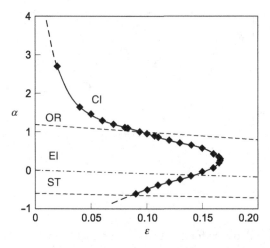

Fig. 26.2. Phase diagram showing the stability of spiral wave solutions to the core instability (CI) of the two-dimensional CGL equation (heavy line with diamonds). The oscillatory range (OR), Eckhaus instability (EI) and strong turbulence (ST) regimes are indicated. Reprinted Figure 13 with permission from Aranson and Kramer (2002). Copyright 2002 by the American Physical Society.

Chapter 24. In contrast to excitable media, for CGL spirals the instability giving rise to the core motion does not saturate, and the core continues to accelerate as it moves.

The stability of CGL spirals to the core instability has been determined numerically, and the results are summarized in the phase diagram in Fig. 26.2. Complex spatiotemporal states with distinctive characteristics exist within the core-unstable region for large β, and are delimited by the lines OR, EI, and ST in this figure.

Apart from the core instability, CGL spirals may undergo other instabilities. Spiral waves may be stable on large bounded domains, even though they are unstable on unbounded domains due to an unstable essential spectrum. The transition from convective to absolute instability causes the spiral to break up. Spiral-wave breakup can occur in either the far field or the core of the spiral wave (Bär and Or-Guil, 1999; Brusch *et al.*, 2000; Sandstede and Scheel, 2000).

26.2 Defect-mediated turbulence

Weakly driven dissipative pattern-forming systems often exhibit spatiotemporal chaos in the form of defect-mediated turbulence. In such turbulent states the dynamics of a pattern is dominated by the rapid motion, nucleation, and annihilation of point defects or vortices (Coullet *et al.*, 1989). Examples can be found in striped patterns in wind-driven sand, electroconvection in liquid crystals (Rehberg *et al.*, 1989), nonlinear optics (Ramazza *et al.*, 1992), fluid convection (Morris *et al.*, 1993;

(a)   (b)

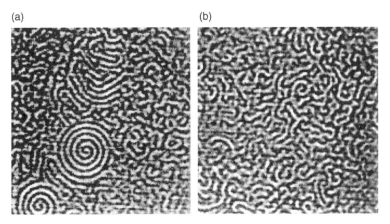

Fig. 26.3. Defect-mediated turbulence in the BZ reaction. (a) Spatial structure close to the instability; (b) fully developed spatiotemporal turbulence. The control parameter is the concentration of H_2SO_4 in the feed reactor. Reprinted by permission from Ouyang and Flesselles (1996). Copyright 1996 Macmillan Publishers Ltd.

Daniels and Bodenshatz, 2002), cardiac tissue (Davidenko *et al.*, 1992), and Langmuir circulation in the oceans (Haeusser and Leibovich, 1997), to name only a few. Defect-mediated turbulence has also been observed in experiments on the BZ reaction (Ouyang and Flesselles, 1996). Figure 26.3 shows a chemical pattern near the onset of the instability giving rise to spatiotemporal turbulence. Note that small, well-defined spirals can still be seen embedded in a sea of turbulent dynamics. Well beyond the instability one sees fully developed turbulence. These results suggest that the dynamics of these very different systems can be characterized by a universal description which is based only on the defect dynamics.

Since the CGL equation provides a generic description of an oscillatory medium, detailed studies of defect-mediated turbulence have been carried out for this system. Figure 26.4 shows the CGL system in the defect-mediated turbulence regime and illustrates distribution of spiral defects in the turbulent dynamics described above.

In the defect-mediated turbulent state the average number of defect pairs is fixed but their instantaneous number fluctuates around a statistically stationary value (Fig. 26.5). Assuming the creation and annihilation of defect pairs occur in a statistically independent fashion, a simple model for the probability distribution of the number of defect pairs at time t, $p(n, t)$, was constructed by Gil *et al.* (1990). The master equation describing the pair-number probability density is

$$\frac{\partial p(n, t)}{\partial t} = c(n - 1)p(n - 1, t) + a(n + 1)p(n + 1, t)$$
$$- (c(n) + a(n))p(n, t), \tag{26.7}$$

(a)   (b)

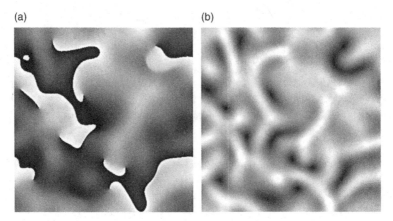

Fig. 26.4. Defect-mediated turbulence in the CGL equation. (a) The phase, $\arg(A)$, as gray shades; (b) the amplitude, $|A|$, with a similar color coding. Topological defects can be identified as points in the phase field around which one finds all shades of gray. The amplitude exhibits a random spatial pattern.

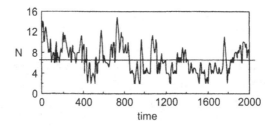

Fig. 26.5. Number of defects as a function of time for the CGL equation with parameters $\alpha = 2.0$ and $\beta = -0.85$. Reprinted Figure 1 with permission from Gil *et al.*, 1990. Copyright 1990 by the American Physical Society.

where $c(n)$ and $a(n)$ are the creation and annihilation rates of defect pairs. The stationary solution of the master equation leads to the recursion relation,

$$p(n) = \frac{c(n-1)}{a(n)} p(n-1). \tag{26.8}$$

Assuming that the creation rate is constant, $c(n) = c$, and that the annihilation rate is determined by a binary collision process so that $a(n) = a_0 n^2$ with a_0 a constant, the solution of this equation is a squared-Poissonian distribution,

$$p(n) \propto \frac{(c/a_0)^n}{(n!)^2}. \tag{26.9}$$

The prediction that the probability density is a squared-Poissonian distribution has been verified both in simulations of the CGL equation and in experiments on systems exhibiting defect-mediated turbulence.

In experimental studies of defect-mediated turbulence, effects arising from defects entering and leaving through the boundaries must be taken into account. Such boundary effects can alter the defect creation and annihilation rates that enter in the master equation model. Boundary effects have been accounted for in the study of defect-mediated turbulence in inclined-layer convection (Daniels and Bodenschatz, 2002). Including these boundary effects in the master equation model led to a modified Poissonian distribution that was able to describe the experimental data.

26.3 Other spatiotemporal states

In addition to defect-mediated turbulence, a number of other spatiotemporal states have been observed in simulations of the CGL equation and related reaction–diffusion models in two spatial dimensions. In particular, the analogs of phase turbulence and spatiotemporal intermittency as well as glassy vortex states, shown in Fig. 26.6, have been studied (Brito *et al.*, 2003; Chaté and Manneville, 1996). In the vortex glass, spirals grow from random initial conditions to fill the entire periodic domain. The spirals occupy cells whose boundaries are shock lines where spiral wave fronts collide. Such glassy states have very long lifetimes. A description of many of these states is given in Aranson and Kramer (2002).

(a)   (b)

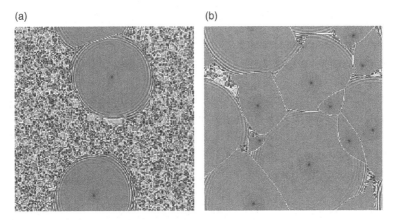

Fig. 26.6. (a) Nucleation of spirals from random initial conditions. (b) After a transient, a vortex glassy state is formed. The amplitude field $|A|$ is shown. Reprinted from Chaté and Manneville (1996). Copyright 1996, with permission from Elsevier.

26.4 Scroll wave solutions

Just as in excitable media, scroll wave solutions of the CGL oscillatory medium exist in three-dimensional systems. Following the development in Chapter 24, it is useful to focus on the dynamics of the singular filament that corresponds to the core of the scroll wave. We again assume that the filament of the scroll wave is described by the space curve $\mathbf{R}(s, t)$, parameterized by the arc length. We adopt a coordinate system similar to that used in Chapter 24, where any point $\mathbf{X}$ in the vicinity of filament can be written as $\mathbf{X}(s, \mathbf{u}, t) = \mathbf{R}(s, t) + u_1 \hat{\mathbf{n}}(s, t) + u_2 \hat{\mathbf{b}}(s, t)$, where $\hat{\mathbf{n}}(s, t)$ and $\hat{\mathbf{b}}(s, t)$ are the normal and binormal, respectively. Here $\mathbf{u} = (u_1, u_2)$. The tangent $d\mathbf{R}(s)/ds = \hat{\mathbf{t}}(s)$, normal, and binormal satisfy the Frenet–Serret equations, $d\hat{\mathbf{t}}/ds = K\hat{\mathbf{n}}$, $d\hat{\mathbf{n}}/ds = -K\hat{\mathbf{t}} + \tau\hat{\mathbf{b}}$, and $d\hat{\mathbf{b}}/ds = -\tau\hat{\mathbf{n}}$, where τ is the torsion.

Employing the methods discussed in Chapter 24, Frenet–Serret equations can be used to reduce the CGL equation to an equation of motion for the filament. Such an analysis was carried out by Gabbay *et al.* (1997), who obtained the filament equations,

$$\hat{\mathbf{n}} \cdot \mathbf{R}_t = -(1 + \beta^2)K, \quad \hat{\mathbf{b}} \cdot \mathbf{R}_t = 0, \tag{26.10}$$

where β is the parameter appearing in the CGL equation. This equation is valid in the regime where the radius of curvature of the filament is much larger than the filament core radius. Apart from a straight filament, one of the simplest geometries to investigate is a closed circular filament with radius R and no phase twist. In this case Eq. (26.10) reduces to

$$\frac{dR(t)}{dt} = -(1 + \beta^2)\frac{1}{R(t)}. \tag{26.11}$$

The solution of this equation is $R^2(t) = R^2(t_0) - 2(1 + \beta^2)(t - t_0)$. Thus, the curvature-driven dynamics causes the square of the radius of the filament to contract linearly with time.

Equation (26.10) breaks down if β is large, and this breakdown was investigated by Aranson *et al.* (1998). By carrying out an asymptotic expansion for $\epsilon \ll 1$ of the CGL equation in the form given in Eq. (26.5) used in the analysis of the core instability, an equation of motion for the filament that includes inertial terms may be obtained, and is given by

$$\frac{d\mathbf{v}}{dt} = -\hat{K}(\epsilon\mathbf{v} - K\mathbf{n}), \tag{26.12}$$

where $\mathbf{v}$ is the velocity of the filament and $\hat{K}$ is a friction term that appears in the analysis of the $2d$ core instability. Since the velocity may vary along the filament,

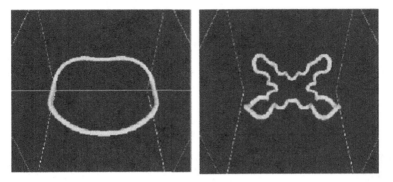

Fig. 26.7. Time evolution of a vortex ring undergoing a filament instability that bends and stretches the ring. Reprinted Figure 6 with permission from Aranson *et al.*, 1998. Copyright 1998 by the American Physical Society.

the acceleration term can cause an instability which will lead to stretching and bending of the filament. An example of such an instability for a vortex ring is shown in Fig. 26.7.

26.5 Twisted filaments

Since the scroll wave filament can be described by a ribbon curve when the phase field is taken into account, we can consider what may happen when a twist is applied to the phase field (Gabbay *et al.*, 1997; Nam *et al.*, 1998; Rousseau *et al.*, 1998). In physical situations twist may arise from gradients or inhomogeneities in the medium. For scroll rings or periodic boundary conditions the ribbon curve is closed, and White's theorem (White, 1969),

$$\text{Lk} = \text{Tw} + \text{Wr} = \text{Tw}_f + \text{Tw}_r + \text{Wr}, \tag{26.13}$$

is applicable. This equation expresses the conservation law relating various topological quantities characterizing the ribbon curve. The (integer) link number Lk represents half of the sum of signed crossings of the two ribbon boundaries. The twist,

$$\text{Tw} = \oint \tau(s)ds + \oint (d\varphi/ds)ds \equiv \text{Tw}_f + \text{Tw}_r, \tag{26.14}$$

which is again decomposed into the twist of the filament Tw_f and the ribbon twist Tw_r, is the number of times the phase field is wrapped around the binormal along the curve. The writhe

$$\text{Wr} = \frac{1}{4\pi} \oint \oint \frac{(\hat{\mathbf{t}}(t) \times \hat{\mathbf{t}}(s)) \cdot (\mathbf{R}(t) - \mathbf{R}(s))}{|\mathbf{R}(t) - \mathbf{R}(s)|^3} dt ds \tag{26.15}$$

quantifies the nonplanarity of the filament curve.

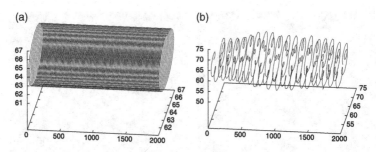

Fig. 26.8. Helical (a) and superhelical (b) vortex filaments in large CGL oscillatory media.

Consider the region of parameter space where the spiral solution is linearly stable, which implies stability of the core and at least absolute stability of the plane wave of wavenumber $k_s(\alpha, \beta)$. In a finite box of height L_z along z which is periodic in z, as long as the filament is unbroken, the link number is conserved, and the link density γ is quantized by the box size,

$$\mathrm{Lk} = 2\pi\, n_t = \gamma L_z. \tag{26.16}$$

Here n_t is the number of times the phase winds around the filament curve. If the filament is straight, the Frenet frame is degenerate. The writhe is zero, and one can choose to have $\mathrm{Tw}_f = 0$ and $\mathrm{Lk} = \mathrm{Tw}_r$, so that γ can also be seen as the (initial) phase twist density. The conservation of the total link requires one to consider γ as an extra parameter of the problem in addition to (α, β), the usual parameters of the CGL equation.

Consider a straight, infinite filament oriented along the z axis, with a constant link per unit length γ. The solution of the three-dimensional CGL equation for this case may be expressed in cylindrical coordinates (r, θ, z) as

$$A_f = \sqrt{1 - \gamma^2}\, F(r') \exp\left(i[\omega_f t \pm (\theta + \gamma z) + \psi(r')]\right) \tag{26.17}$$

where $r' = r\sqrt{1 - \gamma^2}$ and $\omega_f = \omega_s(1 - \gamma^2) - \gamma^2\alpha$, with ω_s, F, and ψ given by the 2d spiral solution (26.1).

We are interested in the stability of this solution. Three-dimensional simulations of the CGL equation in a cylindrical box of height L_z and diameter $L_x = L_y$ which is periodic in z illustrate what happens as the twist on the filament is increased. The initial state corresponds to the phase-twisting the 2d spiral solution n_t times along the z axis. For a small link density, in a large region of the (α, β) plane, the initial state quickly relaxes to solution (26.17): i.e. the system merely takes into account

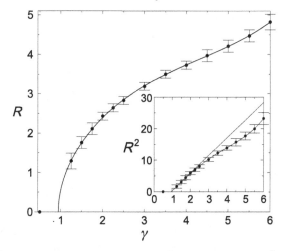

Fig. 26.9. Radius of the helix versus γ. The inset shows that R^2 varies linearly with γ consistent with a supercritical Hopf bifurcation.

the amplitude, wavenumber, and frequency corrections with respect to the $2d$ spiral solution.

Increasing the link per unit length γ, the straight filament remains stable up to some critical value γ_c. For larger γ values a helical deformation of the filament occurs. The radius R of the helix grows and saturates at a finite value forming a helical filament with wavelength λ_h, shown in Fig. 26.8 (Rousseau et $al.$, 2008). Note that $\gamma_h \equiv 2\pi/\lambda_h \neq \gamma$. The radius of the helix may be used as an order parameter for the bifurcation, and is plotted versus γ in Fig. 26.9 (Rousseau et $al.$, 1998). The frequency of the asymptotic (far field) waves also changes from ω_f to ω_h. The helix itself rotates with constant angular velocity ω, and any point on it moves along the z-axis at a constant velocity $v = \omega/\gamma_h$. This velocity is just a phase velocity: a localized perturbation merely diffuses; i.e. the group velocity is zero. These results indicate that the straight filament has undergone a (supercritical) Hopf bifurcation to traveling waves. The direction of propagation and rotation is determined by the sign of γ, which breaks the $z \rightarrow -z$ parity symmetry.

The curvature and torsion of the helical solution are constant along the z axis: $K/(R\gamma_h) = \tau = \gamma_h/(1 + R^2\gamma_h^2)$. We then have

$$\frac{\mathrm{Tw}_f}{\mathrm{Lk}} = \frac{\gamma_h}{\gamma} - \frac{\mathrm{Wr}}{\mathrm{Lk}} = \frac{\gamma_h}{\gamma\sqrt{1 + R^2\gamma_h^2}}, \qquad (26.18)$$

and the ribbon twist is expressed as

$$\frac{\mathrm{Tw_r}}{\mathrm{Lk}} = 1 - \frac{\gamma_h}{\gamma}.$$ (26.19)

A *negative* ribbon twist appears to "compensate" the torsion arising from the instability. This helical bifurcation (termed *sproing*) was seen in excitable media and analyzed in a similar way by Henze *et al.* (1990).

As the twist density increases, the helical filament undergoes a series of secondary bifurcations to more complex coiled forms. A secondary Hopf bifurcation leads to a supercoiled structure shown in Fig. 26.8. Far from the filament defect line, temporal oscillations of the local field are characterized by three independent frequencies. These are associated with a dominant local pulsation and amplitude modulations related to the variation of the distance to the defect line, leading to two extra frequencies for this superhelix solution. A variety of other localized solutions, which involve localized superhelical regions and long helical segments, exist in other regions of the parameter space.

27

Complex oscillatory and chaotic media

In the previous chapter generic features of spiral wave dynamics in oscillatory media were described on the basis of the complex Ginzburg–Landau equation. Spiral waves can also exist in complex oscillatory media where the local dynamics can have period-doubled or even chaotic oscillations. In regimes where complex-oscillatory behavior is found, the new feature that appears in spiral waves is a line defect across which the phase of the oscillation changes by 2π. The presence of line defects leads to spatiotemporal patterns not seen in media with simple local oscillatory dynamics.

Complex periodic or chaotic oscillations do not have simple single-loop trajectories in concentration phase space. For example, a period-n limit cycle is described by a period-n orbit that loops n times in concentration phase space before closing on itself (see Fig. 27.1). In such circumstances no simple single-valued angle variable may be introduced to play the role of the phase. It is often possible to generalize the definition of phase, even for systems whose dynamics is chaotic, and this is related to the phenomenon of phase synchronization (Rosenblum *et al.*, 1997; Pikovsky *et al.*, 2001; Osipov *et al.*, 2003).

A spiral wave is an example of a self-organized structure that is a result of phase synchronization in a medium with complex local dynamics. Reaction–diffusion equation studies (Goryachev and Kapral, 1996a, 1996b; Goryachev *et al.*, 1998, 2000) and experiments (Yoneyama *et al.*, 1995; Park and Lee, 1999, 2002; Guo *et al.*, 2004; Park *et al.*, 2004) have demonstrated that spiral waves with synchronization defect lines exist in spatially distributed systems that undergo period-doubling bifurcations. An example of an experimental spiral wave with a line defect in the Belousov–Zhabotinsky reaction is shown in Fig. 27.2. Spiral waves persist even if the local dynamics is chaotic (Klevecz *et al.*, 1991; Brunnet *et al.*, 1994; Goryachev and Kapral, 1996a; Zhan and Kapral, 2006). Line defects change the structure and symmetry of spiral waves and influence the nature of the spatiotemporal dynamics in media with complex-oscillatory local dynamics.

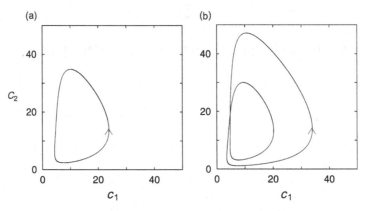

Fig. 27.1. (a) A single-loop period-1 orbit for the Willamowski–Rössler model; (b) the phase space trajectory for a period-2 limit cycle that makes two loops in phase space before closing on itself.

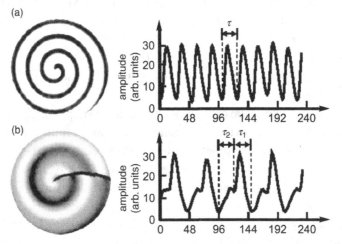

Fig. 27.2. (a) (left) Spiral wave in the Belousov–Zhabotinsky reaction–diffusion system; (right) the local dynamics is period-1. (b) Spiral wave in the period-2 regime with a line defect. Reprinted Figure 1 with permission from Park *et al.*, 2004. Copyright 2004 by the American Physical Society.

27.1 Spiral waves and line defects

In order to investigate the nature of spiral wave patterns in complex oscillatory media we consider reaction–diffusion systems,

$$\frac{\partial}{\partial t}\mathbf{c}(\mathbf{r}, t) = \mathcal{R}(\mathbf{c}(\mathbf{r}, t)) + D\nabla^2\mathbf{c}(\mathbf{r}, t), \tag{27.1}$$

with reactive dynamics that display a period-doubling cascade to chaos. Most simulations have been carried out on the Rössler (Rössler, 1976) and Willamowski–Rössler (Willamowski and Rössler, 1980) models. Both of these models involve three chemical concentrations $\mathbf{c}(\mathbf{r}, t) = (c_X, c_Y, c_Z)$, and the corresponding reaction rates are, respectively,

$$
\begin{aligned}
\mathcal{R}_X^R &= -c_Y - c_Z, \\
\mathcal{R}_Y^R &= c_X + A c_Y, \\
\mathcal{R}_Z^R &= c_X c_Z - C c_Z + B,
\end{aligned} \tag{27.2}
$$

and

$$
\begin{aligned}
\mathcal{R}_X^{WR} &= k_1 c_X - k_{-1} c_X^2 - k_2 c_X c_Y + k_{-2} c_Y^2 - k_4 c_X c_Z + k_{-4}, \\
\mathcal{R}_Y^{WR} &= k_2 c_X c_Y - k_{-2} c_Y^2 - k_3 c_Y + k_{-3}, \\
\mathcal{R}_Z^{WR} &= -k_4 c_X c_Z + k_{-4} + k_5 c_Z - k_{-5} c_Z^2.
\end{aligned} \tag{27.3}
$$

In both cases, as a bifurcation parameter is changed, reaction–diffusion systems with these reaction kinetic terms undergo a period-doubling bifurcation, although the parameter value at which the bifurcation occurs differs in the PDE and ODE systems. Furthermore, in these reaction–diffusion systems, only a finite number of period-doubling bifurcations are seen before more complex spatiotemporal dynamics appears.

Several features associated with period doubling of spiral waves complicate the description of the nature of the bifurcation. A simple spiral wave in a circular domain may be viewed in a frame rotating with the period ω of the spiral. In such a frame the spiral will appear stationary, and it constitutes a relative equilibrium state of the system. As a consequence of the rotational symmetry, only saddle-node and Hopf bifurcations are permitted, and the standard period-doubling bifurcation, which is associated with a Floquet multiplier of -1, cannot occur. The symmetry arguments are akin to those discussed in Chapter 24 for the meandering transition for spiral waves. The most reasonable explanation for the spiral period-doubling bifurcation is that it is a Hopf bifurcation where the Hopf frequency ω_H is in 2:1 resonance with the spiral frequency, so that $\omega_H = \omega/2$. The full explanation of why the 2:1 resonance occurs and is generic was given by Sandstede and Scheel (2007).

In order to understand the nature of the line defect, first consider a spiral wave in a simple oscillatory (period-1) medium. For such a period-1 oscillation the phase variable is just the angle variable $\varphi = \arctan((c_Y - c_Y^0)/(c_X - c_X^0))$, where c_X^0 and c_Y^0 are reference concentrations. The reference concentrations can be chosen to lie near the unstable fixed point that gave rise to the oscillation in a Hopf bifurcation, or in the interior of the planar projection of the limit cycle (see Fig. 27.1). The line integral of the gradient of the phase along the closed curve gives the topological charge $\pm n_t$ of the spiral,

$$\oint \nabla \varphi(\mathbf{r}, t) \cdot d\mathbf{l} = \pm 2\pi n_t. \qquad (27.4)$$

For a stable one-armed spiral $n_t = 1$, and the phase is incremented by 2π along such a closed curve.

Next, suppose a control parameter is changed and the local dynamics signals a bifurcation from period 1 to period 2 where the period of the orbit doubles so that $T_2 \approx 2T_1$. In the period-1 regime the spiral wave has rotational symmetry such that evolution in time through one period T_1 corresponds to rotation through 2π of the entire concentration field of the spiral about its center. Just beyond the bifurcation point, where the period of the orbit doubles, one full rotation of the spiral no longer restores the concentration field to its original value. Instead two spiral rotations are required to restore symmetry. The reaction–diffusion medium is continuous in space, so that when the closed path surrounding the defect is traversed one must arrive at the starting value of the concentration. In view of the fact that it takes two rotations of the spiral through an angle 2π to restore the entire concentration field to its initial value, there must be an additional phase jump of 2π to satisfy the continuity requirements. This phase jump of 2π occurs along a synchronization line defect emanating from the spiral core. Figure 27.3 shows the change in the spiral wave structure on both sides of the period-2 bifurcation point. When the local dynamics is period-1, a simple spiral wave exists. In the period-2 regime, the spiral wave has a synchronization line defect.

The nature of the synchronization line defect can be seen by examining the structure and transformations of local trajectories in phase space as the line defect is crossed. The phase space trajectories at five spatial points along an arc that crosses the line defect are shown in Fig. 27.4. As the arc is traversed, the larger, outer loop of the local orbit constantly shrinks while the smaller, inner loop grows. On the line defect both loops merge and then pass each other, exchanging their positions in phase space. Since the entire loops coincide in phase space on the line defect, the local oscillation is effectively period-1 on this line while it is period-2 elsewhere in the medium.

27.2 Archimedean spiral splay field and line defects

The splay state is a self-organized structure in a spatially distributed medium in which all local oscillators execute identical periodic trajectories but with different phases. In Chapter 25 we saw that the core region that surrounds the topological defect in the complex Ginzburg–Landau equation is very small. Outside

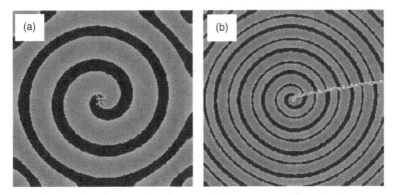

Fig. 27.3. Instantaneous images of the $c_Z(\mathbf{r}, t)$ field showing spiral waves in the Willamowski–Rössler reaction–diffusion system for two parameter values: (a) $k_2 = 1.430$ (period-1 regime) and (b) $k_2 = 1.510$ (period-2 regime). The other parameters are: $k_1 = 31.2$, $k_{-1} = 0.2$, $k_{-2} = 0.1$, $k_3 = 10.8$, $k_{-3} = 0.12$, $k_4 = 1.02$, $k_{-4} = 0.01$, $k_5 = 16.5$, and $k_{-5} = 0.5$. (a) There is a single defect point (spiral tip), indicated by a large star, for the period-1 spiral wave. (b) The period-2 spiral has a line defect, which is superimposed on the pattern. Reprinted Figure 1 with permission from Zhan and Kapral (2005). Copyright 2005 by the American Physical Society.

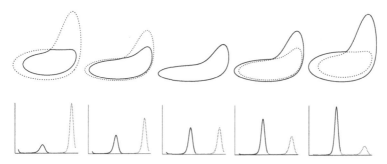

Fig. 27.4. Loop exchange in local orbits in (c_X, c_Y, c_Z) phase space as the line defect curve is crossed. First and second halves of an oscillation period are shown by different line styles. Reprinted with permission from Goryachev and Kapral (2000). Copyright 2000, American Institute of Physics.

the core the local dynamics asymptotically tends to a limit cycle attractor with harmonic character. In this context, the spiral wave may be considered to be a generalized splay state field where the phase space trajectories at all spatial points, except those in the core region, have identical structure but differ in their phases. Similar considerations apply to complex oscillatory media where the local dynamics has a period-doubled or even chaotic character. The spiral waves discussed above are Archimedean spirals, and for such waves the geometrical structure in space and the dynamical behavior in time are related. We can view the system

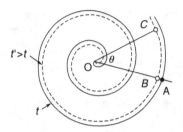

Fig. 27.5. Schematic picture of Archimedean spirals at times t (solid line) and t' (dashed line) with points A, B, and C used in the analysis.

as being in an Archimedean-spiral generalized splay state with the exception of a small core region. The form of line defect can be determined using these ideas (see Zhan and Kapral, 2005).

Given the Archimedean spiral structure, outside the core region the phase of any point, or the lag time between any two spatial points, can be determined. An Archimedean spiral (solid line) at time t is shown in Fig. 27.5. A point A on this spiral is given by

$$r_A = \frac{P}{2\pi}(\theta_A - \theta_{A0}), \tag{27.5}$$

where P is the pitch, r_A and θ_A ($+\infty > \theta_A > 0$) are the polar coordinates of point A, and θ_{A0} is the initial angle with respect to a reference axis that characterizes this Archimedean spiral. Counterclockwise is chosen as the positive direction. Point B lies on another Archimedean spiral (dashed line) at time t' ($t' > t$), and its equation is

$$r_B = \frac{P}{2\pi}(\theta_B - \theta_{B0}), \tag{27.6}$$

where θ_{B0} ($\theta_{B0} \neq \theta_{A0}$) is a different initial phase that characterizes the dashed spiral. Since points A and B lie along the same radial vector, they have identical polar angles ($\theta_B = \theta_A$). The only difference in the local dynamics between points B and A is the lag time,

$$\Delta t_{BA} = (\theta_{B0} - \theta_{A0})\frac{T}{2\pi} = (r_A - r_B)\frac{T}{P}, \tag{27.7}$$

which arises because their initial phases are different. For any point C in the domain with a polar angle θ that is different from that of A, the expression for the lag time between C and A is

$$\Delta t_{CA} = \left(r_A - \left(r_C - \frac{P}{2\pi}\theta\right)\right)\frac{T}{P}, \tag{27.8}$$

where the fact that points C and B have identical phases has been used.

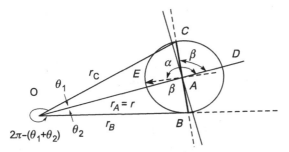

Fig. 27.6. Diagram showing the geometry used in the calculation. The point A lies on the line defect, and the heavy dashed arrow is tangent to the line defect at point A. The perpendicular distances $\overline{AB} = \overline{AC} = \xi$. The angles are $\beta = \angle DAC = \angle BAO$ and $\alpha = \pi/2 + \beta = \angle DAE$.

Figure 27.6 shows a schematic representation of a portion of the line defect for a system with period-2 oscillatory dynamics. At a point A on the line defect the local dynamics is period-1. In the vicinity of the line defect there is a narrow interfacial zone of width 2ξ that connects the period-2 regions on either side of the line that differ in phase by 2π. Without loss of generality, the two points B and C are at the same distance ξ away from the point A. The line BC is perpendicular to the tangent to the line defect at point A. The lengths of the line segments are $\overline{AB} = \overline{AC} = \xi$.

The lag time $T(\xi)$ between points B and C is

$$T(\xi) = \left(\frac{1}{2} + \frac{r_C - r_B}{P} - \frac{\theta_1 + \theta_2}{4\pi}\right)T. \qquad (27.9)$$

Points B and C are taken to lie in the period-2 region where the Archimedean spiral splay field approximation is valid. Assuming that $r_A(r_A = r), r_B, r_C \gg 1$ and $\theta_1, \theta_2 \ll 1$, $r_C - r_B \approx 2\xi \cos\beta$, and $\theta_1 + \theta_2 \approx \frac{2\xi}{r}\sin\beta$, where $\beta = \angle DAC = \angle BAO$. We find

$$T(\xi) = \left(\frac{1}{2} + \frac{2\xi \cos\beta}{P} - \frac{2\xi \sin\beta}{4\pi r}\right)T. \qquad (27.10)$$

Using the fact that the solution X of the equation $a \sin X + b \cos X = c$, where $a, b,$ and c are constants, can be written as $\sin(X + \theta) = c(a^2 + b^2)^{-1/2}$ and $\theta = \arctan(b/a)$, we obtain

$$\beta = \arctan\left(\frac{4\pi r}{P}\right) + \arcsin\left(\frac{T(\xi)/T - 1/2}{2\xi\sqrt{(1/P)^2 + (1/4\pi r)^2}}\right). \qquad (27.11)$$

It follows that $\alpha(r) = \angle DAE$, the angle between the polar and tangent directions of the line defect at A, is $\alpha(r) = \frac{\pi}{2} + \beta = \frac{\pi}{2} + \alpha_1 + \alpha_2$, where

$$\alpha_1 = \arctan\left(\frac{4\pi r}{P}\right), \quad \alpha_2 = \arcsin\left(\frac{\Lambda}{2\sqrt{(1/P)^2 + (1/4\pi r)^2}}\right). \quad (27.12)$$

The equality between α and $\pi/2 + \beta$ is independent of shifts in the direction of the line BC around A (clockwise or counterclockwise). Here $\Lambda = (T(\xi)/T - 1/2)/\xi$ plays the role of an order parameter. While $\Lambda(\xi)$ varies rapidly within the interfacial zone surrounding the line defect, it tends to a nearly constant plateau value outside the narrow zone. Consequently, $\alpha(r)$ does not depend strongly on ξ for values of ξ lying outside the interfacial zone.

An equation for the line defect can be constructed from this information. If α is the angle between the polar and the tangent directions of any curve expressed in polar coordinates (r, θ), we have $\tan\alpha = r/r'$, where $r' = dr/d\theta$. Taking the curve to be the line defect we obtain

$$\theta_b = \theta_a + \int_a^b \frac{1}{r}\tan\alpha(r)dr, \quad (27.13)$$

where θ_a and θ_b are polar angles at two points a and b on the line defect. Given the position of the spiral tip and any point a on the line defect, we can determine the location of any other point b on the line defect by performing the integral in Eq. (27.13) using the knowledge of the functional form of $\alpha(r)$. Figure 27.7 shows

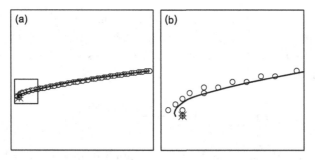

Fig. 27.7. (a) Comparison of the simulation results for the line defect for $k_2 = 1.510$ (overlapping open circles) with the prediction of Eq. (27.13) (solid line, obscured by the open circles). For the chosen parameters, $T = 1.919$, $P = 33.5$, and $\Lambda \approx 0.0036$. (b) A magnification of a portion of the curve in panel (a), indicated by a square, showing the comparison near the spiral tip. Reprinted Figure 7 with permission from Zhan and Kapral (2005). Copyright 2005 by the American Physical Society.

that the analytical formula is able to describe the shape of the line defect, even rather close to the core where the splay field analysis begins to lose its validity.

27.3 Line defect dynamics and spiral core motion

A variety of interesting phenomena exist which are related to the existence of line defects in systems with complex periodic and chaotic dynamics. The presence of line defects in complex oscillatory media breaks the rotational symmetry of the spiral wave. This broken symmetry gives rise to a generic mechanism for core motion (Davidsen *et al.*, 2004) that differs from the instabilities that cause the nonsaturating core instability in simple oscillatory media (Chapter 25), and the meandering instability in simple excitable media (Chapter 24). Such core motion is predicted by the analysis of Sandstede and Scheel (2007).

Consider a single spiral for the Rössler reaction–diffusion system in a disk-shaped domain with no-flux boundary conditions (Fig. 27.8). The spiral core moves, albeit very slowly, taking typically several thousand oscillations to move by one wavelength. After transients, the core moves ballistically at constant speed until it encounters a boundary in a finite system, which influences its motion. The speed v increases continuously and monotonically from the onset of motion at the period-2 bifurcation point, $C_2 \approx 3.03$, for the Rössler system. The velocity has a power law form $v \sim (C-C_2)^{3/2}$. The angle α between the direction of motion and the attached

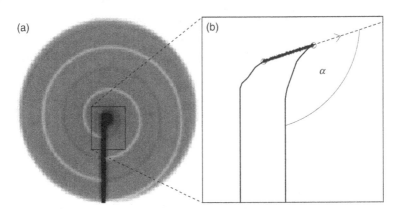

Fig. 27.8. (a) Instantaneous image of the scalar field $\Delta c_z(\mathbf{r}, t) = 1/\tau \int_0^\tau |c_z(\mathbf{r}, t + t') - c_z(\mathbf{r}, t+t'+\tau)| dt'$ with $\tau = 5.95$ in the period-2 regime of the Rössler system for $A = B = 0.2$, $C = 3.5$, and $D = 0.4$. A line defect emanates from the spiral core. (b) Magnification of the rectangular region showing the line defect at two times. The angle α is defined as the angle between the spiral core's trajectory (thick black line) and the attached line defect. Reprinted Figure 1 with permission from Davidsen *et al.* (2004). Copyright 2004 by the American Physical Society.

line defect is constant for a given C. It decreases monotonically from $180°$ at C_2 towards $90°$ with increasing C.

27.4 Turbulence in chaotic systems

When the parameters are changed further beyond the period-2 bifurcation point new phenomena arise. The medium usually undergoes several additional period-doubling bifurcations before the local dynamics becomes chaotic. If the underlying dynamics is period 2^n, 2^{n-1} different types of synchronization defect lines with periodicities 2^k ($k < n$) may exist. Defect lines may persist even when the local dynamics is chaotic, and deep in the chaotic regime turbulent states are found where defect lines are spontaneously created and annihilated. We now consider some of the different scenarios for the onset of turbulent dynamics involving line defects.

Some of these scenarios are related to spiral core motion induced by the presence of line defects. Multi-spiral, spatially disordered configurations occur spontaneously in any sufficiently large domain. For oscillatory media described by the CGL equation, spatially disordered frozen solutions do not exist (Brito *et al.*, 2003), since the weak effective interaction between spirals gives rise to ultra-slow core motion. These results should apply to systems with simple oscillatory dynamics. Figure 27.9 shows the vortex glass phase for the Rössler system in the period-1 regime. This state is analogous to the vortex glass state in the CGL equation shown in Fig. 26.6. Very slow core motion occurs in this regime. In the period-2 regime spirals have a tendency to be grouped in pairs connected by line defects (Fig. 27.9(b)). The spiral pattern rearranges, since a connected pair tends to drift apart and a state that is different from a vortex glass is observed. To see this consider a single spiral pair connected by a straight-line defect. The two cores move with speed v

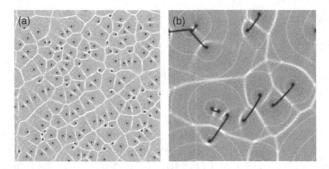

Fig. 27.9. (a) Plot of Δc_z showing a vortex glass state in the period-1 regime ($C = 2.5$). (b) A similar plot for the period-2 regime where line defects connect spiral cores ($C = 3.5$). Reprinted Figure 5 with permission from Davidsen *et al.* (2004). Copyright 2004 by the American Physical Society.

at angles $\pm\alpha$ relative to the line defect and drift away from each other with speed $2v\cos(\pi - \alpha)$, which is positive as long as $|\alpha| > 90°$. The dependence of v and α on C is indistinguishable from that for the single-spiral case discussed earlier. In multi-spiral disordered period-2 regimes, spiral cores move nearly independently, stretching the line defects, until they meet another core, leading to reconnections and creation/annihilation of line defects. In a period-1 medium these cores would annihilate. Here, they repel each other after a new line defect connects them and the annihilation of spiral pairs is prevented. Spatiotemporal chaos in period-2 media preserves the number of spiral cores through the complex dynamics of the line defects connecting them.

Deeper in the chaotic regime the shock lines that form the boundaries of the spiral regions can act as sources for line defects. As system parameters in the Rössler system are changed, chaos first appears on the shock lines for $C \simeq 4.20$. For $C = 4.3$, where most of the medium is still in the period-4 regime, two-banded chaos occurs on the shock lines and leads to local fluctuations which nucleate circular-shaped line defects. For $C \simeq 4.306$, the circular domains are smaller than a certain critical value and collapse shortly after their formation. For larger C the circular-shaped domains proliferate, forming large domains whose growth is limited by collisions with spiral cores or other domains (Fig. 27.10). The transition to line-defect turbulence in media with spiral waves changes the nature of the local dynamics seen in the bulk of the medium. As the line defects propagate, they collide, and these collisions result in an effective band-merging in the orbits of local trajectories so that they take the form of two-banded chaotic trajectories. Although the local trajectories retain their period-4 structure between two passages of line defects, the long-time trajectory cannot be distinguished from that of two-banded chaos. Thus, the global transition of the medium to defect-line turbulence can be characterized locally as intermittent band-merging. The density of line defects grows as a power law beyond the transition (Goryachev *et al.*, 2000) and exhibits properties of a nonequilibrium phase transition. Even more fully developed turbulent states arise when parameters are tuned to lie deeper in the chaotic regime.

The dynamics of the line defects can influence spiral core motion, changing the ballistic core dynamics to Brownian core motion. In the chaotic regime complicated configurations arise, which involve more than one line defect attached to the core, and the core motion is no longer characterized by a simple angle α. Connection and reconnection events among line defects and between line defects and the core continuously occur. Close to the onset of this chaotic regime, the spiral core trajectory often changes its direction abruptly, even though long periods of ballistic motion still persist. As shown in Fig. 27.11, deeper in the turbulent regime the ballistic flights become shorter and shorter, and the core trajectory resembles Brownian motion (see Davidsen *et al.*, 2004). In this regime, line defects are generated rapidly and

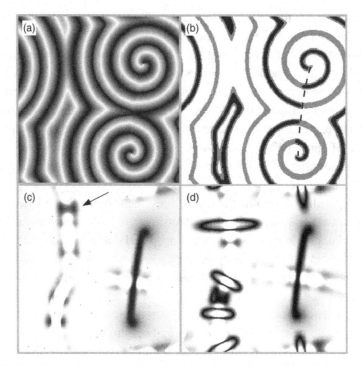

Fig. 27.10. (a) Configuration of a spiral wave pair for the Rössler model with $C = 4.30$. The phase field, which does not resolve the period-4 structure, is shown. (b) The $c_z(\mathbf{r})$ field in gray shades which resolves the period-4 structure and the line defect (dashed line). Panel (c) presents a representation of the concentration field that emphasizes the line defect and shock lines (indicated by an arrow) and suppresses the spiral wave structures. Panel (d) shows the development of circular-shaped line defects on the shock lines for $C = 4.32$. Reprinted with permission from Goryachev *et al.* (2000). Copyright 2000 World Scientific Publishing Company.

homogeneously in the medium. The core motion is characterized by a well-defined diffusion constant.

27.5 Defect-mediated turbulence

Defect-mediated turbulence in oscillatory media was discussed in Chapter 25, and it was shown that a simple stochastic model that accounts for the creation and destruction of defect pairs could capture many aspects of the dynamics of the number of defect pairs in the system. Treating the defect pairs as statistically independent entities, the nucleation rate for pairs of defects was taken to be independent of the

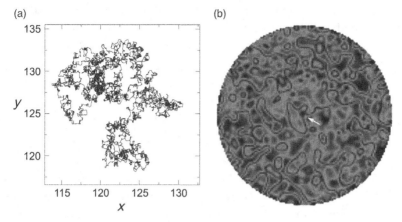

Fig. 27.11. (a) Brownian trajectory of the spiral core in the turbulent regime far from onset of core motion ($C = 5.8$). (b) Turbulent line defect motion responsible for Brownian core dynamics.

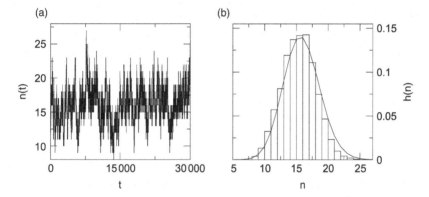

Fig. 27.12. Defect-mediated turbulence for the Willamowski–Rössler reaction–diffusion model in the deterministic chaotic regime. Parameters are: $k_1 = 31.2$, $k_{-1} = 0.2$, $k_2 = 1.45$, $k_{-2} = 0.072$, $k_3 = 10.8$, $k_{-3} = 0.12$, $k_4 = 1.02$, $k_{-4} = 0.01$, $k_5 = 16.5$, $k_{-5} = 0.5$. (a) Number of defect pairs $n(t)$ versus time. (b) Normalized histogram $h(n)$ of the defect pair number. The solid curve is the corresponding squared-Poissonian distribution. Reprinted Figure 2 with permission from Davidsen and Kapral (2003). Copyright 2003 by the American Physical Society.

number of pairs n and, based on the topological nature of the defects, the annihilation rate was taken to be proportional to n^2. This directly led to a squared-Poissonian distribution for the probability distribution function of n.

Defect-mediated turbulence can also exist in systems of the sort considered in this chapter, where the underlying local dynamics is chaotic rather than oscillatory. Consider the Willamowski–Rössler model in a domain with periodic boundary conditions to illustrate the phenomenon. The fluctuations in the number of pairs of

topological defects in this system are shown in Fig. 27.12, and provide evidence
that a defect-mediated turbulent state can exist in chaotic oscillatory media. The
total number of defects in the medium is exactly twice the number of pairs
because the net topological charge is conserved and equal to zero due to the
periodic boundary conditions. Hence, topological defects can be created and annihilated only in pairs of opposite topological charge. One can easily derive a
probability distribution function $p(n)$ for the number of defect pairs provided
that the defects are statistically independent entities following the calculation
given in Chapter 25. If the creation rate $c(n) = c = $ const and the annihilation rate
$a(n) = an^2$, the stationary probability distribution is a squared-Poissonian distribution, $p(n) \propto (c/a)^n/(n!)^2$. Figure 27.12 shows that the stationary probability
distribution function for the Willamowski–Rössler model agrees with the predicted
form of a squared-Poissonian distribution. The assumptions leading to this distribution are justified, since the annihilation rate scales approximately with n^2 and
the creation rate is approximately independent of n.

While the stationary probability distribution has a squared-Poissonian form like
that for simple oscillatory media, differences between defect-mediated turbulence
in oscillatory and chaotic media are seen when the scaling structure of the power
spectrum,

$$S_L(f) = \lim_{T \to \infty} \frac{1}{2T} \left| \int_{-T}^{T} dt\, n(t) e^{-i2\pi ft} \right|^2, \tag{27.14}$$

is examined. For defect-mediated turbulence in the complex Ginzburg–Landau
equation it is found that the power spectrum scales as $S_L(f) \propto 1/f^\gamma$ for intermediate frequencies with an exponent $\gamma = 1.9$, independent of the CGL equation
parameters in the turbulent regime. The power spectrum has the same scaling form
for the Willamowski–Rössler model but with an exponent γ that is far from the
CGL value. In addition, the exponent γ depends on the system parameters. For
$k_{-2} = 0.072$, $\gamma = 1.43$, and for $k_{-2} = 0.075$, $\gamma = 1.60$. Thus, different media can
have different second-order statistics for $n(t)$ depending on the underlying local
dynamics. This implies that a universal description of defect-mediated turbulence
is not possible for chaotic media.

The analysis of the series of waiting times between consecutive creation events
and consecutive annihilation events can provide insight into how the nontrivial
correlations in $n(t)$ at intermediate time scales arise. In both cases the waiting
times are exponentially distributed and statistically uncorrelated, as for a random
walk. This implies that the correlations in $n(t)$ are due to the interaction of creation
and annihilation events. This is further confirmed by the normalized pair correlation

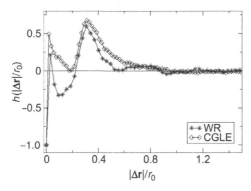

Fig. 27.13. Normalized pair correlation function $h(|\Delta\mathbf{r}|/r_0)$ with $r_0 = \sqrt{\langle n\rangle/L^2}$, where L is the linear dimension of the system.

function for the defects, defined as

$$h(|\Delta\mathbf{r}|) = \frac{\langle n_+(\mathbf{r}, t)n_-(\mathbf{r} + \Delta\mathbf{r}, t)\rangle_{\mathbf{r}, t}}{\langle n\rangle^2} - 1, \qquad (27.15)$$

where $n_\pm(\mathbf{r}, t)$ is the number of defects with $n_t = \pm 1$ at site $\mathbf{r}$ at time t. This function is shown in Fig. 27.13 (see Davidsen and Kapral, 2003). For the CGL equation, the first peak and its decay are very similar to what one would expect for a process where the motion of single defects is unaffected by defects of opposite charge as long as they do not collide: the peak is due mainly to the creation of pairs, and decays to zero approximately as $1/r$ as expected in two dimensions. In contrast, $h(|\Delta\mathbf{r}|)$ for the Willamowski–Rössler model has a pronounced negative value in the vicinity of $|\Delta\mathbf{r}| \approx 0.1 \, r_0$. These strong anti-correlations imply that, with high probability, defects with opposite topological charge annihilate each other directly after their creation, or they separate quickly. This manifests itself in the behavior of $n(t)$ and is probably responsible for the nontrivial exponents in the power spectrum.

28

Resonantly forced oscillatory media

When an oscillatory medium is subjected to external periodic forcing, a variety of spatial structures can form. These spatial patterns are characterized by fronts with distinctive dynamics that separate homogeneous domains in the system. Such self-organized structures arise in a number of different physical contexts, including magnetic field effects on liquid crystals (Frisch *et al.*, 1994), oscillatory grid patterns in liquid crystals (Sano *et al.*, 1993), Rayleigh–Bénard convection (Meyer *et al.*, 1988), and optical parametric oscillators (Longhi, 2001). They also occur in resonantly forced oscillatory reaction–diffusion systems. Spatially resolved experimental studies have been carried out on the Belousov–Zhabotinsky reaction (Petrov *et al.*, 1997; Lin *et al.*, 2000, 2004) and catalytic CO oxidation on Pt surfaces (Bertram *et al.*, 2003) in the presence of external resonant forcing.

In Chapter 25 we saw how the reaction–diffusion equation could be reduced to the complex Ginzburg–Landau equation near the Hopf bifurcation point. This generic CGL equation could then be used to explore the spatiotemporal dynamics in a way that did not rely on specific details of the reaction kinetics. Similarly, for resonantly forced reaction–diffusion systems near the Hopf bifurcation point, it is possible to derive a generic Ginzburg–Landau equation that can be used to explore the properties of the dynamics.

First we consider a spatially homogeneous system whose dynamics is described by the set of ordinary differential equations,

$$\frac{d}{dt}\mathbf{c}(t) = \mathcal{R}(\mathbf{c}; \mu) + a\mathbf{\Phi}(\mathbf{c}, \mu, t), \qquad (28.1)$$

where the external forcing function $\mathbf{\Phi}(\mathbf{c}, \mu, t) = \mathbf{\Phi}(\mathbf{c}, \mu, t + T_f)$ is periodic with period $T_f = 2\pi/\omega_f$. The constant a measures the amplitude of the forcing. The reaction rate and forcing term depend on the parameter μ. In the following we assume that the reactive dynamics $d\mathbf{c}(t)/dt = \mathcal{R}(\mathbf{c})$ is oscillatory with a limit cycle solution $\mathbf{c}_0(t) = \mathbf{c}_0(t + T_0)$ with period $T_0 = 2\pi/\omega_0$. For certain values of

268

the forcing amplitude, when the forcing frequency ω_f is approximately a rational multiple of the natural frequency ω_0, the system may become entrained to the external forcing. When $\omega_f/\omega_0 \approx p/q$, where p and q are coprime integers, the system is said to be entrained at a $p:q$ resonance, and oscillations occur with frequency ω_f/p. The continuous time-translation symmetry of the system is broken, and the system possesses p stable limit cycles $c_0(t + k2\pi/\omega_f)$, $k = 0, \ldots, p-1$. If the system is viewed stroboscopically at the period T_f of the external forcing the system will cycle through p periodic fixed points of the dynamics. An example of the dynamics of a 3:1 resonantly forced system is shown in Fig. 28.1. The three coexisting limit cycle solutions differ in phase by $n2\pi/3$ where $n = 1, 2, 3$. The regions in the $(\omega_f/\omega_0, a)$ parameter plane where the system is entrained are Arnold tongues (Nicolis, 1995). The Arnold tongue for a model of 3:1 resonantly-forced CO oxidation on a Pt surface is shown in Fig. 28.2.

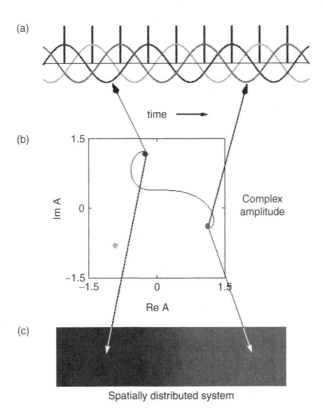

Fig. 28.1. (a) Three coexisting limit cycle solutions for a 3:1 resonantly forced system. The period of the forcing is indicated by vertical lines. (b) Plot in the complex amplitude plane of the three fixed points under stroboscopic dynamics. (c) In a spatially distributed system, domains that differ in phase are connected by interfaces or fronts. The front structure in the complex amplitude plane connecting two fixed points is also shown in the middle panel. Figure provided by C. Hemming.

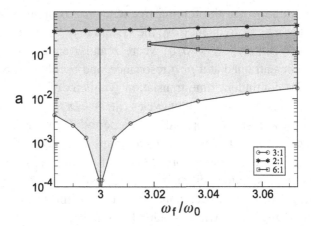

Fig. 28.2. 3:1 Arnold tongue for resonantly forced CO oxidation on Pt(110) as a function of the frequency (ω_f/ω_0) and forcing amplitude a. Also shown are regions of 2:1 and 6:1 entrainment. Reprinted Figure 9 with permission from Davidsen *et al.* (2005). Copyright 2005 by the American Physical Society.

To describe resonantly forced spatially distributed systems we supplement the reaction–diffusion equation with a periodic forcing term:

$$\frac{\partial}{\partial t}\mathbf{c}(\mathbf{r}, t) = \mathcal{R}(\mathbf{c}; \mu) + \mathbf{D}\nabla^2\mathbf{c} + a\mathbf{\Phi}(\mathbf{c}, \mu, t). \tag{28.2}$$

The solutions of this equation can also be entrained to the external driving, but now, because of the existence of spatial degrees of freedom, a much richer structure of resonantly locked solutions can be found. The resonantly locked solutions again lie in Arnold tongues in the ($\omega_f/\omega_0, a$) parameter plane, but different spatiotemporal states can exist within the same tongue. Examples of the types of pattern that can exist are provided by the results of experimental studies on periodic forcing of the Belousov–Zhabotinsky reaction in the oscillatory regime of its dynamics. The ferroin-catalyzed version of this reaction is light sensitive, and when the oscillatory reacting medium is subjected to periodically varying light intensity the entrained states shown in Fig. 28.3 are observed. Although all of the states in the Arnold tongue are resonantly locked at the 2:1 resonance, different spatiotemporal patterns are found as the light intensity and forcing frequency are varied. These patterns are shown in Fig. 28.4, and range from rotating spiral waves to a variety of standing wave patterns.

The resonantly forced reaction–diffusion equation can be reduced to the resonantly forced complex Ginzburg–Landau equation near the Hopf bifurcation point. The analysis is similar to that carried out earlier in Chapter 25 for unforced oscillatory systems but is technically more involved. As in the earlier analysis, we

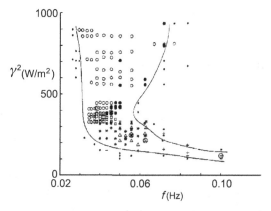

Fig. 28.3. 2:1 resonant tongue for the light-sensitive Belousov–Zhabotinsky reaction indicating the different types of spatial structure seen as a function of the light intensity γ^2 and forcing frequency f. The patterns corresponding to the different symbols are shown in Fig. 28.4. Reprinted Figure 1 with permission from Lin *et al.* (2000). Copyright 2000 by the American Physical Society.

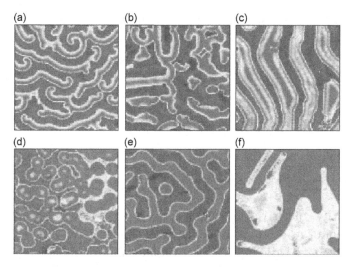

Fig. 28.4. Spatial patterns corresponding to the symbols in the 2:1 Arnold tongue in Fig. 28.3. (a) Rotating spiral wave (+); (b) mixed rotating spiral and standing wave pattern (triangle); (c–f) different standing wave patterns: (c) (*); (d) (square); (e) (filled circle); (f) (open circle). Reprinted Figure 2 with permission from Lin *et al.* (2000). Copyright 2000 by the American Physical Society.

introduce a complex amplitude field $A(\mathbf{r}, t)$ and write the concentration field in terms of this amplitude as

$$\mathbf{c}(\mathbf{r}, t) = \mathbf{c}_0(\mu_0) + A(\mathbf{r}, t)e^{i\omega_f t/p}\mathbf{a} + A^*(\mathbf{r}, t)e^{-i\omega_f t/p}\mathbf{a}^* + \mathcal{O}(|A|^2). \quad (28.3)$$

Here $\mathbf{c}_0(\mu_0)$ is the steady-state concentration at the Hopf bifurcation point $\mu = \mu_0$, and $\mathbf{a}$ is the right eigenvector of the linear stability matrix $\delta\mathcal{R}(\mathbf{c}_0(\mu_0); \mu_0)/\delta\mathbf{c}$. The asterisk denotes the complex conjugate. The reduction of a spatially homogeneous resonantly forced oscillatory system to the resonantly forced Stuart–Landau equation was carried out by Gambaudo (1985), while the corresponding reduction of the resonantly-forced reaction diffusion equation to the resonantly-forced CGL equation can be performed using the methods developed by Elphick *et al.* (1987). Full details of this reduction for a $p:q$ resonance are given in Hemming (2003). The result is

$$\frac{\partial}{\partial t}A(\mathbf{r}, t) = (\mu + i\nu)A - (1 + i\alpha)|A|^2 A + \gamma A^{*(p-1)} + (1 + i\beta)\nabla^2 A, \quad (28.4)$$

where γ gauges the strength of the forcing and the parameter ν measures the detuning from the resonance. The CGL equation parameters α and β are the same as those introduced in Chapter 25. The strong resonances correspond to $p = 1, \ldots, 4$, and we consider these resonances in more detail below.

28.1 2:1 resonance

The 2:1 forced Ginzburg–Landau equation ($p = 2$) in one spatial dimension takes the form

$$\frac{\partial}{\partial t}A(x, t) = (\mu + i\nu)A - (1 + i\alpha)|A|^2 A + \gamma A^* + (1 + i\beta)\frac{\partial^2}{\partial x^2}A. \quad (28.5)$$

This equation does not have a variational form. However, we may gain some insight into the nature of its solutions if we take the parameters ν, α, and β to be zero. In this case we may write Eq. (28.5) in variational form as (Coullet *et al.*, 1990; Coullet and Emilsson, 1992a, 1992b)

$$\frac{\partial}{\partial t}A(x, t) = -\frac{\delta\mathcal{F}[A, A^*]}{\delta A^*}, \quad (28.6)$$

where the free energy functional $\mathcal{F}[A, A^*]$ is given by

$$\mathcal{F}[A, A^*] = \int dx \left(-\mu|A|^2 + \frac{1}{2}|A|^4 - \frac{\gamma}{2}(A^{*2} + A^2) + \frac{1}{2}\left|\frac{dA}{dx}\right|^2 \right). \quad (28.7)$$

The stationary homogeneous solutions of Eq. (28.6) (or Eq. (28.5) with $v = \alpha = \beta = 0$) are $A = A^* = \pm\sqrt{\mu + \gamma}$, while the stationary spatially dependent solutions may be obtained by solving the ordinary differential equation

$$\frac{d^2}{dx^2} A_0 + \mu A_0 - |A_0|^2 A_0 + \gamma A_0^* = 0. \tag{28.8}$$

This equation possesses stationary kink-like solutions connecting the two homogeneous steady states. Letting $A_0 = X_0 + iY_0$, we obtain two types of front solution. Ising fronts have the form

$$X_0^I(x) = \sqrt{\mu + \gamma} \tanh(x\sqrt{(\mu + \gamma)/2}), \quad Y_0^I(x) = 0. \tag{28.9}$$

These fronts are stable if $\gamma > \mu/3$. The amplitude A_0^I of the Ising front is always real, and vanishes at the center of the front at $x = 0$. Bloch fronts have a different form,

$$X_0^B(x) = \sqrt{\mu + \gamma} \tanh\left(x\sqrt{2\gamma}\right), \quad Y_0^B(x) = \pm\frac{\sqrt{\mu - 3\gamma}}{\cosh(x\sqrt{2\gamma})}. \tag{28.10}$$

They are stable if $\gamma < \mu/3$, and have a complex amplitude which does not vanish at $x = 0$.

It is not possible to obtain analytical front solutions for the resonantly forced CGL equation (28.5); however, if v, α, and β are small parameters of order ϵ one may obtain approximate solutions that provide insight into the front structure. In contrast to the variational case, for nonzero v, α, and β traveling front solutions exist. For small values of these parameters the front velocity will also be small and of order ϵ. To describe a traveling front we can transform to a moving frame with velocity v, where $u = x - vt$, so that $A(x, t) = A(x - vt) \equiv A_0(u)$, and write Eq. (28.5) as

$$(1 + i\beta)\frac{d^2 A_0}{du^2} + v\frac{dA_0}{du} + (\mu + iv)A_0 - (1 + i\alpha)|A_0|^2 A_0 + \gamma A_0^* = 0. \tag{28.11}$$

To find a solution we substitute $A_0 = \epsilon \mathcal{A} + \mathcal{O}(\epsilon^2)$ into this equation and obtain, to linear order in ϵ, the equation in $\mathcal{A}$,

$$\left(\frac{d^2}{du^2} + \mu - 2|A_0|^2\right)\mathcal{A} + \left(\gamma - A_0^2\right)\mathcal{A}^*$$

$$= -v\frac{dA_0}{du} - iv A_0 + i\alpha|A_0|^2 A_0 - i\beta\frac{d^2 A_0}{du^2}. \tag{28.12}$$

This equation may be written as a pair of coupled equations for the real and imaginary parts of $\mathcal{A} = (\mathcal{A}_R, \mathcal{A}_I)$. The left eigenvector of the operator on the right-hand side corresponding to the eigenvalue zero is $(dX_0/du, dY_0/du) \equiv (X_{0u}, Y_{0u})$.

Consequently, taking the scalar product with this vector gives the solvability condition,

$$\int du \left[- v(X_{0u}^2 + Y_{0u}^2) + (v - \alpha|A_0|^2)(X_{0u}Y_0 - Y_{0u}X_0) \right.$$

$$\left. + \beta(X_{0u}Y_{0uu} - Y_{0u}X_{0uu}) \right] = 0. \tag{28.13}$$

Inserting the expression for the Bloch front A_0^B yields an expression for the front velocity:

$$v = \pm \left(\frac{(\mu + \gamma)(\mu - 3\gamma)}{2\gamma} \right)^{1/2} \frac{3\pi((\beta - \alpha)\gamma + \alpha\mu - v)}{2(3\mu - \gamma)}. \tag{28.14}$$

From this expression we see that when $\gamma < \mu/3$ the Bloch front velocity is real and can be positive or negative. The velocity vanishes as γ approaches $\mu/3$ from below. For $\gamma > \mu/3$ the Ising front with zero velocity is stable. This nonequilibrium Ising–Bloch bifurcation is sketched in Fig. 28.5. The plot of the front velocity versus the forcing amplitude γ shows the appearance of two branches of positive and negative velocities of the Bloch front that emerge from the zero-velocity Ising front as the forcing amplitude is changed. The figure also shows how the Ising and Bloch fronts connect the two stable states in the complex amplitude plane. The Ising fronts pass through the origin while the Bloch fronts have a nonzero imaginary part that indicates how the imaginary part of the complex amplitude field changes across the interface.

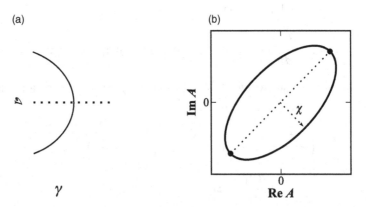

Fig. 28.5. (a) Front velocity versus γ. The bifurcation from an Ising front to Bloch front occurs at $\gamma = \gamma_c = \mu/3$. (b) Sketch of the fronts in the complex amplitude plane. The parameter χ denotes the extremal value $Y_0^B(0)$ and is called the chiral order parameter in applications to spin systems.

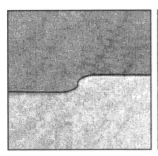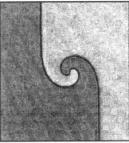

Fig. 28.6. Formation of a Bloch spiral wave in the 2:1 resonantly forced CGL equation. The two homogeneous resonantly locked states are indicated by black and gray colors and are separated by two different Bloch fronts whose velocities have opposite sign and are connected at a Néel point in the center of the domain. As time evolves a Bloch spiral wave forms. Reprinted from Coullet and Emilsson (1992). Copyright 1992, with permission from Elsevier.

In two-dimensional systems the existence of Bloch fronts can lead to pattern formation that is not seen in the unforced CGL equation. Figure 28.6 shows how Bloch spirals can form in these regions of the parameter space. Consider an initial state where the two homogeneous states in the 2:1 system are separated by two different Bloch fronts with positive and negative velocities. They are connected through a defect or Néel point. As the fronts propagate in opposite directions, a two-armed spiral wave is formed with Bloch fronts separating the domains in the spiral. Such fronts can also undergo transverse instabilities similar to those discussed earlier for cubic autocatalytic reactions, giving rise to complex spatiotemporal structures. The unstable fronts can spawn Bloch spirals, leading to spiral wave turbulent states (Hagberg and Meron, 1994).

28.2 Other strong resonances

Systems in the 3:1 resonance tongue have $p = 3$, and can be described by the equation

$$\frac{\partial}{\partial t} A(\mathbf{r}, t) = (\mu + i\nu)A - (1 + i\alpha)|A|^2 A + \gamma A^{*2} + (1 + i\beta)\nabla^2 A. \quad (28.15)$$

In contrast to 2:1 forced systems, Eq. (28.15) possesses three inequivalent fixed points (see Fig. 28.1), and the chiral symmetry of the system is broken. For 2:1

resonantly forced systems, the phase difference between the two states on the right and left sides of a phase front is π. Since there is no difference between phase shifts by π or $-\pi$, the left–right symmetry is preserved. For an interface separating two of the three homogeneous resonantly locked states in a 3:1 resonance, the phase jump across the front may be either $2\pi/3$ or $-2\pi/3$. Since these phase shifts are inequivalent, the left–right symmetry is broken, and fronts in such systems generally have a nonzero velocity. As an example, if the three homogeneous resonantly locked states are placed in contact with a phase defect where the three phases meet, the fronts separating pairs of locked states will move, and the system will evolve to form a three-armed spiral as shown in Fig. 28.7.

For the remaining strong 4:1 resonance additional features appear in the front dynamics as a result of the existence of two different types of front separating the four stable phases (Elphick *et al.*, 1998, 1999). Experimental studies of the resonantly forced Belousov–Zhabotinsky reaction have shown that a wide range of self-organized patterns exist in this system when parameters controlling the forcing amplitude, detuning, and other system parameters are changed.

28.3 Complex front dynamics

The fronts observed in resonantly forced reaction–diffusion systems can adopt more complex structures in different parameter regimes. In Chapter 25 we discussed the Benjamin–Feir instability where phase turbulence exists. If the CGL equation parameters are selected to lie in the Benjamin–Feir unstable regime $(1 + \alpha\beta < 0)$

Fig. 28.7. A three-armed spiral for the 3:1 resonantly forced CGL equation. Reprinted with permission from Hemming and Kapral (2000). Copyright 2000, American Institute of Physics.

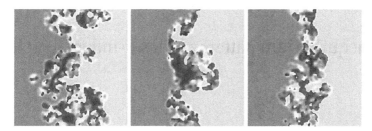

Fig. 28.8. Turbulent fronts separating two homogeneous phases in the 3:1 resonantly forced CGL equation. Reprinted from Hemming and Kapral (2001).
Reproduced by permission of The Royal Society of Chemistry.

then, for sufficiently large forcing amplitudes γ, the instability can be suppressed and the homogeneous state will remain stable. As the forcing amplitude is decreased, a critical amplitude will be reached where the homogeneous phase-locked states are unstable to inhomogeneous perturbations. If two resonantly locked phases are in contact and separated by a phase front that provides the source of the inhomogeneity, the instability manifests itself in complex dynamics in the interfacial zone separating the two homogeneous phases. The complex interfacial structure produced through such a mechanism is shown in Fig. 28.8, which shows turbulent front dynamics for the 3:1 resonantly forced CGL equation. Viewed stroboscopically with the period of the limit cycle, fronts of this type can be described as consisting of an interfacial zone containing a turbulent phase which separates the resonantly locked homogeneous phases. Such fronts may lose their stability for certain values of the forcing amplitude. When this happens the interfacial zone grows without bound to fill the entire domain, with the turbulent phase giving rise to a global spatiotemporally chaotic state. The transverse structures of the two interfaces that separate the turbulent zone from the homogeneous states and scaling properties of the interfacial width have been investigated for both the CGL equation (Hemming and Kapral, 2001, 2002) and coupled map lattice models (Kapral *et al.*, 1994, 1997). Both simulations (Davidsen *et al.*, 2005) and experiments (Bodega *et al.*, 2007) on resonantly forced catalytic CO oxidation on a Pt(110) surface have shown that complex front dynamics and spatiotemporal structures can be observed in this surface reaction. Thus, the addition of periodic forcing to oscillatory reaction–diffusion media can give rise to new classes of self-organized structures whose applications remain to be explored.

29

Nonequilibrium patterns in laser-induced melting

The preceding chapters developed the basic principles needed to describe self-organization and self-assembly in a variety of systems in either initially prepared unstable and metastable states or far-from-equilibrium states. The underlying meso-scopic description involved order parameter fields whose evolution was given in terms of either free energy functionals or amplitude equations. The latter approach is used for systems for which the free-energy-based description is not applicable. However, in many applications to physical and biological problems there is no clear-cut distinction between these two approaches. Often physical systems operate far from equilibrium, and the dynamics may involve both a free energy functional component and a component that cannot be expressed in this form. In this and the following chapters we describe several applications that illustrate how the methods developed in the body of the book may be used to construct models that capture the important aspects of the dynamics. We begin with a discussion of laser-induced melting in this chapter. In the following chapter we consider reactive physical and biological systems where phase segregation and reaction–diffusion dynamics are combined. The last chapter considers active materials where the constituent elements undergo driven or self-propelled motion. The analysis involves the combination of liquid crystal free energy formulations with order parameter dynamics to account for the active motion.

29.1 Laser-induced melting

When a laser with an appropriate intensity is focused onto a solid semiconductor film it can create a variety of ordered and disordered lamellar patterns of coexisting solid and melt regions (Fig. 29.1). The periodicity of the ordered patterns is commensurate with the wavelength of the incident laser radiation (van Driel *et al.*, 1982). Disordered patterns form when the laser beam intensity is low and the beam diameter is large. In this case, the length scales of the disordered patterns are much

(a) (b) (c)

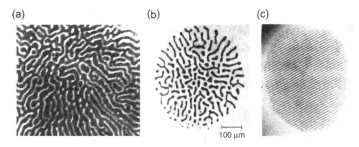

100 µm

Fig. 29.1. Melt morphology of silicon under laser irradiation at $\lambda = 10.6\,\mu\text{m}$. (a) and (b) Disordered structures; (c) ordered structure with the adjacent molten regions separated by λ. (a) Reprinted Figure 1b with permission from Preston *et al.* (1987). Copyright 1987 by the American Physical Society. (b) and (c) Reprinted Figure 1 with permission from van Driel and Dworschak (1992). Copyright 1992 by the American Physical Society.

larger than the wavelength of the incident radiation. These disordered patterns occur because the molten regions have a higher reflectivity than the solid regions. In contrast to ordered patterns, the melt in the disordered pattern is supercooled due to its higher reflectivity and has a temperature lower than the melting temperature T_m while, similarly, the solid regions are superheated with a temperature that is higher than T_m. The film is in a highly nonequilibrium steady state in the presence of the incident laser beam.

A phase-field model with two coupled variables can be constructed to describe such nonequilibrium phenomena (Yeung and Desai, 1994). The phase field $\phi(\mathbf{r}, t)$ is a binary variable with values ± 1 in the liquid and solid regions, respectively. The phase variable ϕ is coupled to a temperature-like field $\mathcal{E}(\mathbf{r}, t)$, which is proportional to the fractional temperature deviation $(T - T_m)/T_m$ of the local temperature $T(\mathbf{r}, t)$ from the melting temperature T_m of the solid film. In dimensionless units the free energy functional is

$$\mathcal{F}[\phi(\mathbf{r}, t)] = \int d^d r \left[-\frac{1}{2}\phi^2 + \frac{1}{4}\phi^4 + \frac{1}{2}(\nabla\phi)^2 - \mathcal{E}\phi \right], \qquad (29.1)$$

where, apart from the usual double well and square gradient terms, one also has the coupling term $(-\mathcal{E}\phi)$. If $T > T_m$ the equilibrium phase is liquid and $\mathcal{E}$ and ϕ are both positive; if $T < T_m$ the equilibrium phase is solid and both $\mathcal{E}$ and ϕ are negative. In either case the coupling term $(-\mathcal{E}\phi)$ is negative and favors the appropriate equilibrium phase. The phase field ϕ obeys a dissipative relaxational equation and the $\mathcal{E}$ field obeys a heat balance equation:

$$\frac{\partial \phi}{\partial t} = -\phi^3 + \phi + \mathcal{E} + \nabla^2 \phi,$$

$$\frac{\partial \mathcal{E}}{\partial t} = -\ell\frac{\partial \phi}{\partial t} + D\nabla^2 \mathcal{E} - r_o\phi + \Delta j. \qquad (29.2)$$

Four dimensionless parameters enter the description: ℓ is a latent-heat-like parameter and D is proportional to the thermal diffusivity of the solid film. The other two terms are proportional to the difference between the incoming heat flux from the laser and the outgoing heat flux through the bottom of the film. Δj is the value of this difference at the interface between the melt and solid phases where $\phi = 0$. If Δj is zero, there is symmetry between the solid and liquid phases: thus, Δj is akin to the off-criticality parameter ψ_o in models A and B. Finally the $-r_o \phi$ term takes into account the difference in the reflectivity between the solid and melt phases. This term is negative since the melt ($\phi > 0$) has a higher reflectivity. The parameters Δj and r_o are also inversely proportional to the film thickness. Although there is no free energy (Lyapunov) functional in general for the coupled set of equations (29.2), such a functional does exist for large latent heat ℓ. From the first of the Eqs. (29.2) one can obtain an expression for $\mathcal{E}$; substituting it in the second equation leads to

$$\ell \frac{\partial \phi}{\partial t} + \frac{\partial^2 \phi}{\partial t^2} + \frac{\partial}{\partial t}(\phi^3 - \phi + \nabla^2 \phi) - D\nabla^2 \frac{\partial \phi}{\partial t}$$
$$= D\nabla^2(\phi^3 - \phi + \nabla^2 \phi) - r_o \phi + \Delta j. \qquad (29.3)$$

In the limit $\ell \gg 1$ and $\ell \gg D$, the left-hand side is dominated by the first term and other terms in it can be neglected. Then the equation reduces to, after dividing by D,

$$\frac{\ell}{D} \frac{\partial \phi}{\partial t} = \nabla^2(\phi^3 - \phi + \nabla^2 \phi) - \frac{r_o}{D}(\phi - \phi_o), \qquad (29.4)$$

where $\phi_o = \Delta j / D$. The resulting Eq. (29.4) is equivalent to Eq. (13.12) that describes the dynamics of microphase separation of block copolymers (Bahiana and Oono, 1990). The correspondence is $D/\ell \to M$, $r_o/D \to B$, $\phi_o \to \psi_o$, and $\phi \to \psi$. Equation (13.12) is obtained using a coarse-grained free energy functional for block copolymers obtained by Leibler (1980) and Ohta and Kawasaki (1986). Therefore in the limit $\ell \gg 1$ and $\ell \gg D$ the Ohta–Kawasaki block copolymer free energy functional serves as a Lyapunov functional for the laser-induced melting model.

Figure 29.2 shows the ϕ field obtained from a numerical integration of Eq. (29.2). There are two regimes with qualitatively different patterns. These are called the weak and strong segregation regimes, in analogy with block copolymers. In the weak segregation regime long parallel stripes straighten out in time. In the strong segregation regime a complicated, interconnected, disordered lamellar structure is found. In this regime, a cross-section of the order parameter profile would show domains of the two phases separated by a thin interfacial region over which ϕ changes from $+1$ to -1. By contrast, in the weak segregation regime the ϕ and $\mathcal{E}$ fields vary approximately sinusoidally with relatively small amplitudes. In both regimes the $\mathcal{E}$ field is out of phase with the ϕ field: $\mathcal{E}$ is negative (positive) where

Fig. 29.2. The ϕ field obtained from a numerical simulation of Eq. (29.2) for $L_x = L_y = 640$, $dx = dy = 1.25$, $\ell = 2.0$, $D = 0.5$, and $\Delta j = 0$. (a) The evolution for $r_0/D = 0.24$ (close to onset; weak segregation) with $t = 1600$, 12800, and 102400, left to right, respectively. The rolls continue to straighten at later times. (b) The evolution for $r_0/D = 0.001875$ (far from the onset; strong segregation) for the same times. The patterns are essentially frozen after this time. Reprinted Figure 1 with permission from Yeung and Desai (1994). Copyright 1994 by the American Physical Society.

ϕ is positive (negative) so that the solid phase ($\phi < 0$) is superheated and the melt phase ($\phi > 0$) is undercooled.

29.2 Static solutions

Consider, for simplicity, the symmetric case where $\Delta j = 0$. In this case, the static solution consists of periodic stripes with wavenumber $k = 2\pi/\lambda$, and their normal points in the x direction. In the pair of Eqs. (29.2), if one sets all time derivatives to zero, eliminates $\mathcal{E}$ using the first equation, and sets $\Delta j = 0$, the resulting equation is the equation that the one-dimensional static solution obeys:

$$0 = \partial_x^2 \left(\phi^3 - \phi - \partial_x^2 \phi \right) - B\phi, \qquad (29.5)$$

where $B = r_0/D$. Although the dynamical analogy with the block copolymer model holds only for large ℓ, the static solutions are the same for all values of ℓ. The static solutions minimize the block copolymer free energy (see Eq. (13.10)). For the one-dimensional static solution, the free energy density is

$$\frac{\mathcal{F}[\phi]}{\lambda} = \frac{1}{\lambda} \int_{-\lambda/2}^{\lambda/2} dx \left[-\frac{1}{2}\phi^2 + \frac{1}{4}\phi^4 + \frac{1}{2}\left(\frac{\partial\phi}{\partial x}\right)^2 \right]$$

$$- \frac{B}{2\lambda} \int_{-\lambda/4}^{\lambda/4} dx \int_{-\lambda/4}^{\lambda/4} dx' \phi(x') |x - x'| \phi(x). \quad (29.6)$$

The equilibrium wavelength is obtained by substituting the static solution with an unspecified λ into the free energy and then minimizing it with respect to λ. The reduced temperature field $\mathcal{E}$ can be obtained using the relation $\partial^2\mathcal{E}/\partial x^2 = B\phi$. We now let $(\phi^3 - \phi) = \mu_B(\phi)$ and consider strong and weak segregation limits.

29.2.1 Strong segregation limit

In the strong segregation regime ($B \ll 1/4$) the domains with $\phi = \pm1$ are separated by sharp interfaces. For small B, Ohta and Kawasaki (1986) obtained the pattern length scale λ^* using a variational method by assuming a form for $\phi(x)$. This length scale can also be obtained, without assuming a form for $\phi(x)$, by using a systematic expansion in the small parameter $1/\lambda$ as follows. We partition the system into four sections consisting of two outer (bulk) regions, corresponding to $\phi = \pm1$, and two inner regions, each consisting of a skin of thickness unity around the two interfaces. In the outer regions we are interested in length scales of order $\lambda \sim k^{-1}$, and this motivates the introduction of the small parameter $\epsilon = \lambda^{-1}$. To extract the large-distance behavior, we rescale the distance in the normal direction as $X = \epsilon x$ in the outer region. The interfacial width is order unity, so x is not rescaled in the inner region. The parameter B must also be rescaled since it determines the wavenumber of the pattern. In the bulk phases $\phi = \pm1$ as $B \to 0$. For $B \to 0$, the first term in Eq. (29.5) scales as ϵ^3. Therefore, one expects $B = \mathcal{O}(\epsilon^3)$ also, and we rescale B as $\tilde{B} = \epsilon^{-3}B$.

In the outer regions the static solution obeys

$$0 = \partial_X^2 \mu(\phi) - \epsilon \tilde{B}\phi, \quad (29.7)$$

where $\mu(\phi) = \mu_B(\phi) - \epsilon^2\partial_X^2\phi$. In the inner region the static solution is given by

$$0 = \partial_x^2 \mu(\phi) - \epsilon^3 \tilde{B}\phi, \quad (29.8)$$

where $\mu(\phi) = \mu_B(\phi) - \partial_x^2\phi$. In each region we expand $\phi = \phi_0 + \epsilon\phi_1 + \mathcal{O}(\epsilon^2)$ and $\mu(\phi) = \mu_0 + \epsilon\mu_1 + \mathcal{O}(\epsilon^2)$. The continuity of μ leads to the matching conditions for ϕ and m^{th} derivatives in the normal direction at the boundary between inner and outer domains: $\phi_i^{inner} = \phi_i^{outer}$ and $\partial_x^m \phi_{i+m}^{inner} = \partial_X^m \phi_i^{outer}$. The inner equation gives

$$\phi(x) = \phi_0(x) + \mathcal{O}(\epsilon^2). \tag{29.9}$$

To zeroth order B does not contribute, and ϕ_0 is the planar interfacial profile discussed earlier for model B in Chapter 7. The higher-order terms, $\phi_{i>0}$, are constrained to be orthogonal to the Goldstone translation mode $d\phi_0/dx$. The outer equation gives

$$\phi(X) = \pm\left(1 + \epsilon \chi \tilde{B} X \left(X \mp \frac{1}{2}\right)\right) + \mathcal{O}(\epsilon^2), \tag{29.10}$$

where $\chi^{-1} = d\mu_B/d\phi|_{\phi=1}$. For $\mu_B = -\phi + \phi^3$ we obtain $\chi = 1/2$. In the outer region the temperature field is

$$\mathcal{E}(X) = \pm\left[\epsilon\frac{\tilde{B}}{2}X\left(X \mp \frac{1}{2}\right) + \epsilon^2\chi\tilde{B}^2\frac{X}{12}\left(X^2(X \mp 1) \pm \frac{1}{8}\right)\right] + \mathcal{O}(\epsilon^3). \tag{29.11}$$

In the inner region $\mathcal{E}(x) = \mathcal{O}(\epsilon^3)$.

Using the above results, we can now calculate the free energy density. There are two interfaces per period so the local term in the free energy density is $2\bar{\sigma}/\lambda + \mathcal{O}(\lambda^{-2})$, where $\bar{\sigma}$ is dimensionless surface tension. The long-range term is $B\lambda^2/96 + \mathcal{O}(\lambda^{-3})$. Thus the free energy density is

$$\frac{\mathcal{F}[\phi]}{\lambda} = \frac{2\bar{\sigma}}{\lambda} + \frac{\lambda^2 B}{96} + \mathcal{O}\left(\frac{1}{\lambda^2}\right). \tag{29.12}$$

Here B is $\mathcal{O}(\lambda^{-3})$. Minimizing $\mathcal{F}$ with respect to λ gives the wavelength of the one-dimensional solution with the lowest free energy as

$$\lambda^* = \left(\frac{96\bar{\sigma}}{B}\right)^{1/3} + \mathcal{O}(B^0). \tag{29.13}$$

Higher-order corrections can also be computed, and we have

$$\lambda^* = \left(\frac{96\bar{\sigma}}{B}\right)^{1/3} + \frac{8\chi\bar{\sigma}}{5} + \mathcal{O}(B). \tag{29.14}$$

Thus, in the strong segregation limit, to leading order in ϵ, the system selects a periodic stripe pattern with $\lambda \sim B^{-1/3}$.

29.2.2 Weak segregation limit

We can also find the wavelength that minimizes the free energy in the weak segregation regime. The homogeneous state first becomes unstable at $B = B_0 = 1/4$ and wavenumber $k = k_0 = 1/\sqrt{2}$. Near onset, we can expand the static solution in orders of ϵ where $\epsilon^2 = B_0 - B$, and let $k = k_0 + \epsilon k_1$. The static solution obeys the equation

$$0 = -\partial_x^2 \left(\phi - \phi^3 + \partial_x^2 \phi \right) - (B_0 - \epsilon^2)\phi. \tag{29.15}$$

We seek a solution for wavelength $\lambda = 2\pi/k$ and write $\phi(x)$ in the form

$$\phi(x) = \sum_{m=1}^{\infty} \epsilon^m f_m(kx), \tag{29.16}$$

where $f_m(kx) \equiv f_m(\theta) = f_m(\theta + 2\pi)$. To fix the position of the interface, we require that f_m be orthogonal to f_m'. We can now write Eq. (29.15) order by order in ϵ. First order in ϵ gives

$$0 = \mathcal{L}_0 f_1(\theta), \tag{29.17}$$

where $\mathcal{L}_0 = - \left(k_0^2 \partial_\theta^2 + k_0^4 \partial_\theta^4 + B_0 \right)$. Assuming odd symmetry, this gives

$$f_1(\theta) = A_k \sin(\theta). \tag{29.18}$$

The order ϵ^2 expression, along with the orthogonality condition, gives $f_2 = 0$. The order ϵ^3 equation is

$$-\mathcal{L}_0 f_3 = - \left(k_1^2 \partial_\theta^2 + 6k_1^2 k_0^2 \partial_\theta^4 - 1 \right) f_1 + k_0^2 \partial_\theta^2 f_1^3. \tag{29.19}$$

The solvability condition requires that the right-hand side of this equation be orthogonal to the zero eigenvector of $\mathcal{L}_0$, which is simply $\sin\theta$. Applying this condition and using Eq. (29.18) gives

$$\frac{3}{8} A_k^3 = k_1^2 A_k - 6k_1^2 k_0^2 A_k + A_k = (1 - 2k_1^2)A_k, \tag{29.20}$$

or $A_k^2 = 8(1 - 2k_1^2)/3$. Substituting this result into Eq. (29.19) and solving the resulting ordinary differential equation gives $f_3(\theta) = A_{3k} \sin(3\theta)$ with $A_{3k} = 9A_k^3/128$. The order ϵ^4 expression gives $f_4(\theta) = 0$. Therefore the static solution is given by

$$\phi(x) = \epsilon A_k \sin(kx) + \epsilon^3 A_{3k} \sin(3kx) + \mathcal{O}(\epsilon^5), \tag{29.21}$$

and

$$\mathcal{E}(x) = \epsilon \frac{A_k}{k^2} \sin(kx) + \epsilon^3 \frac{A_{3k}}{9k^2} \sin(3kx) + \mathcal{O}(\epsilon^5). \qquad (29.22)$$

Using these results the free energy density is

$$\frac{\mathcal{F}[\phi]}{\lambda} = -\frac{\epsilon^2 A_k^2}{4} \left(1 - k^2\right) + \frac{\epsilon^4 3 A_k^4}{32} \left(1 - 9k^2\right)$$
$$+ \frac{\epsilon^2 B A_k^2}{4k^2} + \mathcal{O}(\epsilon^6). \qquad (29.23)$$

The first two terms arise from short-range interactions while the last term is the contribution from long-range interactions. Minimization of this expression with respect to k gives

$$k^{*4} = B + \mathcal{O}(\epsilon^4). \qquad (29.24)$$

If terms of order ϵ^4 in the free energy are retained we obtain

$$k^{*4} = B \left(1 - \frac{5}{16}\epsilon^4\right) + \mathcal{O}(\epsilon^6). \qquad (29.25)$$

The $\mathcal{O}(\epsilon^4)$ correction is fairly small. In the weak segregation limit the selected wavelength behaves as $\lambda^* \sim B^{-1/4}$.

29.3 Linear stability analysis

The homogeneous solution of Eqs. (29.2) is $\phi^* = \Delta j/r_0$, $\mathcal{E}^* = -\phi^* + \phi^{*3}$. To carry out a linear stability analysis around this homogeneous solution, we assume infinitesimal perturbations of the form $\delta\phi_k = \delta_\phi \exp(\omega_k t + i\mathbf{k} \cdot \mathbf{r})$ and $\delta\mathcal{E}_k = \delta_\mathcal{E} \exp(\omega_k t + i\mathbf{k} \cdot \mathbf{r})$. Solving Eqs. (29.2) to first order in δ gives

$$\omega_k = -\frac{\gamma_k'}{2} + \left(\frac{(\gamma_k')^2}{4} - (r_0 + \gamma_k D k^2)\right)^{1/2}, \qquad (29.26)$$

where $\gamma_k = \mu_B' + k^2$ with $\mu_B' = (3\phi^{*2} - 1)$ and $\gamma_k' = \ell + \gamma_k + D k^2 = (\ell + 3\phi^{*2} - 1) + (1 + D)k^2$.

For $\phi^* > 1/\sqrt{3}$, the real part of ω_k is always negative and the homogeneous state is linearly stable. In this case, the initial state lies outside the spinodal curve. For $\phi^* < 1/\sqrt{3}$, the linear dynamics can be divided into several classes. For $\ell < (1 - 3\phi^{*2})$, the homogeneous system is linearly unstable to small-k perturbations. If $4r_0 > (\ell + 3\phi^{*2} - 1)^2$, the instability will be oscillatory. If $\ell > (1 - 3\phi^{*2})$, the system is linearly stable to small-k perturbations. In addition, if $r_0 > D(3\phi^{*2} - 1)^2/4$, the homogeneous state is linearly stable to perturbations of all wavenumbers. For $r_0 < D(3\phi^{*2} - 1)^2/4$ the system is unstable in a finite band of wavenumbers,

$$-\mu'_B - \sqrt{(\mu'_B)^2 - \frac{4r_0}{D}} < 2k^2 < -\mu'_B + \sqrt{(\mu'_B)^2 - \frac{4r_0}{D}}. \qquad (29.27)$$

For $\Delta j = 0$, the stability of the homogeneous state can be characterized as follows: (i) unstable at $k = 0$ if $\ell < 1$, with an oscillatory instability if $4r_0 > (1 - \ell)^2$; (ii) stable for all k if $\ell > 1$ and $r_0/D > 1/4$; and (iii) if $\ell > 1$ and $r_0/D < 1/4$, then the solution is unstable in a band of wavenumbers with $\left(1/2 - \sqrt{1/4 - r_0/D}\right) < k^2 < \left(1/2 + \sqrt{1/4 - r_0/D}\right)$. At threshold, $r_0/D = 1/4$, the band collapses to a point $k = k_0 = 1/\sqrt{2}$.

29.4 Phase diffusion description and lamellar stability

The free energy of the system is a minimum for the inhomogeneous state consisting of globally parallel lamellae. This state is approached via locally parallel lamellar structures whose orientation and wavenumber vary slowly on the length scale of the pattern. The dynamics of the lamellae can be described using the phase diffusion formalism (Pomeau and Manneville, 1979; Cross and Newell, 1984). This formalism allows one to classify the stability of the lamellae and provides a rigorous criterion for wavenumber selection. Below, we restrict the analysis to the symmetric lamellar case and set $\Delta j = 0$ in Eq. (29.2).

The basic idea that underlies the phase diffusion analysis is the recognition that there are two length scales in the problem. The long-distance, long-time behavior on a length scale much larger than the pattern wavelength λ must be separated from the short length scale behavior. To extract this behavior, we note that a local wave vector $\mathbf{k}(\mathbf{r}, t)$ can be defined at each point in the stripe pattern. It is directed normal to the stripes, and its magnitude is the local wavenumber of the stripes. This vector is twofold degenerate: i.e. we can choose either $\mathbf{k}$ or $-\mathbf{k}$. Once this choice is made at one point, it is fixed at all points. We assume that $\mathbf{k}$ varies on length scales $1/\tilde{\epsilon}$, much larger than the pattern wavelength $2\pi/k$. Here we use $\tilde{\epsilon}$ as the small parameter to avoid any confusion with the ϵ that appears in the analysis of the static solutions. Since the ϕ and $\mathcal{E}$ fields describe diffusive dynamics through Eqs. (29.2), we introduce the slow space and time variables $\mathbf{X} = \tilde{\epsilon}\mathbf{r}$ and $T = \tilde{\epsilon}^2 t$, and the fast phase variable $\theta(\mathbf{r}, t)$ such that $\theta = n\pi$ at each solid/melt interface. We also introduce a slow phase variable $\Theta = \tilde{\epsilon}\theta$. The introduction of the separate slow and fast variables allows us to write the evolution equations (29.2) order by order in $\tilde{\epsilon}$. The local wavevector $\mathbf{k}$ of the stripes is related to the fast and slow phase variables by

$$\mathbf{k}(\mathbf{X}, T) = \nabla\theta(\mathbf{r}, t) = \nabla_{\mathbf{X}}\Theta(\mathbf{r}, t). \tag{29.28}$$

The goal of the phase diffusion description is to describe the slow dynamics of $\mathbf{k}$ on the long length scales, $\mathbf{X}$. Expressing the dynamical fields in terms of the phase variables, we have

$$\phi(\mathbf{r}, t) = \phi_k(\theta(\mathbf{r}, t), \mathbf{X}, T), \quad \mathcal{E}(\mathbf{r}, t) = \mathcal{E}_k(\theta(\mathbf{r}, t), \mathbf{X}, T), \tag{29.29}$$

where ϕ_k and $\mathcal{E}_k$ are 2π-periodic functions of θ. Since static solutions exist for a band of wavenumbers, ϕ_k and $\mathcal{E}_k$ depend on the local wavenumber and, hence, on the slow variables $\mathbf{X}$ and T.

Starting from Eqs. (29.2), the dynamical equations in the phase diffusion description are obtained order by order in $\tilde{\epsilon}$. The zeroth-order equations show that ϕ_0 and $\mathcal{E}_0$ correspond to static solutions of the one-dimensional equations. Since the static solutions exist for a band of wavenumbers k, the fields ϕ_0 and $\mathcal{E}_0$ depend on the slow variables through the local wavenumber k. The analysis to first order in $\tilde{\epsilon}$ yields the equation obeyed by the phase variable θ:

$$\tau_k \partial_t \theta = -\nabla \cdot \mathbf{k}G(k) = D_u \partial_u^2 \theta + D_s \nabla_s^2 \theta, \tag{29.30}$$

where

$$D_u = -d(kG(k))/dk, \quad D_s = -G(k), \tag{29.31}$$

with

$$G(k) = \frac{D}{r_0} A_{\mathcal{E}}(k) q_{\mathcal{E},k} - A_\phi(k) q_{\phi,k}, \tag{29.32}$$

and

$$\tau_k = A_\phi(k)\left(q_{\phi,k} + \frac{\ell}{Dk^2}\right) - \frac{1}{r_0} A_{\mathcal{E}}(k) q_{\mathcal{E},k}$$

$$= \frac{D-1}{r_0} A_{\mathcal{E}}(k) q_{\mathcal{E},k} + \frac{\ell}{Dk^2} A_\phi(k) - G(k). \tag{29.33}$$

The A coefficients are related to the static one-dimensional solutions by

$$A_\phi(k) = \frac{1}{2\pi} \int_0^{2\pi} d\theta \, \phi_k(\theta)^2,$$

$$A_\phi(k)q_{\phi,k} = \frac{1}{2\pi} \int_0^{2\pi} d\theta \, [\partial_\theta \phi_k(\theta)]^2,$$

$$A_\mathcal{E}(k) = \frac{1}{2\pi} \int_0^{2\pi} d\theta \, \mathcal{E}_k(\theta)^2, \qquad (29.34)$$

$$A_\mathcal{E}(k)q_{\mathcal{E},k} = \frac{1}{2\pi} \int_0^{2\pi} d\theta \, (\partial_\theta \mathcal{E}_k(\theta))^2.$$

Since ϕ_k and $\mathcal{E}_k$ depend on $k(\mathbf{X}, T)$, these coefficients depend on the slow variables.

If $\tau_k^{-1} D_u \le 0$ there is an Eckhaus instability corresponding to the slow variations of the local wavenumber. If $\tau_k^{-1} D_s \le 0$, then the system is unstable to the zigzag instability which corresponds to variations in the orientation of the lamellae. If $\tau_k > 0$ and one neglects defects,

$$F[k(\mathbf{r}, t)] = \int d\mathbf{r} \int_0^{k(\mathbf{r},t)^2} dq^2 \, G(q) \qquad (29.35)$$

defines a Lyapunov functional with an extremum $G(k^*) = -D_s(k^*) = 0$. Therefore, the selected length scale is exactly that which is marginally stable against the zigzag instability.

For some parameters, τ_k can be negative. If this occurs at a wavenumber larger than that for which either D_s or D_u is negative, the lamellae will be unstable to both the zigzag and Eckhaus instabilities simultaneously. In this case both $\tau_k^{-1} D_s(k)$ and $\tau_k^{-1} D_u(k)$ become negative at the same value of k. From Eq. (29.33) this simultaneous instability supersedes the isolated zigzag instability if the latent heat is sufficiently small,

$$\ell < (1 - D) \frac{Dk^{*2} A_\mathcal{E}(k^*) q_{\mathcal{E},k^*}}{r_0 A_\phi(k^*)}. \qquad (29.36)$$

Here the wavenumber k^* is where the zigzag instability occurs: i.e. $D_s(k^*) = 0$. One cannot determine the final evolution of the unstable state from the phase diffusion dynamics. However, the lamellae are unstable to perturbations of wavenumber k in an entire band of wavenumbers around $k = 0$. This signals a new dynamic instability in the time-dependent dynamics. If the instability arises from negative τ_k, the lamellae are unstable to perturbations with an arbitrary ratio of normal and tangential components.

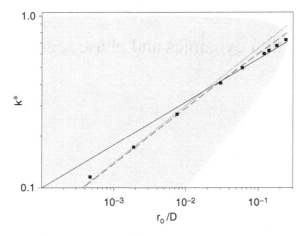

Fig. 29.3. Graphical summary of the stability analysis for the symmetric case, $\Delta j = 0$. The homogeneous state is linearly unstable within the shaded area. The solid line is $k^* = (r_0/D)^{1/4}$: i.e. the lowest-order result for the wavenumber near onset (weak segregation). The dotted line is the lowest-order wavenumber far from onset (strong segregation), $k^* = 2\pi(96D\bar{\sigma}/r_0)^{-1/3}$. The dashed line is the far-from-onset result including next-order corrections. The squares are the length scales obtained from simulations of Eq. (29.2) with $\ell = 2$ and $D = 0.5$. Reprinted Figure 2 with permission from Yeung and Desai (1994). Copyright 1994 by the American Physical Society.

In general, the coefficients $A_\phi(k)$, $A_\varepsilon(k)$, and $q_{\phi,k}$, $q_{\varepsilon,k}$ must be obtained numerically. However, they can be obtained explicitly in the strong and weak segregation limits. In particular, since these coefficients depend only on the static $1d$ solutions, the results obtained earlier in this chapter can be used for this purpose. In the strong segregation limit, the lamellae are unstable to the zigzag instability for $\lambda > \lambda^*$ where

$$\lambda^* = \frac{2\pi}{k^*} = \left(\frac{96D\bar{\sigma}}{r_0}\right)^{1/3} + \frac{8\chi\bar{\sigma}}{5} + \mathcal{O}(\lambda^{-1}). \qquad (29.37)$$

In the weak segregation limit this instability occurs at

$$k^{*4} = \frac{r_0}{D} - \frac{5}{64}\epsilon^4 + \mathcal{O}(\epsilon^6), \qquad (29.38)$$

which is in agreement with the result for the equilibrium wavelength.

Figure 29.3 summarizes the results of the stability analysis in the large ℓ limit. It shows the neutral stability curve $|k - k_0| = \sqrt{r_0/D}$ (shaded region boundary) as well as the boundaries of the zigzag instability. The numerical results agree quite well with the theoretical predictions: in particular, both the strong and weak segregation regimes are observed in the numerical results of Fig. 29.3.

30

Reaction dynamics and phase segregation

Throughout this book we have described the mechanisms underlying phase segregation processes in a variety of systems, as well as the pattern formation processes that occur in reactive systems driven far from chemical equilibrium. There are many similarities in the morphologies of the patterns that arise in these two contexts. But there are also important differences, many of which stem from the fact that a description in terms of free energy functionals is not usually possible for far-from-equilibrium conditions. In this chapter we consider some examples of systems where both chemical reaction and phase segregation interact to change the character of the self-organization process.

30.1 Phase segregation in reacting systems

We begin by investigating one of the simplest situations where a phase-separating mixture undergoes chemical reaction, and examine the nature of the chemical patterns seen in such systems. In particular, we consider a binary mixture of A and B species undergoing the interconversion chemical reaction

$$A \underset{k_r}{\overset{k_f}{\rightleftharpoons}} B, \tag{30.1}$$

where k_f and k_r are the forward and reverse rate coefficients. The mass action rate law is

$$\frac{dn_A(t)}{dt} = -k_f n_A(t) + k_r n_B(t). \tag{30.2}$$

Since the total number of A and B molecules is constant, $n_A(t) + n_B(t) = n_0$, we can express this rate law in terms of the density of either species. Expressing the rate law in terms of $\phi(t) = n_A(t)/n_0$, we find

$$\frac{d\phi(t)}{dt} = -k_{AB}(\phi(t) - \phi_o),\tag{30.3}$$

where $k_{AB} = k_f + k_r$ and $\phi_o = k_r/k_{AB}$.

We also assume that the A and B species are partially miscible, so that the A-B binary mixture can undergo phase segregation, as discussed in Chapter 2. Consequently, ϕ serves as the order parameter field, and we may combine the model B equation of motion for a system with a conserved order parameter (see Chapters 6 and 13) with the mass action rate law to find the kinetic equation for the local order parameter field,

$$\frac{\partial\phi(\mathbf{r},t)}{\partial t} = \nabla^2\frac{\delta\mathcal{F}_{GLW}[\phi]}{\delta\phi} - k_{AB}(\phi - \phi_o).\tag{30.4}$$

Here we have assumed that the local total density field is constant, so that $n_A(\mathbf{r},t) + n_B(\mathbf{r},t) = n_0$. The free energy functional is taken to have the Ginzburg–Landau–Wilson form (see Eq. (3.2)),

$$\mathcal{F}_{GLW}[\phi(\mathbf{r},t)] = \int d^dr\left[f(\phi) + \frac{\kappa}{2}(\nabla\phi)^2\right].\tag{30.5}$$

We can write the kinetic equation for the order parameter in a form that closely resembles that for model B with long-range repulsive interactions (Motoyama and Ohta, 1997). For this purpose we define the Green function $G(\mathbf{r} - \mathbf{r}')$ that satisfies the equation $\nabla^2 G(\mathbf{r} - \mathbf{r}') = -\delta(\mathbf{r} - \mathbf{r}')$. Its Fourier transform is $\hat{G}(\mathbf{k}) = 1/k^2$. Thus, if we rewrite the free energy functional as

$$\begin{aligned}\mathcal{F}[\phi] = \mathcal{F}_{GLW}[\phi] + \mathcal{F}_{LR}[\phi] = \mathcal{F}_{GLW}[\phi]\\ + \frac{k_{AB}}{2}\int d^dr\int d^dr'\,(\phi(\mathbf{r}) - \phi_o)\,G(\mathbf{r} - \mathbf{r}')\,(\phi(\mathbf{r}') - \phi_o)\end{aligned}\tag{30.6}$$

we can rewrite Eq. (30.4) and cast it into the form of the model B order parameter equation,

$$\frac{\partial\phi(\mathbf{r},t)}{\partial t} = \nabla^2\frac{\delta\mathcal{F}[\phi]}{\delta\phi}.\tag{30.7}$$

Written in this form, we see that the phase separation kinetics of the binary reacting mixture shares many features in common with phase segregation in block copolymer and other systems with long-range repulsive interactions and a conserved order parameter. Linearizing the order parameter equation about the initial average value

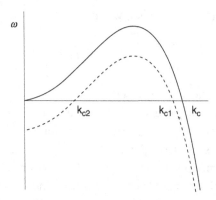

Fig. 30.1. Sketch of the dispersion relation for the binary mixture with no reaction (solid line) and with reaction (dashed line). Reprinted Figure 1 with permission from Glotzer *et al.* (1995). Copyright 1995 by the American Physical Society.

of the order parameter $\bar{\phi}$ and taking the Fourier transform yields the growth rate of small perturbations from $\phi = \bar{\phi}$ as (Glotzer *et al.*, 1995)

$$\omega(k) = \kappa k^2 (k_c^2 - k^2) - k_{AB}, \tag{30.8}$$

where $k_c = (f''(\phi_0)/\kappa)^{1/2}$. This dispersion relation is plotted in Fig. 30.1 for the binary mixture with no reaction ($k_{AB} = 0$) and with reaction. When reaction is present we see that the dispersion relation is modified by reaction to yield a band of unstable wave vectors between k_{c2} and k_{c1}, signaling the formation of patterns with a characteristic wavelength that is independent of the system size.

As suggested by the linear stability analysis, one of the most interesting consequences of the existence of chemical interconversion between the A and B species is the fact that the domain-coarsening process can be arrested, leading to time-independent patterned states with a characteristic spatial scale. This is confirmed by simulations of the kinetic equation for ϕ. Taking the free energy functional to be given by

$$f(\phi) = \phi \ln \phi + (1 - \phi) \ln(1 - \phi) + \chi \phi (1 - \phi), \tag{30.9}$$

where the parameter χ characterizes the interaction energy between the A and B species, the simulation results in Fig. 30.2 show that coarsening stops when reactions between the two species take place. A labyrinthine pattern with a characteristic wavelength, similar to those found earlier for systems with competing interactions, is obtained.

Competition between phase segregation and chemical reaction has been observed experimentally in polymer blends subjected to ultraviolet light that induces polymerization in one component during the segregation process (Tran-Cong and

(a)   (b)

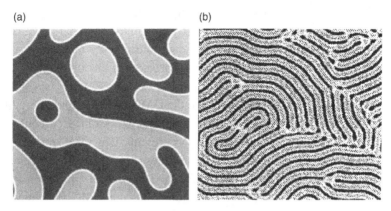

Fig. 30.2. Concentration field $\phi(\mathbf{r}, t_1)$ for a long-time t_1. (a) $\phi(\mathbf{r}, t_1)$ for the system without reaction. The system will continue to coarsen as time evolves. (b) $\phi(r, t_1)$ when reaction is present. One sees domains with a smaller characteristic wavelength. No further coarsening will take place as time increases. Reprinted Figures 2 and 3 with permission from Glotzer *et al.* (1995). Copyright 1995 by the American Physical Society.

Harada, 1996) and in surface-supported Au–Pd adlayers on Rh(110) surfaces where spinodal decomposition and chemical reaction combine to produce lamellar structures (Locatelli *et al.*, 2006).

30.2 Protein-induced pattern formation in biomembranes

The above results showed how a simple chemical reaction between two species can influence the phase segregation process in a binary system. If the reactive dynamics is more complex and is carried out under far-from-equilibrium conditions, its coupling to the phase segregation process can lead to more interesting self-organized structures. Such effects have been considered in the context of surface catalytic reactions (Hildebrand *et al.*, 2003). Biological systems are especially good candidates for the observations of such dynamics. Biochemical network reactions in living systems take place under far-from-equilibrium conditions and often involve positive or negative feedback loops that lead to complex pattern-formation dynamics. In addition, specific molecular interactions can give rise to segregation processes that can couple to the reaction dynamics. We now examine protein-induced pattern formation in biomembranes as an illustration of the self-organized structures that can arise from the competition between these two mechanisms for pattern formation.

In the cell membrane, lipids and proteins lie in proximity and interact strongly. As a result of these interactions, biomembranes can often segregate into different lipid microdomains. It is believed that such domain structure is important for some of the processes at the cellular level: for example, domains can serve as signaling platforms in living cells (Simons and Ikonen, 1997). It has been postulated that the nature of the microdomains can be affected by both phase segregation and chemical pattern formation (John and Bär, 2005a), and we outline the construction of a model that incorporates these mechanisms for GMC proteins interacting with acidic lipids in the membranes of eukaryotic cells.

The chemical mechanisms that underlie the GMC protein–lipid interactions are believed to occur as follows. The cell membrane is assumed to contain acidic and neutral lipid species as well as adsorbed proteins. The GMC proteins interact with and bind only to the acidic lipids, and these interactions can induce phase separation leading to microdomain formation. The protein–lipid interactions are regulated by a protein kinase C (PKC). In quiescent cells GMC proteins are bound to the membrane. When the membrane-bound proteins are phosphorylated by PKC, the proteins unbind and move into the cytosol. In the cytosol dephosphorylation can take place, and the proteins can rebind with the membrane. This mechanism, called the myristoyl–electrostatic (ME) switch, is shown in Fig. 30.3.

With these considerations in mind, John and Bär (2005a, 2005b) formulated the model for protein-induced phase segregation in quantitative terms. The membrane consists of acidic and neutral lipids and the area fraction of the membrane covered by acidic lipids is ϕ_ℓ, so that the area fraction of neutral lipids is $(1 - \phi_\ell)$. The GMC protein is assumed to exist in a membrane-bound form (M) as well as in unphosphorylated (U) and phosphorylated (P) forms in the cytosol. The fraction of the total membrane surface area covered by membrane-bound GMC proteins

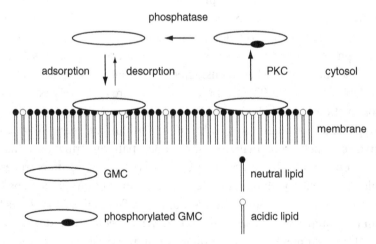

Fig. 30.3. Schematic representation of the mechanism for the ME switch, showing reaction-induced GMC protein binding and unbinding processes with the acidic lipids in the membrane. Reprinted from John and Bär (2005b). Copyright 2005, with permission of the Institute of Physics.

will be denoted by ϕ_M. The concentrations of the protein in the cytosol in the unphosphorylated and phosphorylated forms are c_U and c_P, respectively.

First, consider the dynamics of the lipid and membrane-bound protein species. The kinetic equations determining the evolutions of these species are given by

$$\frac{\partial \phi_\ell(\mathbf{r},t)}{\partial t} = -\nabla \cdot \mathbf{j}_\ell(\mathbf{r},t), \tag{30.10}$$

$$\frac{\partial \phi_M(\mathbf{r},t)}{\partial t} = \mathcal{R}_M(\mathbf{c}(\mathbf{r},t)) - \nabla \cdot \mathbf{j}_M(\mathbf{r},t). \tag{30.11}$$

Equation (30.10) is simply the continuity equation for the local acidic lipid area fraction. The flux of acidic lipids, $\mathbf{j}_\ell(\mathbf{r},t)$ is given by

$$\mathbf{j}_\ell = -D_\ell(\phi_\ell)\nabla\frac{\delta\mathcal{F}}{\delta\phi_\ell}. \tag{30.12}$$

The free energy functional

$$\mathcal{F}[\phi_\ell,\phi_M] = \int d^d r \left[f(\phi_\ell,\phi_M) + \frac{\kappa}{2}(\nabla\phi_\ell)^2 \right], \tag{30.13}$$

contains a local free energy that depends on both the acidic lipid and membrane-bound protein area fractions and a square gradient term that accounts for the interfaces between the lipid microdomains. The local free energy has the usual entropic terms corresponding to the two species discussed in earlier chapters, as well as an interaction term between the lipid and membrane-bound protein with strength characterized by a parameter u:

$$\begin{aligned} f(\phi_\ell,\phi_M) = {}& \phi_\ell \ln \phi_\ell + (1-\phi_\ell)\ln(1-\phi_\ell) \\ & + \phi_M \ln \phi_M + (1-\phi_M)\ln(1-\phi_M) - u\phi_\ell\phi_M, \end{aligned} \tag{30.14}$$

The area-fraction dependence of the diffusion coefficient of the acidic lipids is assumed to take the form $D_\ell(\phi_\ell) = D_\ell \phi_\ell(1-\phi_\ell)$.

The kinetic equation for the membrane-bound protein in Eq. (30.11) has a similar structure. However, in addition to the flux term,

$$\mathbf{j}_M = -D_M(\phi_M)\nabla\frac{\delta\mathcal{F}}{\delta\phi_M}, \tag{30.15}$$

where $D_M(\phi_M) = D_M\phi_M(1-\phi_M)$, the chemical reactions that the proteins undergo are accounted for by the reaction rate $\mathcal{R}_M$. The proteins can adsorb at sites of the membrane which are free of protein with a rate constant k_{ad}. In order to desorb from the membrane, the proteins must overcome an activation barrier determined by the interaction energy $Nu\phi_\ell$ (in units of k_BT) of the proteins with the acidic lipids. Here N measures the size ratio between lipids and proteins. Thus, the desorption rate coefficient is given by $k_{de}e^{-Nu\phi_\ell}$. Once membrane-bound proteins are activated by PKC, they are phosphorylated through reactions that follow Michaelis–Menten kinetics. The PKC activation, characterized by the rate coefficient k_{PKC}, requires lipid proteins as effector molecules, and the rate coefficient is given explicitly by $k^*_{PKC} = k_{PKC}(1 - \phi_M)c^n_P/(K^n_P + c^n_P)$, where n is a parameter. Taking these considerations into account, the reaction rate $\mathcal{R}_M$ is given by

$$\mathcal{R}_M = k_{ad}(1 - \phi_M)c_U - k_{de}\phi_M e^{N\chi c_P} - k^*_{PKC}\frac{\phi_M}{K_M + \phi_M}, \quad (30.16)$$

where the first two terms account for the adsorption and desorption processes and the last term accounts for the PKC activation and phosphorylation kinetics governed by Michaelis–Menten kinetics.

The evolution of the concentrations of the unphosphorylated and phosphorylated GMC protein species in the cytosol is given by reaction–diffusion kinetics. The reaction–diffusion equations are

$$\frac{\partial c_U(\mathbf{r}, t)}{\partial t} = \mathcal{R}_U(\mathbf{c}(\mathbf{r}, t)) + D_c\nabla^2 c_U(\mathbf{r}, t)$$

$$\frac{\partial c_P(\mathbf{r}, t)}{\partial t} = \mathcal{R}_P(\mathbf{c}(\mathbf{r}, t)) + D_c\nabla^2 c_P(\mathbf{r}, t), \quad (30.17)$$

where D_c is the diffusion coefficient of the protein, assumed to be the same for either form in the cytosol. The reaction rates follow from considerations similar to those described above. The concentration of the unphosphorylated form can change by adsorption and desorption on the membrane. Also, in the cytosol, phosphorylated proteins are dephosphorylated by a phosphatase with rate constant k_{Ph}. Thus,

$$\mathcal{R}_U = -k_{ad}(1 - \phi_M)c_U + k_{de}\phi_M e^{-Nuc_P} + k_{Ph}c_P. \quad (30.18)$$

The rate for the phosphorylated form must account for the production due to PKC activation processes as well as depletion due to dephosphorylation processes,

$$\mathcal{R}_P = k^*_{PKC}\frac{\phi_M}{K_M + \phi_M} - k_{Ph}c_P. \quad (30.19)$$

Using physiologically reasonable values for the parameters entering this description, the model yields a variety of interesting structures. These are summarized in the phase diagram in Fig. 30.4 obtained from a study of the linear stability of the uniform steady state of the model. Depending on the system parameters, the uniform

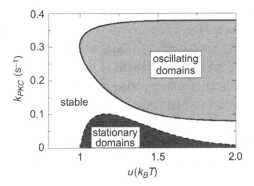

Fig. 30.4. Phase diagram showing regions in parameter space where stationary and oscillating domains can exist. Reprinted from John and Bär (2005b). Copyright 2005, with permission of the Institute of Physics.

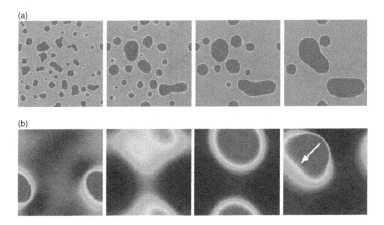

Fig. 30.5. (a) Evolution from the unstable homogeneous steady state for parameters lying in the stationary domain region. One observes domain formation and coarsening. (b) Evolution from the unstable homogeneous steady state for parameters in the oscillating domain region, showing the formation of traveling domains. In both (a) and (b) time increases from left to right. Reprinted from John and Bär (2005b). Copyright 2005. Copyright 2005, with permission of the Institute of Physics.

steady state can lose its stability and form stationary domains or even oscillating domains. Figure 30.5 shows the results of simulations starting from small perturbations about the uniform steady state for parameters corresponding to stationary domains. The unstable uniform steady state evolves into stationary domains of high acidic lipid and protein concentrations that coarsen as time increases.

In contrast, if parameters are chosen to lie in the oscillating domain region, different dynamical behavior is observed. The unstable uniform state initially evolves into a standing wave with growing amplitude (not shown in the figure) until a traveling wave whose wavelength is comparable to that of the cell size develops. Such dynamical structures arise from interactions between the nonlinear far-from-equilibrium biochemical reaction kinetics and the phase segregation process, and are outside the scope of descriptions based solely on free energy functionals. Consequently, in far-from-equilibrium systems we expect a rich variety of spatiotemporal self-organized structures that can be analyzed using the techniques discussed in this book.

31

Active materials

Self-organization and self-assembly take different forms, and their description involves the consideration of different principles when the elements comprising the system undergo active motion. The active motion may arise either because the elements are self-propelled or because external forces or fluxes are applied to the system to induce motion. Biology provides many examples of such active media. Microorganisms such as *Escherichia coli* move by using molecular motors to drive flagella that propel the organism. The amoeba *Dictyostelium discoideum* moves by using pseudopods that change shape through actin polymerization and depolymerization processes. Large numbers of such active agents often display collective behavior: for example, *Dictyostelium discoideum* colonies form streaming patterns, and rippling patterns are seen in myxobacteria such as *Myxococcus xanthus*. Myxobacteria patterns have been modeled using reaction–diffusion-like descriptions (Börner *et al.*, 2002; Igoshin and Oster, 2004). Flocking behavior is also exhibited by birds, fish, mammals, and a variety of microorganisms (Reynolds, 1987). The spatial patterns seen in these systems span a large range of length scales, from kilometers for herds of wildebeest to micrometers for colonies of the amoeba *Dictyostelium discoideum*.

Often particle-based models are employed to describe the nonequilibrium dynamics of active media. Studies based on such models show that collective motion arises as a result of the emergence of orientational long-range order in a system with many degrees of freedom. Simple discrete models are able to capture many of the essential features of the collective behavior (Vicsek *et al.*, 1995; Grégoire *et al.*, 2003). Figure 31.1 shows configurations depicting cohesive flocks obtained from simulations of a discrete model. The model depends on two parameters: α measures the force tending to align the particle velocities, while β measures the cohesive forces between the particles. As β increases, the system evolves from gas to liquid to solid phases.

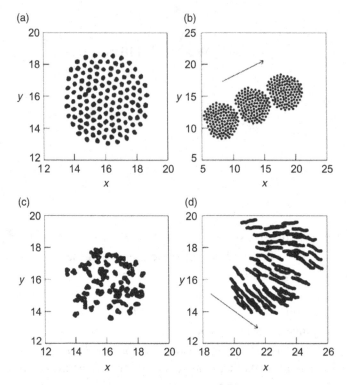

Fig. 31.1. Configurations depicting four phases of systems containing active particles observed in simulations of the modified Vicsek model: (a) solid, $\alpha = 1$,
$\beta = 100$; (b) moving solid, $\alpha = 3$, $\beta = 100$; (c) liquid, $\alpha = 1$, $\beta = 2$; and (d)
moving liquid, $\alpha = 3$, $\beta = 3$. In (a), (c), and (d) 20 consecutive time intervals are
superimposed. In (b) three configurations separated by time intervals of 120 are
shown. In (b) and (d) the arrows indicate the direction of motion. Reprinted from
Grégoire *et al.* (2003). Copyright 2003, with permission from Elsevier.

Flocking is not restricted to living organisms; it is also seen in physical systems
such as agitated granular-rod monolayers (Mishra and Ramaswamy, 2006). While
particle-based schemes are often used to describe the dynamics and collective
behavior of such systems, the pattern-formation processes that occur on mesoscopic
scales may be modeled by free energy functional approaches and their extensions
that account for nonequilibrium effects. Below we consider several examples that
illustrate how the dynamics of active materials can be modeled using the concepts
developed in this book.

31.1 Active nematics

The orientation of anisotropic particles determines the direction of their motion.
The simulation results shown in Fig. 31.1 were obtained for polar self-propelled

particles with a vector order parameter where there is a nonzero average velocity in ordered moving phases. Continuum hydrodynamic models of vector order parameter systems have been constructed (Toner and Tu, 1998). In such models number conservation is assumed, which leads to the coupling between the local density and the local velocity through the nonlinear continuity equation. The model predicts the existence of an ordered phase in which all members of the flock move together with the same mean nonzero velocity. The Goldstone modes associated with the spontaneously broken rotational symmetry are fluctuations in the velocity direction. These Goldstone modes couple with the modes associated with the number conservation law to produce propagating sound modes. The special character of these sound modes leads to anomalously large fluctuations of the flock's number density at long wavelengths.

Here we consider the apolar case where there is no preferred orientation of the rod-like particles, and the system behaves like an active nematic liquid crystal. An active nematic is a collection of active particles with inversion (head–tail) symmetry whose axes are aligned, on average, in a direction $\hat{\mathbf{n}}$. We shall show that active nematics behave in a manner that is qualitatively different from their equilibrium counterparts.

The appropriate order parameters for a system of active nematic particles are the tensor nematic order parameter $\mathbf{Q}(\mathbf{r})$ defined in Eq. (11.2) and the local concentration of active particles $c(\mathbf{r})$. A continuum hydrodynamic model for active nematics on a substrate that provides dissipation was constructed by Ramaswamy *et al.* (2003). The concentration field satisfies a continuity equation,

$$\frac{\partial c(\mathbf{r}, t)}{\partial t} = -\nabla \cdot \mathbf{j}(\mathbf{r}, t) = -\nabla \cdot (c(\mathbf{r}, t)\mathbf{v}(\mathbf{r}, t)), \tag{31.1}$$

where $\mathbf{v}$ is the local velocity field. The local momentum density $\mathbf{g} = mc\mathbf{v}$, where m is the mass of a particle, obeys the dissipative equation of motion,

$$\frac{\partial \mathbf{g}(\mathbf{r}, t)}{\partial t} = -\zeta \mathbf{v} - \nabla \cdot \boldsymbol{\sigma} + \mathbf{f}_R, \tag{31.2}$$

where ζ is a friction constant, $\mathbf{f}_R$ is a Gaussian white-noise random force, and the stress tensor is given by $\boldsymbol{\sigma} = w_0 c \mathbf{I} + w_1 \mathbf{Q} c$, where w_0 and w_1 are constants. The important part of the expression for the stress tensor is the term proportional to $\mathbf{Q}c$. It cannot be derived from a free energy, and is a reflection that nonequilibrium active nematics are being considered. We shall be interested in situations where the inertial term in Eq. (31.2) can be neglected. In this case we can solve for the velocity field in terms of the order parameters:

$$\mathbf{v} = -\frac{1}{\zeta} \left(w_0 \nabla c + w_1 \nabla \cdot (\mathbf{Q}c) - \mathbf{f}_R \right). \tag{31.3}$$

The equation of motion for the director field follows from the Frank free energy functional in Eq. (11.6) with an additional term that accounts for the velocity field:

$$\frac{\partial \hat{\mathbf{n}}}{\partial t} = \lambda \mathbf{A} \cdot \hat{\mathbf{n}} + \boldsymbol{\omega} \times \hat{\mathbf{n}} - \frac{\delta \mathcal{F}_n[\hat{\mathbf{n}}(\mathbf{r})]}{\delta \hat{\mathbf{n}}(\mathbf{r})} + \mathbf{f}_n, \tag{31.4}$$

with

$$-\frac{\delta \mathcal{F}_n[\hat{\mathbf{n}}(\mathbf{r})]}{\delta \hat{\mathbf{n}}(\mathbf{r})}$$
$$= K_1 \nabla(\nabla \cdot \hat{\mathbf{n}}) - K_2[(\hat{\mathbf{n}} \cdot (\nabla \times \hat{\mathbf{n}}))(\nabla \times \hat{\mathbf{n}}) + \nabla \times (\hat{\mathbf{n}}(\hat{\mathbf{n}} \cdot (\nabla \times \hat{\mathbf{n}})))]$$
$$+ K_3[(\hat{\mathbf{n}} \times (\nabla \times \hat{\mathbf{n}})) \times (\nabla \times \hat{\mathbf{n}}) + \nabla \times (\hat{\mathbf{n}} \times (\hat{\mathbf{n}} \times (\nabla \times \hat{\mathbf{n}})))].$$

In these equations λ is the flow alignment parameter, $A_{ij} = (\nabla_i v_j + \nabla_j v_i)/2$, $\boldsymbol{\omega} = (\nabla \times \mathbf{v})/2$, and $\mathbf{f}_n$ is the random force. We are interested in small deviations of both the director and the concentration fields from their equilibrium values, $\delta \hat{\mathbf{n}}_\perp = \hat{\mathbf{n}} - \hat{\mathbf{z}}$ and $\delta c = c - c_0$, respectively. We have selected the equilibrium value of the director to lie along the z axis and used the fact that $\delta \hat{\mathbf{n}}$ is perpendicular to $\hat{\mathbf{z}}$ to linear order. The linearized form of Eq. (31.4) is

$$\frac{\partial \delta \hat{\mathbf{n}}_\perp}{\partial t} = \frac{1}{2}(\lambda - 1)\nabla_\perp v_z + \frac{1}{2}(\lambda + 1)\nabla_z \mathbf{v}_\perp + K_1' \nabla_\perp (\nabla_\perp \cdot \delta \hat{\mathbf{n}}_\perp)$$
$$+ K_2 \nabla_\perp^2 \cdot \delta \hat{\mathbf{n}}_\perp + K_3 \nabla_z^2 \cdot \delta \hat{\mathbf{n}}_\perp + \mathbf{f}_{n\perp}, \tag{31.5}$$

where $K_1' = K_1 - K_2$, and $\mathbf{f}_{n\perp}$ is the part of the Gaussian random force that is perpendicular to $\hat{\mathbf{z}}$. It has correlations given by

$$\langle f_{n\perp i}(\mathbf{r}, t) f_{n\perp j}(\mathbf{r}', t') \rangle = \delta ij \Delta_n \delta(\mathbf{r} - \mathbf{r}')\delta(t - t'). \tag{31.6}$$

To obtain a closed set of equations for $\delta \hat{\mathbf{n}}_\perp$ and δc, we can linearize Eq. (31.3) and use the result to eliminate the velocity in Eqs. (31.1) and (31.5). The linearized form of Eq. (31.3) is

$$v_z = -a_0 \nabla_\perp \cdot \delta \hat{\mathbf{n}}_\perp - (a_1 + a_2)\nabla_z \delta c + \frac{1}{\zeta} f_{Rz},$$
$$\mathbf{v}_\perp = -a_0 \nabla_z \delta \hat{\mathbf{n}}_\perp - a_1 \nabla_\perp \delta c + \frac{1}{\zeta} \mathbf{f}_{R\perp}, \tag{31.7}$$

where $a_0 = w_1 S c_0/\zeta$, $a_1 = (w_0 - w_1 S/3)/\zeta$, $a_2 = w_1 S/\zeta$, and S is the scalar order parameter defined in Eq. (11.2). This result may be substituted into Eqs. (31.1) and (31.5) to obtain the linearized equations of motion for the order parameters. The equation of motion for the concentration field is

$$\frac{\partial \delta c}{\partial t} = D_\perp \nabla_\perp^2 \delta c + D_z \nabla_z^2 \delta c$$
$$+ 2 c_0 a_0 \nabla_z (\nabla_\perp \cdot \delta \hat{\mathbf{n}}_\perp) - \frac{c_0}{\zeta} \nabla \cdot \mathbf{f}_R, \tag{31.8}$$

where $D_\perp = c_0 a_1$ and $D_z = c_0 (a_1 + a_2)$. The linearized equation for the deviation of the director field from its equilibrium value is

$$\frac{\partial \delta \hat{\mathbf{n}}_\perp}{\partial t} = \left(K_z \nabla_z^2 + K_\perp \nabla_\perp^2 + K_L \nabla_\perp \nabla_\perp \right) \cdot \delta \hat{\mathbf{n}}_\perp$$
$$+ D_{cn} \nabla_z \nabla_\perp \delta c + \mathbf{f}_\perp, \tag{31.9}$$

where we have defined $K_z = K_3 - (\lambda+1)a_0/2$, $K_\perp = K_2$, $K_L = K_1' - (\lambda-1)a_0/2$, and $D_{cn} = -(\lambda - 1)(a_1 + a_2)/2 - (\lambda + 1)a_1/2$. The random force is given by $\mathbf{f}_\perp = \mathbf{f}_{n\perp} + \frac{1}{2}(\lambda - 1)\nabla_\perp f_{Rz}/\zeta + \frac{1}{2}(\lambda + 1)\nabla_z f_{R\perp}/\zeta$. These linear equations can be solved by Fourier transformation, and the correlation functions of the order parameters can be computed. Using these solutions, the Fourier transform of the concentration autocorrelation function can be shown to vary as

$$\langle \delta c(\mathbf{k}, t) \delta c(-\mathbf{k}, t) \rangle \propto \frac{1}{k^2}. \tag{31.10}$$

Thus, we see that there are large fluctuations that diverge as k^{-2} as $k \to 0$. This result implies that an active nematic system containing on average N particles should show large number fluctuations with a standard deviation proportional to N in $2d$ systems. This result should be compared to that for systems in thermal equilibrium away from a critical point where number fluctuations are proportional to $\sqrt{N}$. One may also show that the velocity autocorrelation of a tagged particle should decay slowly with time as t^{-1} in $2d$ systems. Active nematics behave qualitatively differently from their equilibrium counterparts. The behavior of the anomalous density fluctuations that appear to characterize flocking phenomena is also captured by phase ordering in a model of hard-core particles sliding downwards under a gravitational field on a one-dimensional fluctuating surface (Das and Barma, 2000; Chatterjee and Barma, 2006).

31.2 Active Langmuir monolayers

As a second example, we consider the dynamics of Langmuir monolayers composed of rod-like liquid crystal molecules containing chiral propellers. The presence of chiral molecules in the monolayer, in combination with nonequilibrium fluxes of material through the layer, leads to cooperative motion and self-organization. Tabe and Yokoyama (2003) carried out experiments on systems of this type where chiral R-(OPOB) molecules were spread on a glycerol surface to form a monolayer membrane (Fig. 31.2). Under appropriate coverage conditions the liquid crystal molecules are tilted relative to the surface, as shown in the figure. Since the glycerol contains water, and the water vapor pressure above the monolayer can be controlled, there is in general a flux of water molecules across the membrane. The transfer of water across the membrane causes the chiral propeller molecules to undergo rotational motion. The molecular motion results in the synchronized collective orientational precession of groups of molecules giving rise to self-organized patterns that could be observed with a reflection-type polarizing microscope. An example of the patterns seen in such experiments is given in Fig. 31.3, which shows oscillating concentric rings of synchronized orientational order in the liquid crystal monolayer. The resulting patterns are very similar to the target patterns observed in excitable media, as discussed in Chapter 24.

The dynamics of such systems can be modeled by choosing a suitable set of order parameters and extending the equations of motion for liquid crystal systems to account for the presence of chiral molecules (Shibata and Mikhailov, 2006). As

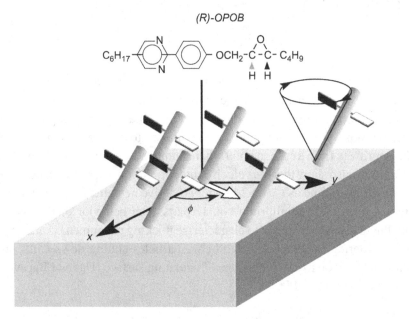

Fig. 31.2. Schematic representation of oriented chiral R-(OPOB) molecules spread on a glycerol surface. The chiral propellers are represented by small horizontal wings. The chemical structure of the R-(OPOB) molecules is also shown in the figure. Reprinted by permission from Tabe and Yokoyama (2003). Copyright 2003 Macmillan Publishers Ltd.

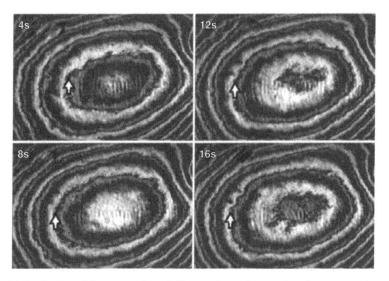

Fig. 31.3. Series of frames at four different times, increasing from top to bottom and left to right, showing the reflectivity of the monolayer. Reprinted by permission from Tabe and Yokoyama (2003). Copyright 2003 Macmillan Publishers Ltd.

in Chapter 11, we let $\hat{\mathbf{n}}(\mathbf{r})$ be the local director field for the liquid crystal. Also, we let $c(\mathbf{r})$ be the local concentration of chiral molecules. Then, the evolution equation for the fraction of chiral molecules satisfies a diffusion equation with a concentration-dependent diffusion coefficient. It can be written in the form

$$\frac{\partial}{\partial t}c(\mathbf{r},t) = \frac{D}{k_B T}\nabla[c(\mathbf{r},t)(1 - c(\mathbf{r},t))\nabla\mu(\mathbf{r},t)]. \tag{31.11}$$

The chemical potential is given by the functional derivative of the free energy, $\mu(\mathbf{r},t) = \delta\mathcal{F}[c(\mathbf{r},t),\hat{\mathbf{n}}(\mathbf{r},t)]/\delta c(\mathbf{r},t)$, where the free energy functional is

$$\mathcal{F}[c,\hat{\mathbf{n}}] = \int d^2 r \left[\frac{1}{2}K_1(\nabla\cdot\hat{\mathbf{n}})^2 + k_B T c \ln c + k_B T(1 - c)\ln(1 - c)\right.$$
$$\left. + \frac{1}{2}G|\nabla c|^2 + \Lambda c\nabla\cdot\hat{\mathbf{n}}\right]. \tag{31.12}$$

This free energy functional contains contributions that arise from the liquid crystal director field and Ginzburg–Landau-like terms from the chiral molecule concentration field. The liquid crystal contribution is just the splay contribution to the Frank free energy density given in Eq. (11.6). The chiral molecule contribution is analogous to that for a binary mixture discussed in Chapter 2. Finally, it is assumed that the splay field couples to the concentration field, and this coupling is given by the term $\Lambda c\nabla\cdot\hat{\mathbf{n}}$ in Eq. (31.12).

The evolution equations for the components of the orientation field can also be derived from the free energy functional, and are

$$\frac{\partial}{\partial t}\hat{n}_x(\mathbf{r}, t) = -\Gamma \frac{\delta \mathcal{F}[c(\mathbf{r}, t), \hat{\mathbf{n}}(\mathbf{r}, t)]}{\delta \hat{n}_x(\mathbf{r}, t)} + \Omega \hat{n}_y(\mathbf{r}, t)$$

$$\frac{\partial}{\partial t}\hat{n}_y(\mathbf{r}, t) = -\Gamma \frac{\delta \mathcal{F}[c(\mathbf{r}, t), \hat{\mathbf{n}}(\mathbf{r}, t)]}{\delta \hat{n}_y(\mathbf{r}, t)} - \Omega \hat{n}_x(\mathbf{r}, t). \qquad (31.13)$$

The new feature is the appearance of the added terms involving the frequency Ω. These terms account for the assumption that the nonequilibrium water flux across the membrane induces rotational motion of the director field. This flux prevents the system from relaxing to equilibrium and, thus, more complex oscillatory states may arise in this system. The precession depends on the presence of chiral molecules in local spatial regions, and the precession frequency could also vary with spatial position in the monolayer. To model this effect the precession frequency is assumed to vary linearly with the local chiral molecule concentration, $\Omega \sim c$. Numerical solutions of this model yield a variety of different spatiotemporal states, many of which are similar to those observed in the experiments on this system (Shibata and Mikhailov, 2006).

The examples chosen for illustration in the last several chapters have served to show how elements that enter in the description of pattern formation and self-assembly processes in equilibrium and far-from-equilibrium systems, which were discussed in detail in the book, can be combined. The focus of much current research is on the behavior of systems in the nonequilibrium domain. In biological systems self-assembly and self-organization almost always occur under such conditions. Physical systems are often studied under conditions that force them out of equilibrium, giving rise to a variety of pattern formation processes not seen in equilibrium systems. The techniques described in this book should provide the background needed to investigate the development of order in such systems.

References

F. F. Abraham, 1979. On the thermodynamics, structure and phase stability of the nonuniform fluid state, *Phys. Repts.* **53**, 93–156.

K. I. Agladze, A. V. Panfilov and A. N. Rudenko, 1988. Nonstationary rotation of spiral waves: three-dimensional effect, *Physica D* **29**, 409–415.

S. M. Allen and J. W. Cahn, 1979. A microscopic theory for antiphase boundary motion and its application to antiphase domain coarsening, *Acta Metall.* **27**, 1085–1095.

S. Alonso, R. Kähler, A. S. Mikhailov and F. Sagués, 2004. Expanding scroll rings and negative tension, turbulence in a model of excitable media, *Phys. Rev. E* **70**, 056201.

D. Andelman, F. Brochard and J.-F. Joanny, 1987. Phase transitions in Langmuir monolayers of polar molecules, *J. Chem. Phys.* **86**, 3673–3681.

I. S. Aranson and L. Kramer, 2002. The world of the complex Ginzburg–Landau equation, *Rev. Mod. Phys.* **74**, 99–143.

I. S. Aranson, K. A. Gorshkov, A. S. Lomov and M. I. Rabinovich, 1990. Stable particle-like solutions of multidimensional nonlinear fields, *Physica D* **43**, 435–453.

I. S. Aranson, A. R. Bishop and L. Kramer, 1998. Dynamics of vortex lines in the three-dimensional complex Ginzburg–Landau equation, *Phys. Rev. E* **57**, 5276–5286.

D. G. Aronson and H. F. Weinberger, 1975. Nonlinear diffusion in population genetics, combustion, and nerve propagation, in *Partial Differential Equations and Related Topics* [Lecture Notes in Mathematics no. 446], J. A. Goldstein, ed., Springer-Verlag, Berlin, pp. 5–49.

M. Asai, H. Ueba and C. Tatsuyama, 1985. Heteroepitaxial growth of Ge films on the Si(100)-2x1 surface, *J. Appl. Phys.* **58**, 2577–2583.

R. J. Asaro and W. A. Tiller, 1972. Interface morphology development during stress corrosion cracking: Part I. Via surface diffusion, *Metall. Trans.* **3**, 1789–1796.

M. Bahiana and Y. Oono, 1990. Cell dynamical system approach to block copolymers, *Phys. Rev. A* **41**, 6763–6771.

P. Bak, C. Tang and K. Wiesenfeld, 1987. Self-organized criticality: An explanation of the 1/f noise, *Phys. Rev. Lett.* **59**, 381–384.

T. Bánsági and O. Steinbock, 2007. Negative filament tension of scroll rings in an excitable system, *Phys. Rev. E* **76**, 045202(R).

M. Bär and M. Or-Guil, 1999. Alternative scenarios of spiral breakup in a reaction–diffusion model with excitable and oscillatory dynamics, *Phys. Rev. Lett.* **82**, 1160–1163.

A.-L. Barabási and H. E. Stanley, 1995. *Fractal Concepts in Surface Growth*, Cambridge University Press, Cambridge.

D. Barkley, 1994. Euclidean symmetry and the dynamics of rotating spiral waves, *Phys. Rev. Lett.* **72**, 164–167.

F. S. Bates, 2005. Network phases in block copolymer melts, *MRS Bull.* **30**, 525–532.

R. Becker and W. Doring, 1935. Kinetische behandlung der keimbildung in ubersattigten dampfen, *Annalen der Physik* **416**, 719–752.

A. L. Belmonte, Q. Ouyang and J.-M. Flesselles, 1997. Experimental survey of spiral dynamics in the Belousov–Zhabotinsky reaction, *J. Phys. II France* **7**, 1425–1468.

B. P. Belousov, 1958. A periodic reaction and its mechanism, in *Sbornik Referatov po Radiatsionni Meditsine Moscow*, Medgiz, Moscow. [English translation in R. J. Field and M. Burger, eds., 1985, *Oscillations and Travelling Waves in Chemical Systems*, Wiley, New York, pp. 605–613]

A. J. Bernhoff, 1991. Spiral wave solutions for reaction–diffusion equations in a fast reaction/slow diffusion limit, *Physica D* **53**, 125–150.

M. Bertram, C. Beta, A. S. Mikhailov, H. H. Rotermund and G. Ertl, 2003. Complex patterns in a periodically forced surface reaction, *J. Phys. Chem. B* **107**, 9610–9615.

H. A. Bethe, 1935. Statistical theory of superlattices, *Proc. R. Soc. London Ser. A* **150**, 552–575.

V. N. Biktashev, A. V. Holden and H. Zhang, 1994. Tension of organizing filaments of scroll waves, *Philos. Trans. R. Soc. London Ser. A* **347**, 611–630.

J. Billingham and D. J. Needham, 1991. The development of traveling waves in quadratic and cubic autocatalysis with unequal diffusion rates. 1. Permanent form traveling waves, *Phil. Trans. R. Soc. London Ser. A* **334**, 1–24.

K. Binder, 1984. Nucleation barriers, spinodals, and the Ginzburg criterion, *Phys. Rev. A* **29**, 341–349.

K. Binder, 1987. Theory of 1st-order phase transitions, *Rep. Progr. Phys.* **50**, 783–859.

C. T. Black, C. B. Murray, R. L. Sandstrom and S. Sun, 2000. Spin-dependent tunneling in self-assembled cobalt-nanocrystal superlattices, *Science* **290**, 1131–1134.

P. S. Bodega, P. Kaira, C. Beta, D. Krefting, D. Bauer, B. Mirwald-Schulz, C. Punckt and H. H. Rotermund, 2007. High frequency periodic forcing of the oscillatory catalytic CO oxidation on Pt(110), *New J. Phys.* **9**, 61.1–61.18.

P. Borckmans, A. De Wit and G. Dewel, 1992. Competition in ramped Turing structures, *Physica A* **188**, 137–157.

U. Börner, A. Deutsch, H. Reichenbach and M. Bär, 2002. Rippling patterns in aggregates of myxobacteria arise from cell–cell collisions, *Phys. Rev. Lett.* **89**, 078101.

W. L. Bragg and E. J. Williams, 1934. The effect of thermal agitation on atomic arrangement in alloys, *Proc. R. Soc. London Ser. A* **145**, 699–730.

M. Bramson, 1983. Convergence of solutions of the Kolmogorov equation to travelling waves, *Mem. Am. Math. Soc.* 285, 1–190.

M. Bramson, P. Calderoni, A. De Masi, P. Ferrari, J. Lebowitz and R. H. Schonmann, 1986. Microscopic selection principle for a diffusion-reaction equation, *J. Stat. Phys.* **45**, 905–920.

A. J. Bray, 1994. Theory of phase-ordering kinetics, *Adv. Phys.* **43**, 357–459.

A. J. Bray, 1997. Velocity distribution of topological defects in phase-ordering systems, *Phys. Rev. E* **55**, 5297–5301.

A. J. Bray and K. Humayun, 1992. Scaling and multiscaling in the ordering kinetics of a conserved order parameter, *Phys. Rev. Lett.* **68**, 1559–1562.

A. J. Bray and K. Humayun, 1993. Towards a systematic calculation of the scaling functions for the ordering kinetics of nonconserved fields, *Phys. Rev. E* **48**, 1609–1612.

A. J. Bray and S. Puri, 1991. Asymptotic structure factor and power-law tails for phase ordering in systems with continuous symmetry, *Phys. Rev. Lett.* **67**, 2670–2673.

A. J. Bray, S. Puri, R. E. Blundell and A. M. Somoza, 1993. Structure factor for phase ordering in nematic liquid crystals, *Phys. Rev. E* **47**, R2261–R2264.

S. A. Brazovskii, 1975. Phase-transitions of an isotropic system to an inhomogeneous state, *Zh. Eksp. Teor. Fiz.* **68**, 175–185 [*Sov. Phys. JETP* **41**, 85 (1975)].

C. J. Brinker, 2004. Evaporation induced self assembly: functional nanostructures made easy, *MRS Bull.* **29**, 631–640.

C. Brito, I. S. Aranson and H. Chaté, 2003. Vortex glass and vortex liquid in oscillatory media, *Phys. Rev. Lett.* **90**, 068301.

E. Brunet and B. Derrida, 1997. Shift in the velocity of a front due to a cutoff, *Phys. Rev. E* **56**, 2597–2604.

L. Brunnet, H. Chaté and P. Manneville, 1994. Long-range order with local chaos in lattices of diffusively coupled ODEs, *Physica D* **78**, 141–154.

L. Brusch, M. G. Zimmermann, M. van Hecke, M. Bär and A. Torcini, 2000. Modulated amplitude waves and the transition from phase to defect chaos, *Phys. Rev. Lett.* **85**, 86–89.

V. M. Burlakov, 2006. Ostwald ripening in rarefied systems, *Phys. Rev. Lett.* **97**, 155703.

J. W. Cahn, 1961. On spinodal decomposition, *Acta Metall.* **9**, 795–801.

J. W. Cahn, 1966. The later stages of spinodal decomposition and the beginnings of particle coarsening, *Acta Metall.* **14**, 1685–1692.

J. W. Cahn, 1968. 1967 Institute of Metals lecture spinodal decomposition, *Trans. Metall. Soc. AIME* **242**, 166–180.

J. W. Cahn and J. E. Hilliard, 1958. Free energy of a nonuniform system. I. Interfacial free energy, *J. Chem. Phys.* **28**, 258–267.

J. W. Cahn and J. E. Hilliard, 1959. Free energy of a nonuniform system. III. Nucleation in a two-component incompressible fluid, *J. Chem. Phys.* **31**, 688–699.

V. Castets, E. Dulos, J. Boissonade and P. De Kepper, 1990. Experimental evidence of a sustained standing Turing-type nonequilibrium chemical pattern, *Phys. Rev. Lett.* **64**, 2953–2956.

P. M. Chaikin and T. C. Lubensky, 1995. *Principles of Condensed Matter Physics*, Cambridge University Press, New York.

A. Chakrabarti and J. D. Gunton, 1993. Lamellar phase in a model for block copolymers, *Phys. Rev. E* **47**, R792–R795.

S. Chandrasekhar, 1992. *Liquid Crystals*, Cambridge University Press, Cambridge.

H. Chaté, 1994. Spatiotemporal intermittency regimes of the one-dimensional complex Ginzburg–Landau equation, *Nonlinearity* **7**, 185–204.

H. Chaté and P. Manneville, 1992. Stability of the Bekki–Nozaki hole solutions to the one-dimensional complex Ginzburg–Landau equation, *Phys. Lett. A* **171**, 183–188.

H. Chaté and P. Manneville, 1996. Phase diagram of the two-dimensional complex Ginzburg–Landau equation, *Physica A* **224**, 348–368.

S. Chatterjee and M. Barma, 2006. Dynamics of fluctuation-dominated phase ordering: Hard-core passive sliders on a fluctuating surface, *Phys. Rev. E* **73**, 011107.

F. Chavez, R. Kapral, G. Rousseau and L. Glass, 2001. Scroll waves in spherical shell geometries, *Chaos* **11**, 757–765.

Y. C. Chou and W. I. Goldburg, 1979. Phase separation and coalescence in critically quenched isobutyric-acid–water and 2,6-lutidine–water mixtures, *Phys. Rev. A* **20**, 2105–2113.

Y. C. Chou and W. I. Goldburg, 1981. Angular distribution of light scattered from critically quenched liquid mixtures, *Phys. Rev. A* **23**, 858–864.

R. H. Clayton and A. V. Holden, 2004. Filament behavior in a computational model of ventricular fibrillation in the canine heart, *IEEE Trans. Biomed. Eng.* **51**, 28–34.

R. H. Clayton, E. A. Zuchkova and A. V. Panfilov, 2006. Phase singularities and filaments: Simplifying complexity in computational models of ventricular fibrillation, *Prog. Biophys. Molec. Biol.* **90**, 378–398.

A. Coniglio and M. Zannetti, 1989. Multiscaling in growth kinetics, *Europhys. Lett.* **10**, 575–580.

P. Coullet and K. Emilsson, 1992a. Strong resonances of spatially distributed oscillators: a laboratory to study patterns and defects, *Physica D* **61**, 119–131.

P. Coullet and K. Emilsson, 1992b. Pattern formation in the strong resonant forcing of spatially distributed oscillators, *Physica A* **188**, 190–200.

P. Coullet, L. Gil and J. Lega, 1989. Defect-mediated turbulence, *Phys. Rev. Lett.* **62**, 1619–1622.

P. Coullet, J. Lega, B. Houchmanzadeh and J. Lajzerwoicz, 1990. Breaking chirality in nonequilibrium systems, *Phys. Rev. Lett.* **65**, 1352–1355.

A. Craievich and J. M. Sanchez, 1981. Dynamical scaling in the glass system $B_2O_3-PbO-Al_2O_3$, *Phys. Rev. Lett.* **47**, 1308–1311.

M. C. Cross and P. Hohenberg, 1993. Pattern formation outside of equilibrium, *Rev. Mod. Phys.* **65**, 851–1112.

M. C. Cross and A. C. Newell, 1984. Convection patterns in large aspect ratio systems, *Physica D* **10**, 299–328.

K. E. Daniels and E. Bodenschatz, 2002. Defect turbulence in inclined layer convection, *Phys. Rev. Lett.* **88**, 034501.

D. Das and M. Barma, 2000. Particles sliding on a fluctuating surface: Phase separation and power laws, *Phys. Rev. Lett.* **85**, 1602–1605.

S. Dattagupta and S. Puri, 2004. *Dissipative Phenomena in Condensed Matter*, Springer, Berlin.

J. M. Davidenko, A. V. Pertsov, R. Salomonsz, W. Baxter and J. Jalife, 1992. Stationary and drifting spiral waves of excitation in isolated cardiac muscle, *Nature (London)* **355**, 349–351.

J. Davidsen and R. Kapral, 2003. Defect-mediated turbulence in systems with local deterministic chaos, *Phys. Rev. Lett.* **91**, 058303.

J. Davidsen, R. Erichsen, R. Kapral and H. Chaté, 2004. From ballistic to Brownian vortex motion in complex oscillatory media, *Phys. Rev. Lett.* **93**, 018305.

J. Davidsen, A. S. Mikhailov and R. Kapral, 2005. Front explosion in a periodically forced surface reaction, *Phys. Rev. E* **72**, 046214.

P. G. Debenedetti, 1996. *Metastable Liquids: Concepts and Principles*, Princeton University Press, Princeton, NJ.

P. Debye, H. R. Anderson and H. Brumberger, 1957. Scattering by an inhomogeneous solid, 2. The correlation function and its application, *J. Appl. Phys.* **28**, 679–683.

G. Dee and J. S. Langer, 1983. Propagating pattern selection, *Phys. Rev. Lett.* **50**, 383–386.

P. G. de Gennes and J. Prost, 1993. *The Physics of Liquid Crystals*, 2nd ed., Clarendon Press, Oxford.

C. Denniston and M. O. Robbins, 2004. Mapping molecular models to continuum theories for partially miscible fluids, *Phys. Rev. E* **69**, 021505.

G. Dewel, P. Borckmans, A. DeWit, B. Rudovics, J.-J. Perraud, E. Dulos, J. Boissonade and P. De Kepper, 1995. Pattern selection and localized structures in reaction-diffusion systems, *Physica A* **213**, 181–198.

S. Dushman, 1904. The rate of the reaction between iodic and hydriodic acids, *J. Phys. Chem.* **8**, 453–482.

D. J. Eaglesham and M. Cerullo, 1990. Dislocation-free Stranski– Krastanow growth of Ge on Si(100), *Phys. Rev. Lett.* **64**, 1943–1946.

B. Echebarria, V. Hakim and H. Henry, 2006. Nonequilibrium ribbon model of twisted scroll waves, *Phys. Rev. Lett.* **96**, 098301.

E. C. Edblom, M. Orbán and I. R. Epstein, 1986. A new iodate oscillator: the Landolt reaction with ferrocyanide in a CSTR, *J. Am. Chem. Soc.* **108**, 2826–2830.

S. F. Edwards and D. R. Wilkinson, 1982. The surface statistics of a granular aggregate, *Proc. R. Soc. London Ser. A* **381**, 17–31.

D. A. Egolf and H. S. Greenside, 1995. Characterization of the transition from defect to phase turbulence, *Phys. Rev. Lett.* **74**, 1751–1754.

G. Ehrlich and F. G. Huda, 1966. Atomic view of surface self-diffusion: tungsten on tungsten, *J. Chem. Phys.* **44**, 1039–1049.

K. R. Elder and R. C. Desai, 1989. Role of nonlinearities in off-critical quenches as described by the Cahn–Hilliard model of phase separation, *Phys. Rev. B* **40**, 243–254.

K. R. Elder, T. M. Rogers and R. C. Desai, 1988. Early stages of spinodal decomposition for the Cahn–Hilliard–Cook model of phase separation, *Phys. Rev. B* **38**, 4725–4739.

C. Elphick, G. Iooss and E. Tirapegui, 1987. Normal form reduction for time-periodically driven differential equations, *Phys. Lett. A* **120**, 459–463.

C. Elphick, A. Hagberg and E. Meron, 1998. Phase front instability in periodically forced oscillatory systems, *Phys. Rev. Lett.* **80**, 5007–5010.

C. Elphick, A. Hagberg and E. Meron, 1999. Multiphase patterns in periodically forced oscillatory systems, *Phys. Rev. E* **59**, 5285–5291.

G. Ertl, 2000. Dynamics of reactions at surfaces, *Adv. Catal.* **45**, 1–69.

F. H. Fenton, E. M. Cherry, H. M. Hastings and S. J. Evans, 2002. Multiple mechanisms of spiral wave breakup in a model of cardiac electrical activity, *Chaos* **12**, 852–892.

P. Fife, 1984. Propagator–controller systems and chemical patterns, in *Non-Equilibrium Dynamics in Chemical Systems*, C. Vidal and A. Pacault, eds., Springer-Verlag, Berlin, pp. 76–88.

R. A. Fisher, 1937. The wave of advantageous genes, *Ann. Eugenics* **7**, 355–369.

R. FitzHugh, 1961. Impulses and physiological states in theoretical models of nerve membrane, *Biophys. J.* **1**, 445–466.

M. Flicker and J. Ross, 1974. Mechanism of chemical instability for periodic precipitation phenomena, *J. Chem. Phys.* **60**, 3458–3465.

D. Forster, 1975. *Hydrodynamic Fluctuations, Broken Symmetry, and Correlation Functions*, Benjamin, Reading, MA.

F. C. Frank, 1958. Liquid crystals: On the theory of liquid crystals, *Disc. Faraday Soc.* **25**, 19–28.

P. Fratzl, J. L. Lebowitz, O. Penrose and J. Amar, 1991. Scaling functions, self-similarity, and the morphology of phase-separating systems, *Phys. Rev. B* **44**, 4794–4811.

G. H. Fredrickson and F. S. Bates, 1996. Dynamics of block copolymers: Theory and experiment, *Ann. Rev. Mater. Sci.* **26**, 501–550.

D. Frenkel and J. P. McTague, 1979. Evidence for an orientationally ordered two-dimensional fluid phase from molecular-dynamics calculations, *Phys. Rev. Lett.* **42**, 1632–1635.

T. Frisch, S. Rica, P. Coullet and J. M. Gilli, 1994. Spiral waves in liquid crystal, *Phys. Rev. Lett.* **72**, 1471–1474.

U. Frisch, 1995. *Turbulence*, Cambridge University Press, Cambridge.

H. Furukawa, 1989. Numerical study of multitime scaling in a solid system undergoing phase separation, *Phys. Rev. B* **40**, 2341–2347.

M. Gabbay, E. Ott and P. N. Guzdar, 1997. Motion of scroll wave filaments in the complex Ginzburg–Landau equation, *Phys. Rev. Lett.* **78**, 2012–2015.

J. M. Gambaudo, 1985. Perturbation of a Hopf bifurcation by an external time-periodic forcing, *J. Diff. Eqns.* **57**, 172–199.

M. Gameiro, K. Mischaikow and T. Wanner, 2005. Evolution of pattern complexity in the Cahn–Hilliard theory of phases separation, *Acta Mater.* **53**, 693–704.

N. Ganapathisubramanian and K. Showalter, 1985. A new iodate driven nonperiodic oscillatory reaction in a continuously stirred tank reactor, *J. Phys. Chem.* **89**, 2118–2119.

T. Garel and S. Doniach, 1982. Phase transitions with spontaneous modulation: The dipolar Ising ferromagnet, *Phys. Rev. B* **26**, 325–329.

V. Gáspár and K. Showalter, 1990. Simple model for the oscillatory iodate oxidation of sulfite and ferrocyanide, *J. Phys. Chem.* **94**, 4973–4979.

B. D. Gaulin, S. Spooner and Y. Morii, 1987. Kinetics of phase separation in $Mn_{0.67}Cu_{0.33}$, *Phys. Rev. Lett.* **59**, 668–671.

W. M. Gelbart, A. Ben-Shaul and D. Roux, eds., 1994. *Micelles, Membranes, Microemulsions, and Monolayers*, Springer-Verlag, New York.

L. Gil, J. Lega and J. L. Meunier, 1990. Statistical properties of defect-mediated turbulence, *Phys. Rev. A* **41**, 1138–1141.

V. L. Ginzburg and L. D. Landau, 1950. On the theory of superconductivity, *Zh. Eksp. Teor. Fiz.* **20**, 1064–1082 [translation in *Collected Papers of L. D. Landau*, Pergamon, Oxford, 1965, pp. 546–568].

B. Giron, B. Meerson and P. V. Sasorov, 1998. Weak selection and stability of localized distributions in Ostwald ripening, *Phys. Rev. E* **58**, 4213–4216.

S. C. Glotzer, 1995. Spinodal decomposition in polymer blends, in *Annual Reviews in Computational Physics* II, D. Stauffer, ed., World Scientific, Singapore, pp. 1–46.

S. C. Glotzer, E. A. Di Marzio and M. Muthukumar, 1995. Reaction-controlled morphology of phase-separating mixtures, *Phys. Rev. Lett.* **74**, 2034–2037.

A. Goldbeter, 1996. *Biochemical Oscillations and Cellular Rhythms: The Molecular Bases of Periodic and Chaotic Behaviour*, Cambridge University Press, Cambridge.

N. Goldenfeld, 1992. *Lectures on Phase Transitions and the Renormalization Group*, Addison-Wesley, Reading, MA.

H. Goldstein, 1950. *Classical Mechanics*, Addison-Wesley, Reading, MA.

R. E. Goldstein, D. J. Muraki and D. M. Petrich, 1996. Interface proliferation and the growth of labyrinths in a reaction–diffusion system, *Phys. Rev. E* **53**, 3933–3957.

M. Golubitsky, V. G. LeBlanc and I. Melbourne, 1997. Meandering of the spiral tip: An alternative approach, *J. Nonlinear Sci.* **7**, 557–586.

K. A. Gorshkov, A. S. Lomov and M. I. Rabinovich, 1996. Three-dimensional particle-like solutions of coupled nonlinear fields, *Phys. Lett. A* **137**, 250–254.

A. Goryachev and R. Kapral, 1996a. Spiral waves in chaotic systems, *Phys. Rev. Lett.* **76**, 1619–1622.

A. Goryachev and R. Kapral, 1996b. Structure of complex-periodic and chaotic media with spiral waves, *Phys. Rev. E* **54**, 5469–5481.

A. Goryachev and R. Kapral, 2000. Synchronization line defects in oscillatory and excitable media, in *Stochastic Dynamics and Pattern Formation in Biological and Complex Systems*, S. Kim, K. J. Lee and W. Sung, eds., American Institute of Physics, Melville, NY, pp. 23–35.

A. Goryachev, H. Chaté and R. Kapral, 1998. Synchronization defects and broken symmetry in spiral waves, *Phys. Rev. Lett.* **80**, 873–876.

A. Goryachev, R. Kapral and H. Chaté, 2000. Synchronization defect lines, *Int. J. Bifurcat. Chaos* **10**, 1537–1564.

G. Grégoire, H. Chaté and Y. Tu, 2003. Moving and staying together without a leader, *Physica D* **181**, 157–170.

M. Ya. Grinfeld, 1986. Instability of the separation boundary between a nonhydrostatically stressed elastic body and a melt, *Dokl Akad. Nauk SSSR* **290**, 1358; *Sov. Phys. Dokl.* **31**, 831–834.

A. Yu. Grosberg and A. R. Khokhlov, 1997. *Giant Molecules*, Academic Press, San Diego, CA.

B. Grossmann, H. Guo and M. Grant, 1991. Kinetic roughening of interfaces in driven systems, *Phys. Rev. A* **43**, 1727–1743.

D. Gruner, R. Kapral and A. Lawniczak, 1993. Nucleation, domain growth, and fluctuations in a bistable chemical system, *J. Chem. Phys.* **99**, 3938–3945.

J. D. Gunton and M. Droz, 1983. *Introduction to the Theory of Metastable and Unstable States* [Lecture Notes in Physics no. 183], Springer-Verlag, Berlin.

J. D. Gunton, M. San Miguel and P. S. Sahni, 1983. The dynamics of first order phase transitions, in *Phase Transitions and Critical Phenomena*, vol. 8, C. Domb and J. L. Lebowitz, eds., Academic Press, London, pp. 267–482.

H. Y. Guo, L. Li, H. L. Wang and Q. Ouyang, 2004. Chemical waves with line defects in the Belousov–Zhabotinsky reaction, *Phys. Rev. E* **69**, 056203.

J. E. Guyer and P. W. Voorhees, 1995. Morphological stability of alloy thin films, *Phys. Rev. Lett.* **74**, 4031–4034.

T. M. Haeusser and S. Leibovich, 1997. Amplitude and mean drift equations for the oceanic Ekman layer, *Phys. Rev. Lett.* **79**, 329–332.

P. S. Hagan, 1982. Spiral waves in reaction–diffusion equations, *SIAM J. Appl. Math.* **42**, 762–786.

A. Hagberg and E. Meron, 1993. Domain-walls in nonequilibrium systems and emergence of persistent patterns, *Phys. Rev. E* **48**, 705–708.

A. Hagberg and E. Meron, 1994. From labyrinthine patterns to spiral turbulence, *Phys. Rev. Lett.* **72**, 2494–2497.

V. Hakim and A. Karma, 1999. Theory of spiral wave dynamics in weakly excitable media: Asymptotic reduction to a kinematic model and applications, *Phys. Rev. E* **60**, 5073–5105.

C. Harrison, D. H. Adamson, Z. Cheng, J. M. Sebastian, S. Sethuraman, D. A. Huse, R. A. Register and P. M. Chaikin, 2000. Mechanisms of ordering in striped patterns, *Science* **290**, 1558–1560.

C. Harrison, Z. Cheng, S. Sethuraman, P. M. Chaikin, D. A. Vega, J. M. Sebastian, R. A. Register and D. H. Adamson, 2002. Dynamics of pattern coarsening in a two-dimensional smectic system, *Phys. Rev. E* **66**, 011706.

T. Hashimoto, M. Shibayama and H. Kawai, 1980. Domain-boundary structure of styrene-isoprene block copolymer films cast from solution. 4. Molecular-weight dependence of lamellar microdomains, *Macromolecules* **13**, 1237–1247.

T. Hashimoto, J. Kumaki and H. Kawai, 1983. Time-resolved light scattering studies on kinetics of phase separation and phase dissolution of polymer blends. 1. Kinetics of phase separation of a binary mixture of polystyrene and poly(vinyl methyl ether), *Macromolecules* **16**, 641–648.

T. Hashimoto, M. Itakura and H. Hasegawa, 1986a. Late stage spinodal decomposition of a binary polymer mixture. I. Critical test of dynamical scaling on scattering function, *J. Chem. Phys.* **85**, 6118–6128.

T. Hashimoto, M. Itakura and N. Shimidzu, 1986b. Late stage spinodal decomposition of a binary polymer mixture. II. Scaling analyses on $Q_m(\tau)$ and $l_m(\tau)$, *J. Chem. Phys.* **85**, 6773–6786.

C. Hemming, 2003. *Resonantly Forced Oscillatory Reaction-Diffusion Systems*, Ph.D. Thesis, University of Toronto.

C. Hemming and R. Kapral, 2000. Resonantly forced inhomogeneous reaction-diffusion systems, *Chaos* **10**, 720–730.

C. Hemming and R. Kapral, 2001. Turbulent fronts in resonantly forced oscillatory systems, *Faraday Discuss. Chem. Soc.* **120**, 371–382.

C. Hemming and R. Kapral, 2002. Front explosion in a resonantly forced complex Ginzburg–Landau system, *Physica D* **168–169**, 10–22.

H. Henry and V. Hakim, 2000. Linear stability of scroll waves, *Phys. Rev. Lett.* **85**, 5328–5331.

H. Henry and V. Hakim, 2002. Scroll waves in isotropic excitable media: Linear instabilities, bifurcations, and restabilized states, *Phys. Rev. E* **65**, 046235.

C. Henze, E. Lugosi and A. T. Winfree, 1990. Helical organizing centers in excitable media, *Can. J. Phys.* **68**, 683–710.

M. Hildebrand, M. Ipsen, A. S. Mikhailov and G. Ertl, 2003. Localized nonequilibrium nanostructures in surface chemical reactions, *New J. Phys.* **5**, 61.1–61.28.

A. L. Hodgkin and A. F. Huxley, 1952. A quantitative description of membrane current and its application to conduction and excitation in nerve, *J. Physiol. London* **117**, 500–544.

A. L. Hodgkin, A. F. Huxley and B. Katz, 1952. Measurement of current–voltage relations in the membrane of the giant axon of loligo, *J. Physiol. London* **116**, 424–448.

P. C. Hohenberg and B. I. Halperin, 1977. Theory of dynamical critical phenomena, *Rev. Mod. Phys.* **49**, 435–479.

D. Horváth and K. Showalter, 1995. Instabilities in propagating reaction–diffusion fronts of the iodate–arsenous acid reaction, *J. Chem. Phys.* **102**, 2471–2478.

D. Horváth, V. Petrov, S. K. Scott and K. Showalter, 1993. Instabilities in propagating reaction–diffusion fronts, *J. Chem. Phys.* **98**, 6332–6343.

R. B Hoyle, 2006. *Pattern Formation: An Introduction to Methods*, Cambridge University Press, Cambridge.

K. Huang, 1987. *Statistical Mechanics*, Wiley, New York.

Z. F. Huang and R. C. Desai, 2002. Epitaxial growth in dislocation-free strained alloy films: Morphological and compositional instabilities, *Phys. Rev. B* **65**, 205419.

O. A. Igoshin and G. Oster, 2004. Rippling of myxobacteria. *Math. Biosci.* **188**, 221–233.

W. Janke, W. E. Skaggs and A. T. Winfree, 1989. Chemical vortex dynamics in the Belousov–Zhabotinskii reaction and in the two-variable oregonator model, *J. Phys. Chem.* **93**, 740–749.

H. Jinnai, Y. Nishikawa, R. J. Spontak, S. D. Smith, D. A. Agard and T. Hashimoto, 2000. Direct measurement of interfacial curvature distributions in a bicontinuous block copolymer morphology, *Phys. Rev. Lett.* **84**, 518–521.

K. John and M. Bär, 2005a. Alternative mechanisms of structuring biomembranes: Self-assembly versus self-organization, *Phys. Rev. Lett.* **95**, 198101.

K. John and M. Bär, 2005b. Travelling lipid domains in a dynamic model for protein-induced pattern formation in biomembranes, *Phys. Biol.* **2**, 123–132.

R. Kapral and K. Showalter, eds., 1995. *Chemical Waves and Patterns*, Kluwer, Dordrecht.

R. Kapral, R. Livi, G.-L. Oppo and A. Politi, 1994. Dynamics of complex interfaces, *Phys. Rev. E* **49**, 2009–2022.

R. Kapral, R. Livi and A. Politi, 1997. Critical behavior of complex interfaces, *Phys. Rev. Lett.* **79**, 2277–2280.

M. Kardar, G. Parisi and Y.-C. Zhang, 1986. Dynamic scaling of growing interfaces, *Phys. Rev. Lett.* **56**, 889–892.

M. Karttunen, I. Vattulainen and A. Lukkarinen, eds., 2004. *Novel Methods in Soft Matter Simulations* [Lecture Notes in Physics no. S640], Springer-Verlag, Berlin.

K. Kawasaki, M. C. Yalabik and J. D. Gunton, 1978. Growth of fluctuations in quenched time-dependent Ginzburg–Landau model systems, *Phys. Rev. A* **17**, 455–470.

J. P. Keener, 1988. The dynamics of three-dimensional scroll waves in excitable media, *Physica D* **31**, 269–276.

D. A. Kessler, Z. Ner and L. M. Sander, 1998. Front propagation: Precursors, cutoffs and structural stability, *Phys. Rev. E* **58**, 107–114.

C. Kittel and H. Kroemer, 1980. *Thermal Physics*, Freeman, San Francisco.

M. Kléman, 1983. *Points, Lines and Walls: In Liquid Crystals, Magnetic Systems and Various Disordered Media*, Wiley, New York.

R. R. Klevecz, J. Pilliod and J. Bolen, 1991. Autogenous formation of spiral waves by coupled chaotic attractors, *Chronobiol. Int.* **8**, 6–13.

A. N. Kolmogorov, I. G. Petrovskii and N. S. Piskunov, 1937. Study of the diffusion equation with growth of the quantity of matter and its application to a biology problem, *Moscow Univ. Bull. Math., Ser. Int., Section A* **1**, 1.

C. Kooy and U. Enz, 1960. Experimental and theoretical study of the domain configuration in thin layers of $BaFe_{12}O_{19}$, *Philips Res. Repts.* **15**, 7–29.

J. M. Kosterlitz and D. J. Thouless, 1973. Self-consistent theory of localization, *J. Phys. C* **6**, 1181–1203.

J. M. Kosterlitz and D. J. Thouless, 1978. Two-dimensional physics, in *Progress in Low Temperature Physics*, vol. VIIB, D. F. Brewer, ed., North-Holland, Amsterdam, pp. 371–433.

J. Krug, 1997. Origins of scale invariance in growth processes, *Adv. Phys.* **46**, 139–282.

J. Krug, M. Plischke and M. Siegert, 1993. Surface diffusion currents and the universality classes of growth, *Phys. Rev. Lett.* **70**, 3271–3274.

Y. Kuramoto, 1984. *Chemical Oscillations, Waves and Turbulence*, Springer-Verlag, Berlin.

Y. Kuramoto and T. Tsuzuki, 1976. Persistent propagation of concentration waves in dissipative media far from thermal equilibrium, *Prog. Theor. Phys.* **55**, 356–369.

Y. Kwon, K. Thornton and P. W. Voorhees, 2007. Coarsening of bicontinuous structures via nonconserved and conserved dynamics, *Phys. Rev. E* **75**, 021120.

J. Lahann and R. Langer, 2005. Smart materials with dynamically controllable surfaces, *MRS Bull.* **30**, 185–188.

L. D. Landau, 1937. On the theory of phase transitions, *Zh. Eksp. Teor. Fiz.* **7**, 19–32; On the theory of phase transitions II, *Zh. Eksp. Teor. Fiz.* **7**, 627–632 [translation in *Collected Papers of L. D. Landau*, Pergamon, Oxford, 1965, pp. 193–216].

L. D. Landau and E. M. Lifshitz, 1986. *Theory of Elasticity*, 3rd ed., Pergamon, New York.

L. D. Landau, E. M. Lifshitz and L. P. Pitaevskii, 1980. *Statistical Physics*, 3rd ed., Part 1, Oxford, New York.

J. S. Langer, 1969. Statistical theory of the decay of metastable states, *Ann. Phys.* **54**, 258–275.

J. S. Langer, 1971. Theory of spinodal decomposition in alloys, *Ann. Phys.* **65**, 53–86.

K. J. Lee and H. L. Swinney, 1995. Lamellar structures and self-replicating spots in a reaction-diffusion system, *Phys. Rev. E* **51**, 1899–1915.

K. J. Lee, W. D. McCormick, Q. Ouyang and H. L. Swinney, 1993. Pattern-formation by interacting chemical fronts, *Science* **261**, 192–194.

F. K. LeGoues, M. Copel and R. M. Tromp, 1990. Microstructure and strain relief of Ge films grown layer by layer on Si(001), *Phys. Rev. B* **42**, 11690–11700.

L. Leibler, 1980. Domain-boundary structure of styrene–isoprene block copolymer films cast from solution. 4. Molecular-weight dependence of lamellar microdomains, *Macromolecules* **13**, 1602–1617.

I. Lengyel and I. R. Epstein, 1991. Modelling of Turing structures in the chlorite–iodide–malonic acid–starch reaction system, *Science* **251**, 650–652.

F. Léonard and R. C. Desai, 1998. Alloy decomposition and surface instabilities in thin films, *Phys. Rev. B* **57**, 4805–4815.

G. Li, O. Qi and H. L. Swinney, 1996. Transitions in two-dimensional patterns in a ferrocyanide–iodate–sulfite reaction, *J. Chem. Phys.* **105**, 10830–10837.

I. M. Lifshitz, 1962. Kinetics of ordering during second-order phase transitions, *Sov. Phys. JETP* **15**, 939–942.

I. M. Lifshitz and V. V. Slyozov, 1961. The kinetics of precipitation from supersaturated solid solutions, *J. Phys. Chem. Solids* **19**, 35–50.

A. L. Lin, M. Bertram, K. Martinez and H. L. Swinney, 2000. Resonant phase patterns in a reaction–diffusion system, *Phys. Rev. Lett.* **84**, 4240–4243.

A. L. Lin, A. Hagberg, E. Meron and H. L. Swiney, 2004. Resonance tongues and patterns in periodically forced reaction–diffusion systems, *Phys. Rev. E* **69**, 066217.

F. Liu and N. Goldenfeld, 1989. Dynamics of phase separation in block copolymer melts, *Phys. Rev. A* **39**, 4805–4810.

A. Locatelli, T. O. Mentes, L. Aballe, A. S. Mikhailov and M. Kiskinova, 2006. Formation of regular surface-supported mesostructures with periodicity controlled by chemical reaction rate, *J. Phys. Chem. B*, 19108–19111.

S. Longhi, 2001. Spiral waves in a class of optical parametric oscillators, *Phys. Rev. E* **63**, 055202.

C. Luengviriya, U. Storb, G. Lindner, S. C. Müller, M. Bär and M. J. B. Hauser, 2008. Scroll wave instabilities in an excitable medium, *Phys. Rev. Lett.* **100**, 148302.

R. Luther, 1906. Räumliche fortpflanzung chemischer reaktionen, *Z. Elektrochem.* **12**, 596–600.

A. Malevanets and R. Kapral, 1996. Links, knots, and knotted labyrinths in bistable systems, *Phys. Rev. Lett.* **77**, 767–770.

A. Malevanets and R. Kapral, 1997. Microscopic model for FitzHugh–Nagumo dynamics, *Phys. Rev. E* **55**, 5657–5670.

A. Malevanets and R. Kapral, 1998. Knots in bistable reacting systems, in *Ideal Knots*, A. Stasiak, V. Katritch and L. H. Kauffman, eds., World Scientific, Singapore, pp. 234–254.

A. Malevanets, A. Careta and R. Kapral, 1995. Biscale chaos in propagating fronts, *Phys. Rev. E* **52**, 4724–4735.

P. Manneville, 1990. *Dissipative Structures and Weak Turbulence*, Academic Press, Boston, MA.

S. C. Manrubia, A. S. Mikhailov and D. H. Zanette, 2004. *Emergence of Dynamical Order*, World Scientific, Singapore.

A. Maritan, F. Toigo, J. Koplik and J. Banavar, 1992. Dynamics of phase separation in block copolymer melts, *Phys. Rev. Lett.* **69**, 3193–3195.

G. F. Mazenko, 1989. Theory of unstable thermodynamic systems, *Phys. Rev. Lett.* **63**, 1605–1608.

G. F. Mazenko, 1990. Theory of unstable growth, *Phys. Rev. B* **42**, 4487–4505.

G. F. Mazenko, 1991. Theory of unstable growth. II. Conserved order parameter, *Phys. Rev. B* **43**, 5747–5763.

G. F. Mazenko, 2006. *Nonequilibrium Statistical Mechanics*, Wiley, New York.

H. M. McConnell, 1989. Theory of hexagonal and stripe phases in monolayers, *Proc. Nat. Acad. Sci. USA* **86**, 3452–3455.

H. M. McConnell, 1990. Harmonic shape transitions in lipid monolayer domains, *J. Phys. Chem.* **94**, 4728–4731.

B. Meerson, 1999. Fluctuations provide strong selection in Ostwald ripening, *Phys. Rev. E* **60**, 3072–3075.

B. Meerson, L. M. Sander and P. Smereka, 2005. The role of discrete-particle noise in the Ostwald ripening, *Europhys. Lett.* **72**, 604–610.

N. D. Mermin, 1979. Topological theory of defects in ordered media, *Rev. Mod. Phys.* **51**, 591–648.

C. W. Meyer, D. S. Cannell, G. Ahlers, J. B. Swift and P. C. Hohenberg, 1988. Pattern competition in temporally modulated Rayleigh-Bénard convection, *Phys. Rev. Lett.* **61**, 947–950.

A. S. Mikhailov, 1994. *Foundations of Synergetics I. Distributed Active Systems*, Springer, Berlin.

A. S. Mikhailov, V. A. Davydov and V. S. Zykov, 1994. Complex dynamics of spiral waves and motions of curves, *Physica D* **70**, 1–39.

R. A. Milton and S. K. Scott, 1995. Instabilities of cubic autocatalytic waves on two and three dimensional domains, *J. Chem. Phys.* **102**, 5271–5277.

S. Mishra and S. Ramaswamy, 2006. Active nematics are intrinsically phase separated, *Phys. Rev. Lett.* **97**, 090602.

S. W. Morris, E. Bodenschatz, D. S. Cannell and G. Ahlers, 1993. Spiral defect chaos in large aspect ratio Rayleigh–Bénard convection, *Phys. Rev. Lett.* **71**, 2026–2029.

M. Motoyama and T. Ohta, 1997. Morphology of phase-separating binary mixtures with chemical reaction, *J. Phys. Soc. Jpn.* **66**, 2715–2725.

J. D. Murray, 1989. *Mathematical Biology*, Springer-Verlag, Berlin.

S. E. Nagler, R. F. Shannon, Jr., C. R. Harkless, M. A. Singh and R. M. Nicklow, 1988. Time-resolved x-ray scattering study of ordering and coarsening in Cu_3Au, *Phys. Rev. Lett.* **61**, 718–721.

J. Nagumo, S. Animoto and S. Yoshizawa, 1962. An active pulse transmission line simulating nerve axon, *Proc. Inst. Radio Eng.* **50**, 2061–2070.

K. Nam, E. Ott, P. N. Guzdar and M. Gabbay, 1998. Stability of spiral wave vortex filaments with phase twists, *Phys. Rev. E* **58**, 2580–2585.

D. R. Nelson, 1983. Defect mediated phase transitions, in *Phase Transitions and Critical Phenomena*, vol. 7, C. Domb and J. L. Lebowitz, eds., Academic Press, London, pp. 1–99.

S. Nettesheim, A. van Oertsen, H. H. Rotermund and G. Ertl, 1993. Reaction-diffusion patterns in the catalytic CO-oxidation on Pt(100): front propagation and spiral waves, *J. Chem. Phys.* **98**, 9977–9985.

A. C. Newell and J. A. Whitehead, 1971. Review of the finite bandwidth concept, in *Instability of Continuous Systems*, H. H. E. Leipholz, ed., Springer-Verlag, Berlin, pp. 284–289.

A. C. Newell, T. Passot and J. Lega, 1993. Ordered parameter equations for patterns, *Ann. Rev. Fluid Mech.* **23**, 399–453.

G. Nicolis, 1995. *Introduction to Nonlinear Science*, Cambridge University Press, Cambridge.

G. Nicolis and I. Prigogine, 1977. *Self-Organization in Nonequilibrium Systems: From Dissipative Structures to Order Through Fluctuations*, Wiley, New York.

S. O. Nielsen, C. F. Lopez, G. Srinivas and M. L. Klein, 2004. Coarse grain models and the computer simulation of soft materials, *J. Phys. Condens. Mater.* **16**, R481–R512.

K. Nozaki and N. Bekki, 1984. Exact solutions of the generalized Ginzburg–Landau equation, *J. Phys. Soc. Jpn.* **53**, 1581–1582.

T. Ohta and K. Kawasaki, 1986. Equilibrium morphology of block copolymer melts, *Macromolecules* **19**, 2621–2632.

T. Ohta, D. Jasnow and K. Kawasaki, 1982. Universal scaling in the motion of random interfaces, *Phys. Rev. Lett.* **49**, 1223–1226.

T. Ohta, R. Mimura and M. Kobayashi, 1989. Higher-dimensional localized patterns in excitable media, *Physica D* **34**, 115–144.

T. Ohta, A. Ito and A. Tetsuka, 1990. Self-organization in an excitable reaction-diffusion system: Synchronization of oscillatory domains in one dimension, *Phys. Rev. A* **42**, 3225–3232.

K. Oki, H. Sagane and T. Eguchi, 1977. Separation and domain structure of $\alpha + B_2$ phase in Fe-Al alloys, *J. Phys. Colloq. (Paris)* **38**, Ser. C7, 414–417.

L. Onsager, 1949. The effects of shape on the interaction of colloidal particles, *Ann. N.Y. Acad. Sci.* **51**, 627–659.

C. W. Oseen, 1933. The theory of liquid crystals, *Trans. Faraday Soc.* **29**, 883–899.

G. V. Osipov, B. Hu, C. S. Zhou, M. V. Ivanchenko and J. Kurths, 2003. Three types of transitions to phase synchronization in coupled chaotic oscillators, *Phys. Rev. Lett.* **91**, 024101.

Q. Ouyang and J.-M. Flesselles, 1996. Transition from spirals to defect turbulence driven by a convective instability, *Nature* **379**, 143–146.

Q. Ouyang and H. L. Swinney, 1991. Transition from a uniform state to hexagonal and striped Turing patterns, *Nature* **352**, 610–612.

G. A. Ozin and A. C. Arsenault, 2005. *Nanochemistry: A Chemical Approach to Nanomaterials*, Royal Society of Chemistry, Cambridge.

P. Palffy-Muhoray, 2007. The diverse world of liquid crystals, *Physics Today*, **60**, no. 9, 54–60.

A. V. Panfilov, 1999. Three-dimensional organization of electrical turbulence in the heart, *Phys. Rev. E* **59**, R6251–R6254.

G. C. Paquette and Y. Oono, 1994. Structural stability and selection of propagating fronts in semilinear parabolic partial differential equations, *Phys. Rev. E* **49**, 2368–2388.

J. S. Park and K. J. Lee, 1999. Complex periodic spirals and line-defect turbulence in a chemical system, *Phys. Rev. Lett.* **83**, 5393–5396.

J. S. Park and K. J. Lee, 2002. Formation of a spiraling line defect and its meandering transition in a period-2 medium, *Phys. Rev. Lett.* **88**, 224501.

J. S. Park, S. J. Woo and K. J. Lee, 2004. Transverse instability of line defects of period-2 spiral waves, *Phys. Rev. Lett.* **93**, 098302.

M. Park, C. Harrison, P. M. Chaikin, R. A. Register and D. H. Adamson, 1997. Block copolymer lithography: Periodic arrays of similar to 10(11) holes in 1 square centimeter, *Science* **276**, 1401–1404.

R. K. Pathria, 1972. *Statistical Mechanics*, Pergamon, Oxford.

R. L. Pego, 1989. Front migration in the nonlinear Cahn–Hilliard equation, *Proc. R. Soc. London Ser. A* **422**, 261–278.

J. A. Pelesko, 2007. *Self Assembly*, Chapman and Hall/CRC, New York.

L. S. Penrose, 1965. Dermatoglyphic topology, *Nature* **205**, 544–546.

D. M. Petrich and R. E. Goldstein, 1994. Nonlocal contour dynamics model for chemical front motion, *Phys. Rev. Lett.* **72**, 1120–1123.

V. Petrov, Q. Ouyang and H. L. Swinney, 1997. Resonant pattern formation in a chemical system, *Nature* **388**, 655–657.

H. R. Petty, R. G. Worth and A. L. Kindzelskii, 2000. Imaging sustained dissipative patterns in the metabolism of individual living cells, *Phys. Rev. Lett.* **84**, 2754–2757.

A. S. Pikovsky, M. Rosenblum and J. Kurths, 2001. *Synchronization: A Universal Concept in Nonlinear Sciences*, Cambridge University Press, Cambridge.

A. Pimpinelli and J. Villain, 1998. *Physics of Crystal Growth*, Cambridge University Press, Cambridge.

L. M. Pismen, 1980. Pattern selection at the bifurcation point, *J. Chem. Phys.* **72**, 1900–1907.

L. M. Pismen, 1999. *Vortices in Nonlinear Fields: From Liquid Crystals to Superfluids, From Non-Equilibrium Patterns to Cosmic Strings*, Oxford University Press, New York.

L. M. Pismen, 2006. *Patterns and Interfaces in Dissipative Dynamics*, Springer, Berlin.

Y. Pomeau and P. Manneville, 1979. Stability and fluctuations of a spatially periodic convective flow, *J. Phys. Lett. Paris* **40**, L609–L612.

G. Porod, 1952. Die rontgenkleinwinkelstreuung von dichtgepakten kolloiden systemen 2, *Kolloid Zeit. Zeit. Polymere* **125**, 51–57.

G. Porod, 1982. General theory, in *Small Angle X-Ray Scattering*, O. Glatter and O. Kratky, eds., Academic Press, London, pp. 17–51.

J. S. Preston, H. M. van Driel and J. E. Sipe, 1987. Order-disorder transitions in the melt morphology of laser-irradiated silicon, *Phys. Rev. Lett.* **58**, 69–72.

S. Puri, 2004. Kinetics of phase transitions, *Phase Trans.* **77**, 407–431.

S. Puri, A. J. Bray and F. Rojas, 1995. Ordering kinetics of conserved XY models, *Phys. Rev. E* **52**, 4699–4703.

S. Puri, K. R. Elder and R. C. Desai, 1989. Approximate asymptotic solutions to the d-dimensional Fisher equation, *Phys. Lett. A* **142**, 357–360.

H. Qian and G. F. Mazenko, 2004. Vortex kinetics of conserved and nonconserved $O(n)$ models, *Phys. Rev. E* **70**, 031104/1–7.

S. Ramaswamy, R. A. Simha and J. Toner, 2003. Active nematics on a substrate: Giant number fluctuations and long-time tails, *Europhys. Lett.* **62**, 196–202.

P. L. Ramazza, S. Residori, G. Giacomelli and F. Arecchi, 1992. Statistics of topological defects in linear and nonlinear optics, *Europhys. Lett.* **19**, 475–480.

I. Rehberg, S. Rasenat and V. Steinberg, 1989. Traveling waves and defect-initiated turbulence in electroconvecting nematics, *Phys. Rev. Lett.* **62**, 756–759.

C. W. Reynolds, 1987. Flocks, herds, and schools: a distributed behavioral model. *Computer Graphics* **21**, 25–34.

J. Roebuck, 1902. The rate of the reaction between arsenious acid and iodine in acid solutions; the rate of the reverse reaction; and the equilibrium between them, *J. Phys. Chem.* **6**, 365–398.

M. C. Rogers and S. W. Morris, 2005. Buoyant plumes and vortex rings in an autocatalytic chemical reaction, *Phys. Rev. Lett.* **95**, 024505.

T. M. Rogers, 1989. *Domain Growth and Dynamical Scaling During the Last Stages of Phase Separation*, Ph.D. Thesis, University of Toronto.

T. M. Rogers and R. C. Desai, 1989. Numerical study of late-stage coarsening for off-critical quenches in the Cahn–Hilliard equation of phase separation, *Phys. Rev. B* **39**, 11956–11964.

T. M. Rogers, K. R. Elder and R. C. Desai, 1988. Numerical study of the late stages of spinodal decomposition, *Phys. Rev. B* **37**, 9638–9649.

F. Rojas, S. Puri and A. J. Bray, 2001. Kinetics of phase ordering in the $O(n)$ model with a conserved order parameter, *J. Phys. A* **34**, 3985–4002.

C. Roland and R. C. Desai, 1990. Kinetics of quenched systems with long-range repulsive interactions, *Phys. Rev. B* **42**, 6658–6669.

M. G. Rosenblum, A. S. Pikovsky and J. Kurths, 1997. From phase to lag synchronization in coupled chaotic oscillators, *Phys. Rev. Lett.* **78**, 4193–4196.

O. Rössler, 1976. An equation for continuous chaos, *Phys. Lett. A* **57**, 397–398.

G. Rousseau, H. Chaté and R. Kapral, 1998. Coiling and supercoiling of vortex filaments in oscillatory media, *Phys. Rev. Lett.* **80**, 5671–5674.

G. Rousseau, H. Chaté and R. Kapral, 2008. Twisted vortex filaments in the three-dimensional complex Ginzburg–Landau equation, *Chaos* **18**, 026103.

J. S. Rowlinson and B. Widom, 1982. *Molecular Theory of Capillarity*, Clarendon Press, Oxford.

S. A. Safran, 2003. *Statistical Thermodynamics of Surfaces, Interfaces, and Membranes*, Westview Press, Cambridge, MA.

H. Sagane, K. Oki and T. Eguchi, 1977. Observation of phase separation in Fe_3Al alloys, *Trans. Jpn. Inst. Metals* **18**, 488–496.

C. Sagui and R. C. Desai, 1994. Kinetics of phase separation in two-dimensional systems with competing interactions, *Phys. Rev. E* **49**, 2225–2244.

C. Sagui and R. C. Desai, 1995a. Late-stage kinetics of systems with competing interactions quenched into the hexagonal phase, *Phys. Rev. E* **52**, 2807–2821.

C. Sagui and R. C. Desai, 1995b. Effects of long-range repulsive interactions on Ostwald ripening, *Phys. Rev. E* **52**, 2822–2840.

C. Sagui, A. M. Somoza and R. C. Desai, 1994. Spinodal decomposition in an order-disorder phase transition with elastic fields, *Phys. Rev. E* **50**, 4865–4879.

H. Sakaguchi, 1990. Breakdown of the phase dynamics, *Prog. Theor. Phys.* **84**, 792–800.

E. Sander and T. Wanner, 1999. Monte Carlo simulations for spinodal decomposition, *J. Stat. Phys.* **95**, 925–948.

B. Sandstede and A. Scheel, 2000. Absolute and convective instabilities of waves on unbounded and large bounded domains, *Physica D* **145**, 233–277.

B. Sandstede and A. Scheel, 2001. Superspiral structures of meandering and drifting spiral waves, *Phys. Rev. Lett.* **86**, 171–174.

B. Sandstede and A. Scheel, 2007. Period doubling of spiral waves and defects, *SIAM J. Appl. Dynam. Systems* **6**, 494–547.

B. Sandstede, A. Scheel and C. Wulff, 1997. Dynamics of spiral waves on unbounded domains using center-manifold reductions, *J. Differ. Equ.* **141**, 122–149.

M. Sano, H. Kokubo, B. Janiaud and K. Sato, 1993. Phase wave in a cellular structure, *Progr. Theor. Phys.* **90**, 1–34.

A. Saul and K. Showalter, 1985. Propagating reaction–diffusion fronts, in *Oscillations and Traveling Waves in Chemical Systems*, R. J. Field and M. Burger, eds., Wiley, New York, pp. 419–439.

M. Schick, 1998. Avatars of the gyroid, *Physica A* **251**, 1–11.

F. Schlögl, 1972. Chemical reaction models for nonequilibrium phase transitions, *Z. Phys.* **253**, 147–161.

V. A. Schukin and D. Bimberg, 1999. Spontaneous ordering of nanostructures on crystal surfaces, *Rev. Mod. Phys.* **71**, 1125–1171.

R. L. Schwoebel and E. J. Shipsey, 1966. Step motion on crystal surfaces, *J. Appl. Phys.* **37**, 3682–3686.

S. K. Scott and K. Showalter, 1992. Simple and complex propagating reaction–diffusion fronts, *J. Phys. Chem.* **96**, 8702–8711.

M. Seul and D. Andelman, 1995. Domain shapes and patterns: The phenomenology of modulated phases, *Science* **267**, 476–483.

M. Seul and V. S. Chen, 1993. Isotropic and aligned stripe phases in a monomolecular organic film, *Phys. Rev. Lett.* **70**, 1658–1661.

M. Seul and M. J. Sammon, 1990. Competing interactions and domain-shape instabilities in a monomolecular film at an air–water interface, *Phys. Rev. Lett.* **64**, 1903–1906.

M. Seul, L. R. Monar, L. O'Gormann and R. Wolfe, 1991. Morphology and local-structure in labyrinthine stripe domain phases, *Science* **254**, 1616–1618.

M. Seul, N. Y. Morgan and C. Sire, 1994. Domain coarsening in a two-dimensional binary mixture: Growth dynamics and spatial correlations, *Phys. Rev. Lett.* **73**, 2284–2287.

T. Shibata and A. S. Mikhailov, 2006. Nonequilibrium self-organization phenomena in active Langmuir monolayers, *Chaos* **16**, 037108.

A. Shinozaki and Y. Oono, 1991. Asymptotic form factor for spinodal decomposition in three-space, *Phys. Rev. Lett.* **66**, 173–176.

A. Shinozaki and Y. Oono, 1993. Spinodal decomposition in 3-space, *Phys. Rev. E* **48**, 2622–2654.

K. Showalter and J. J. Tyson, 1987. Luther's 1906 discovery and analysis of chemical waves, *J. Chem. Ed.* **64**, 742–744.

B. I. Shraiman, A. Pumir, W. van Saarloos, P. C. Hohenberg, H. Chaté and M. Holen, 1992. Spatiotemporal chaos in the one-dimensional complex Ginzburg-Landau equation, *Physica D* **57**, 241–248.

E. D. Siggia, 1979. Late stages of spinodal decomposition in binary mixtures, *Phys. Rev. A* **20**, 595–605.

K. Simons and E. Ikonen, 1997. Functional rafts in cell membranes, *Nature* **387**, 569–572.

G. I. Sivashinsky, 1977. Diffusional thermal theory of cellular flames, *Combust. Sci. Tech.* **15**, 137–146.

G. I. Sivashinsky, 1983. Instabilities, pattern formation, and turbulence in flames, in *Ann. Rev. Fluid Mech.* **15**, 179–199.

G. S. Skinner and H. L. Swinney, 1991. Periodic to quasiperiodic transition of chemical spiral rotation, *Physica D* **48**, 1–16.

B. J. Spencer, P. W. Voorhees and S. H. Davis, 1993. Morphological instability in epitaxially strained dislocation-free solid films: Linear stability theory, *J. Appl. Phys.* **73**, 4955–4970.

D. J. Srolovitz, 1989. On the stability of surfaces of stressed solids, *Acta Metall.* **37**, 621–625.

O. Stiller, S. Popp, I. Aranson and L. Kramer, 1995a. All we know about hole solutions in the CGLE, *Physica D* **87**, 361–370.

O. Stiller, S. Popp and L. Kramer, 1995b. From dark solitons in the defocusing nonlinear Schrödinger to holes in the complex Ginzburg–Landau equation, *Physica D* **84**, 424–436.

U. Storb, C. R. Neto, M. Bär and S. C. Müller, 2003. A tomographic study of desynchronization and complex dynamics of scroll waves in an excitable chemical reaction with a gradient, *Phys. Chem. Chem. Phys.* **5**, 2344–2353.

J. T. Stuart, 1960. On the non-linear mechanics of wave disturbances in stable and unstable parallel flows. 1. The basic behaviour in plane Poiseuille flow, *J. Fluid Mech.* **9**, 353–370.

M. Suzuki, 1976a. Scaling theory of nonequilibrium systems near instability point. 1. General aspects of transient phenomena, *Prog. Theor. Phys. Jpn.* **56**, 77–94.

M. Suzuki, 1976b. Scaling theory of nonequilibrium systems near instability point. 2. Anomalous fluctuation theorems in extensive region, *Progr. Theor. Phys. Jpn.* **56**, 477–493.

Y. Tabe and H. Yokoyama, 2003. Coherent collective precession of molecular rotors with chiral propellers, *Nature Materials* **2**, 806–809.

M. Tabor and I. Klapper, 1994. The dynamics of knots and curves, *Nonlinear Science Today* **4**, 7–13.

H. Tanaka, 1994. New coarsening mechanisms for spinodal decomposition having droplet pattern in binary fluid mixture: Collision-induced collisions, *Phys. Rev. Lett.* **72**, 1702–1705.

H. Tanaka, T. Yokokawa, H. Abe, T. Hayashi and T. Nishi, 1990. Transition from metastability to instability in a binary-liquid mixture, *Phys. Rev. Lett.* **65**, 3136–3139.

H. Tomita, 1984. Sum rules for small angle scattering by random interface, *Prog. Theor. Phys. Jpn.* **72**, 656–658.

J. Toner and Y. Tu, 1998. Flocks, herds, and schools: A quantitative theory of flocking, *Phys. Rev. E* **58**, 4828–4858.

H. Toyoki, 1992. Structure factors of vector-order-parameter systems containing random topological defects, *Phys. Rev. B* **45**, 1965–1970.

Q. Tran-Cong and A. Harada, 1996. Reaction-induced ordering phenomena in binary polymer mixtures, *Phys. Rev. Lett.* **76**, 1162–1165.

A. M. Turing, 1952. The chemical basis of morphogenesis, *Phil. Trans. R. Soc. London Ser. B* **237**, 37–72.

J. J. Tyson and J. P. Keener, 1988. Singular perturbation theory of traveling waves in excitable media, *Physica D*, **32**, 327–361.

V. Vanag and I. Epstein, 2001. Pattern formation in a tunable medium: The Belousov–Zhabotinsky reaction in an aerosol OT microemulsion, *Phys. Rev. Lett.* **87**, 228301.

H. M. van Driel and K. Dworschak, 1992. Locking of optical and thermodynamic length scales in laser-induced melt-solid patterns on silicon, *Phys. Rev. Lett.* **69**, 3487–3490.

H. M. van Driel, J. E. Sipe and J. F. Young, 1982. Laser-induced periodic surface structure on solids: a universal phenomenon, *Phys. Rev. Lett.* **49**, 1955–1958.

W. van Saarloos, 2003. Front propagation into unstable states, *Phys. Repts.* **386**, 29–222.

M. Venturoli, M. M. Sperotto, M. Kranenburg and B. Smit, 2006. Mesoscopic models of biological membranes, *Phys. Repts.* **437**, 1–54.

T. Vicsek, A. Czirók, E. Ben-Jacob, I. Cohen and O. Shochet, 1995. Novel type of phase transition in a system of self-driven particles, *Phys. Rev. Lett.* **75**, 1226–1229.

J. Villain, 1991. Continuum models for crystal growth from atomic beams with and without deposition, *J. Phys. I (France)* **1**, 19–42.

C. Wagner, 1961. Theorie der alterung von niederschlagen durch umlosen (Ostwald-reifung), *Z. Elektrochem.* **65**, 581–591.

D. Walgraef, 1997. *Spatio-Temporal Pattern Formation*, Springer, New York.

J. H. White, 1969. Self-linking and Gauss integral in higher dimentions, *Am. J. Math.* **91**, 693–728.

K. D. Willamowski and O. E. Rössler, 1980. Irregular oscillations in a realistic abstract quadratic mass-action system, *Z. Naturforsch. A* **35**, 317–318.

K. G. Wilson, 1975. Renormalization group-critical phenomena and kondo problem, *Rev. Mod. Phys.* **47**, 773–840.

K. G. Wilson and J. Kogut, 1974. The renormalization group and the ε expansion, *Phys. Repts.* **12**, 75–199.

A. T. Winfree, 1972. Spiral waves of chemical activity, *Science*, **175**, 634–636.

A. T. Winfree, 1973. Scroll-shaped waves of chemical activity in three dimensions, *Science* **181**, 937–939.

A. T. Winfree, 1987. *When Time Breaks Down: The Three-Dimensional Dynamics of Electrochemical Waves and Cardiac Arrhythmias*, Princeton University Press, Princeton, NJ.

A. T. Winfree, 2001. *The Geometry of Biological Time*, Springer-Verlag, New York.

N.-C. Wong and C. M. Knobler, 1978. Light scattering studies of phase separation in isobutyric acid + water mixtures, *J. Chem. Phys.* **69**, 725–735.

Y. N. Xia and G. M. Whitesides, 1998. Soft lithography, *Ann. Rev. Mater. Sci.* **28**, 153–184.

J. H. Yao, K. R. Elder, H. Guo and M. Grant, 1993. Theory and simulation of Ostwald ripening, *Phys. Rev. B* **47**, 14110–14125.

J. H. Yao, K. R. Elder, H. Guo and M. Grant, 1994. Late stage droplet growth, *Physica A* **204**, 770–788.

C. Yeung, 1988. Scaling and the small-wave-vector limit of the form factor in phase-ordering dynamics, *Phys. Rev. Lett.* **61**, 1135–1138.

C. Yeung and R. C. Desai, 1994. Pattern formation in laser-induced melting, *Phys. Rev. E* **49**, 2096–2114.

C. Yeung, Y. Oono and A. Shinozaki, 1994. Possibilities and limitations of Gaussian-closure approximations for phase-ordering dynamics, *Phys. Rev. E* **49**, 2693–2699.

M. Yoneyama, A. Fujii and S. Maeda, 1995. Wavelength-doubled spiral fragments in photosensitive monolayers, *J. Am. Chem. Soc.* **117**, 8188–8191.

B. Yurke, A. N. Pargellis, I. Chuang and N. Turok, 1992. Coarsening dynamics in nematic liquid crystals, *Physica B* **178**, 56–72.

A. N. Zaikin and A. M. Zhabotinsky, 1970. Concentration wave propagation in two-dimensional liquid-phase self-oscillating system, *Nature* **225**, 535–537.

A. M. Zhabotinsky, 1964. Periodic processes of the oxidation of malonic acid in solution, *Biofizika* **9**, 306–311.

M. Zhan and R. Kapral, 2005. Model for line defects in complex-oscillatory spiral waves, *Phys. Rev. E* **72**, 046221.

M. Zhan and R. Kapral, 2006. Destruction of spiral waves in chaotic media, *Phys. Rev. E* **73**, 026244.

Z. Q. Zhang and S. A. E. G. Falle, 1994. Stability of reaction–diffusion fronts, *Proc. R. Soc. London Ser. A* **446**, 517–528.

R. K. P. Zia, 1985. Normal coordinates and curvature terms in an interface Hamiltonian, *Nucl. Phys. B* **251**, 676–690.

V. S. Zykov, 1987. *Simulation of Wave Processes in Excitable Media*, Manchester University Press, Manchester.

Index

Printed in the United States
by Baker & Taylor Publisher Services